"GOOD HONEST BEER"
FOR BODILY HEALTH
A glass or two of beer at the end of the day's work is almost a tradition with the English working man. It refreshes him after his labours: it builds strength for the morrow.
Here in New Zealand this age-old English custom also prevails, for "Beer is the Wine of the British Race." In Auckland the 5 o'clock reviver is "LION"—the wholesome product of an Auckland industry dating back to the year 1862.
The Khyber Pass Breweries not only constitute an important Auckland industry in themselves, but their products also give strength and energy to thousands of workers in other industries throughout the Province.
LION
PALE ALE

"HE'S CONCERNED ABOUT HIS IMAGE — HE SAYS BEER'S GETTING TOO REAL AND HE CAN'T DECIDE WHETHER IT'S TO BE 'RUGBY, RACING AND GIN' OR 'RUGBY, RACING AND SHERRY'"

Drinks DOMINION BITTER
EVERY
For the past eleven years, Mr. Storrie of Oban Road, Browns Bay, Auckland has enjoyed three, sometimes four, bottles of Dominion Bitter every day. He says, "Since I commenced drinking Waitemata regularly I have not had a days illness, sleep soundly, and feel a hundred per cent fit." You too, will find that Dominion Bitter contributes to good health and long life. Always insist on Dominion Bitter — a Waitemata Product.
DOMINION BITTER
A product of DOMINION BREWERIES LIMITED, QUEEN'S ARCADE, Auckland

OVERHEARD DURING THE GREAT NORTHERN
HARD WORK TAKES IT OUT OF A MAN
YES – BUT 'LION' PUTS IT BACK!
Lion is a winner — and a favourite with everyone — yet it pays big dividends in health and well-being. Every win should be truly celebrated with a "Lion" — every loss made but another excuse to sample its golden goodness. Racing is thirsty work — without "Lion"!
LION ALE
BREWED AT THE "LION" BREWERY — KHYBER PASS — AUCKLAND

That's worth a GOLD TOP
THE ALE WITH
PURITY
BODY
FLAVOUR
1-56

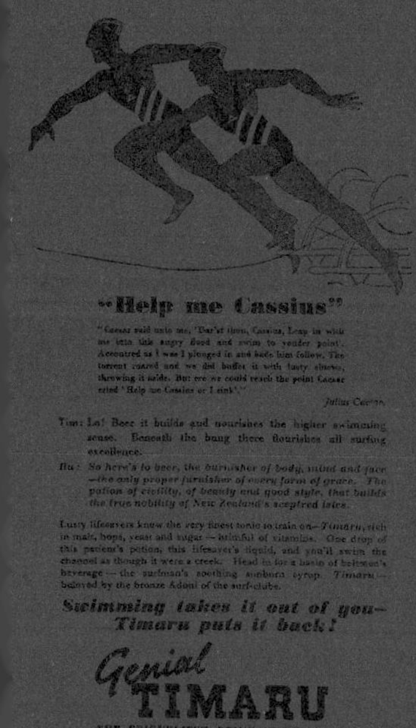
"Help me Cassius"
"Caesar said unto me, 'Dar'st thou, Cassius, Leap in with me into this angry flood and swim to yonder point.' Accoutred as I was I plunged in and bade him follow. The torrent roared and we did buffet it with lusty sinews, throwing it aside. But ere we could reach the point Caesar cried 'Help me Cassius or I sink'."
Julius Caesar.
Tim: Lo! Beer it builds and nourishes the higher swimming sense. Beneath the bung there flourishes all surfing excellence.
Ru: So here's to beer, the burnisher of body, mind and face —the only proper furnisher of every form of grace. The potion of civility, of beauty and good style, that builds the true nobility of New Zealand's sceptred isles.
Lusty lifesavers know the very finest tonic to train on—Timaru, rich in malt, hops, yeast and sugar — brimful of vitamins. One drop of this patient's potion, this lifesaver's liquid, and you'll swim the channel as though it were a creek. Head in for a basin of beltmen's beverage — the surfman's soothing sunburn syrup. Timaru — beloved by the bronze Adoni of the surf-clubs.
Swimming takes it out of you— Timaru puts it back!
Genial TIMARU
THE FRIENDLIEST DRINK IN THE WORLD

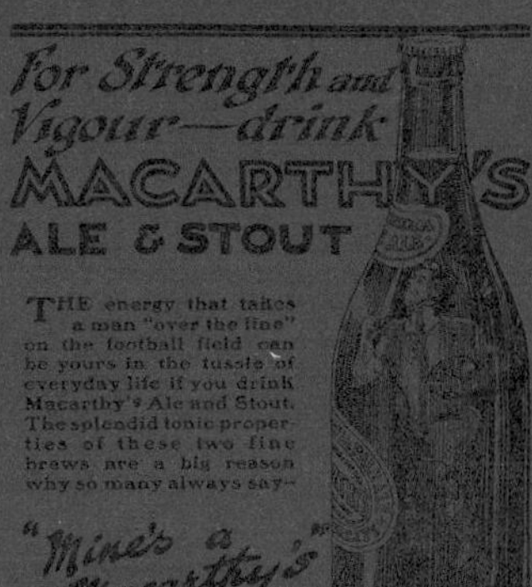
For Strength and Vigour—drink MACARTHY'S ALE & STOUT
THE energy that takes a man "over the line" on the football field can be yours in the tussle of everyday life if you drink Macarthy's Ale and Stout. The splendid tonic properties of these two fine brews are a big reason why so many always say—
"Mine's a Macarthy's"

That's worth a GOLD TOP
THE ALE WITH
PURITY
BODY
FLAVOUR
2-56

A Supper Tonic for Middle Age!
FROM ALL HOTELS AND BOTTLE STORES
MELLO
NATURALLY MATURED STOUT
Mello in Name · Mellow in Taste

CONTINUOUS FERMENT

CONTINUOUS FERMENT

A History of Beer and Brewing in New Zealand

GREG RYAN

AUCKLAND UNIVERSITY PRESS

First published 2023
Auckland University Press
University of Auckland
Private Bag 92019
Auckland 1142
New Zealand
www.aucklanduniversitypress.co.nz

ISBN 978 1 86940 987 6

A catalogue record for this book is available from the National Library of New Zealand

Book and cover design by Duncan Munro
Illustration input by Tim Nolan / Blackant Mapping Solutions

This book was printed on FSC® certified paper

Printed in China by 1010 Printing International Ltd

Beer labels on cover: (1)–(3) 'Natural XXXX Oatmalt Stout', 'Tui', 'Ballins Sparkling Beer' – all from the Collection of beer and drink labels (1940–1969), (Ref. MSX-10203), Patricia Edwards-Sextus Collection, Alexander Turnbull Library, Wellington, New Zealand; (4) 'Waikato Breweries: Waikato Brew 22' (Ref. Eph-A-ALCOHOL-Edwards-Waikato-1-07) and (5) 'Robert Ford and Company Ltd (Hokitika): Ford's Pale XXXX Ale' (Ref: Eph-A-ALCOHOL-Edwards-Westland-1-01), Patricia Edwards-Sextus Collection, Alexander Turnbull Library, Wellington, New Zealand; (6) 'W Burridge & Son: Brown Jug Beer' (Ref: Eph-A-ALCOHOL-Beer-Labels-1930s-01) and (7) 'Robert Ford and Company Ltd (Hokitika): Stout' (Ref. Eph-A-ALCOHOL-Beer-Labels-1940s-01), Alexander Turnbull Library, Wellington, New Zealand. All others from the private collection of the author.

Contents

Acknowledgements

I must begin by thanking Belgium and specifically a bottle of Rodenbach Grand Cru in the railway station café in Kortrijk that was present when the seed for this book was planted in May 2002. Thereafter, Lincoln University provided me with internal research funding, technical support and periods of leave as the vague ideas took shape. I am also very grateful to the staff of the Alexander Turnbull Library and National Library, Wellington; the Hocken Collections – Uare Taoka o Hākena, University of Otago; Canterbury Museum Library; the Macmillan Brown Library, University of Canterbury; Nelson Provincial Museum; Parliamentary Library, Wellington; Auckland City Libraries; the State libraries of New South Wales and Victoria; and especially Edna Carson at the Lion Technical Library, Auckland, who answered all manner of obscure requests and provided things I did not know I needed – including on her last day of work before a much-deserved retirement. DB Breweries, John Duncan, Jeremy Finn, Doug Munro, the late Basil Poff, Geoff Rice and Lindley Turnbull all provided valuable references and ideas.

I could not have had a better local and source of excellent beer than the University of Canterbury Club under Charles Swettenham's management. The beer tastings with Craig Bowen and Ralph Bungard, and the constant reminders from Charles to finish the book, provided much inspiration. In this and other settings, Brent Giblin, Steve Dunn and Neil Macbeth have been friends and drinking companions for more than forty years, mostly within the parameters of the licensing laws. Alyson Gardner, Tracey Shields, Emma Stewart and especially Roslyn Kerr have kept me sane and mostly out of trouble as friends and colleagues at Lincoln University. Jim McAloon joined me during the early 1990s to explore the second generation of New Zealand Guinness and has influenced me as a historian, brewer and friend ever since. At Auckland University Press, Sam Elworthy has always been very encouraging and trusted me despite my liberties with deadlines. Caren Wilton has once again been an exemplary and informative copyeditor who understands the importance of cats to productivity. Kate Stone and Katharina Bauer also refined the manuscript. I am also very grateful to two anonymous readers for their enthusiasm, many helpful suggestions and corrections.

My greatest debt is to the non-beer drinkers in my world. Hugh Ryan has waited his entire life for this book, provided a great deal of entertainment and distraction along the way, and negotiated an entirely fair wage to assist in keeping the bibliography and brewery database up to date. Liz Martyn was at the table in Kortrijk. She has done an extraordinary amount of research for this and related projects, has read every word – and quite a few more that were politely deemed surplus to requirements – and has a remarkable eye for interesting beer. This book would not exist without her.

How Beer is Made

The four basic ingredients of beer are grain (usually barley, but also wheat, rye, rice and others), water, hops and yeast. The basic brewing process is to extract sugars from the grain so that the yeast can turn it into alcohol and carbon dioxide, creating beer.

Grains are harvested and processed by malting. They are soaked in water to germinate and form the enzymes needed for starch separation. The grain is then dried in a kiln. The extent of drying and roasting influences the colour, taste and style of the beer – such as lighter malt for ale, and darker for stout. The finished malt is then milled to break apart the grain kernels so that they dissolve better in water.

Milled malt is then steeped in near-boiling water for about an hour. The starch in the grist dissolves, and sugar, protein and tannin are released. This mash is then filtered to separate a liquid wort from the spent grains. The wort is boiled for about an hour while hops and any other desired additives, such as spices, are added. Hops provide bitterness, to balance the sugar in the wort and provide flavour. They also act as a natural preservative.

At the end of the boil, the wort is cooled and put in a vessel where yeast is added to allow fermentation to occur. The yeast turns the malt sugar into alcohol and carbon dioxide. Depending on the type of yeast, the wort preparation used and the fermenting temperature, a top-fermented (ale, stout, etc.) or bottom-fermented (lager) beer is produced. The beer may then be filtered before being bottled or casked, and may be left to age.

The alcoholic strength of beer is measured by specific gravity – a measurement of the density of liquid relative to pure water. As yeast converts sugars to alcohol and carbon dioxide, the wort becomes less dense and the specific gravity drops. By comparing the specific gravity before fermentation (known as original gravity) to that after fermentation (final gravity) and applying a formula, the alcohol content can be determined.

Introduction

This book is informed by an appreciation for beer. It is not essential to enjoy one's research topic. In many cases, this is impossible; in others, illegal or at least immoral. But, to understand the multiple roles of beer in both shaping and reflecting important themes in New Zealand history, it helps to have some appreciation of the diverse possibilities of the beverage, the settings in which it is consumed and the perspective of the drinker.

Beer is much more than a cold amber liquid for the relief of parched throats, or a putty in the hands of marketers seeking to create images without reference to taste. It is important to consider the culture of beer consumption in its own right. Why did people drink, and did they do so excessively by contemporary international standards? What did people drink and in what circumstances? How did tastes change over time? What role did brewers and publicans play in the community other than as dispensers of alcohol? To determine the impact of beer on New Zealand society is to distinguish between those who drank to excess, those who drank in calm moderation, those who abstained quietly and those who abstained noisily. Migrants to nineteenth-century New Zealand came from societies in which beer drinking was regarded as a physical and social necessity.

Yet in the writing of New Zealand history, the complexity of beer and its contribution to the social fabric have been swamped by the campaign to prohibit it. The dominance of a prohibitionist perspective is not entirely surprising. In the international historical literature on alcohol, there has been a strong focus on the emergence of temperance and prohibition in Britain and North America during the nineteenth century, and in particular the rise and fall of national prohibition in the United States. Much of this history places alcohol within the machinations of local and national politics and legislative processes. Its deleterious effects on colonised indigenous people have also been a recurring theme. Work emphasising sociability, the business of production, or sites of alcohol consumption is more the exception than the rule, although some of the more substantial examinations of these realms do tend to focus on beer more than other alcoholic beverages.

Likewise, where alcohol has appeared in the writing of New Zealand history, aside from its influence on some of the practitioners, the accounts almost entirely concentrate on its negative effects. As Caroline Daley notes: 'New Zealand social historians have tended to focus on the wowsers of our past, the prescribers who preached a message of personal temperance if not prohibition.'[1] Indeed, the prohibition movement, at its peak from the mid-1880s to the late 1920s, tends to be portrayed as an inevitable and justifiable response to widespread drunkenness and related instability in colonial society. It would be foolish to deny that some people in the nineteenth and early twentieth centuries drank more than was good for them or for those around them. Reliable evidence of drink-induced financial, physical and psychological damage is inescapable. However, an account of alcohol-induced catastrophe need not dominate to the exclusion of all other possibilities. Too much of the existing literature is uncritical, too inclined to take prohibitionist rhetoric at face value, and is largely devoid of meaningful international comparisons.[2]

In 1968, P. F. McKimmey traced the rise of the temperance and prohibition movement to 1893. In 1977, A. R. Grigg continued the story up to 1914, while Richard Newman provided a brief treatment of the crucial 1911 election in which support for prohibition peaked at 55.8 per cent.[3] These works have become the standard sources for those few historians who have mentioned prohibition, let alone alcohol more generally, over the past half-century, including Jock Phillips's discussion of 'the boozer and the decent bloke' in *A Man's Country? The Image of the Pakeha Male – A History* and chapters by Stevan Eldred-Grigg in *Pleasures of the Flesh: Sex and Drugs in Colonial New Zealand, 1840–1915*.[4] McKimmey and Grigg worked from the premise that there was a severe alcohol problem in pre-First World War New Zealand that needed to be addressed. Both found much of their confirmation in the opinions of prohibitionists, and neither gave much attention to the variety and complexity of those who opposed prohibition. In 2006, Paul Christoffel provided a detailed and valuable analysis of New Zealand alcohol law and policy since 1881, but his work is very much concerned with the motives and methods of those who sought to restrict rather than those who wished to enjoy alcohol.[5]

The following traces the history of beer and the brewing industry in New Zealand from the first strategic foray at Dusky Sound in 1773 to the establishment of permanent breweries alongside early Pākehā settlement amid a sense that beer was an important barometer of the quality and prosperity of a developing colonial society. Thereafter, the growth of the industry during and beyond the gold discoveries of the 1860s, and an accompanying bureaucracy for brewers, sellers and drinkers, necessarily addresses an enduring mythology about nineteenth-century consumption habits. In this context, especially as powerful players in the later brewing industry began to take shape, prohibition is seen less as a mounting crusade that failed by fewer than two thousand votes to rescue a drink-blighted country from itself in 1919, and more as a strange sort of cargo-cult movement in which progress was contested every step of the way by articulate antagonists who constantly questioned the validity of its claims.

The consolidation of New Zealand Breweries and the unlikely emergence of Dominion Breweries during the inter-war decades, and the creative strategies they used to achieve prosperity, provide some balance to the competing moral, political and industrial forces that eventually culminated in the post-war six o'clock swilling of weak beer. That misguided attempt to circumscribe drinking has assumed rather mythical proportions that tend to distract from the more nuanced beer and brewing culture either side of it, and sometimes in parallel with it. If the introduction of later closing from 1967 was only a partial victory for more civilised drinking conditions, and indeed created problems of its own, it certainly heralded some degree of liberalisation. Tastes sometimes followed international trends, although at other times they reflected what was available rather than what was desired. The account concludes with the surge of 'craft' brewing during the early twenty-first century, but equally the contested terrain as to what that term means and whether it matters. Caution must prevail with the notion that craft beer rescued New Zealand beer from an abyss of blandness, as the limited market share for craft beer suggests that the majority of beer drinkers have not yet shown an inclination to be rescued.

The scope must always be much broader than beer. Patterns of settlement and infrastructure, population increases and shifts, were mirrored in the appearance, and later disappearance, of breweries. The need for raw materials always linked brewing to the progress of agriculture, horticulture and the science of production. The sale of beer bound it to the complexities of licensing legislation, customs and excise, taxation, and later the subtleties and oddities of branding and advertising.[6]

There was also no question that beer, as the most widely consumed alcoholic beverage by the late nineteenth century, has always been close to the centre of national politics, where its fortunes have shifted with the economic well-being of the country and the mood of politicians who periodically saw it as a convenient source of revenue. Events that had their genesis far beyond beer, such as the 1958 'Black Budget', created actual and perceived impacts for beer. Beer also intersected with the politics of race and gender in ways that were frequently paternalistic and seldom liberating. This is also a history of business practices and workplace cultures – some successful, some unsuccessful, by turns lucky and unlucky. Some, despite their best efforts, were eventually outmanoeuvred by the powerful duopoly of the later twentieth century, which itself was swallowed by global brewing empires – a development that was certainly not unique to New Zealand but perhaps had greater impact in a small market.

Above all, this is a history of consumption and to some extent places of consumption. Without those wishing to drink beer in its various forms and settings, there would be no brewing industry.

'View, possibly of Dusky Bay, New Zealand', watercolour by William Hodges, probably made during Captain Cook's Second Voyage (1772–1775). Mitchell Library, State Library of New South Wales

ONE

'Inspissated Juice of Wort' and Other Early Alcoholic Forays Before 1850

New Zealand's history with alcohol is interwoven with European 'discovery' and colonisation, Māori having left no record of producing indigenous intoxicating liquor. Beer, which would become the national drink, appeared early, preceding permanent European settlement in New Zealand by nearly twenty years. The first brew was by Captain James Cook and the crew of the *Resolution* at Dusky Sound in April 1773. In practical terms, it left no legacy for the later brewing industry – other than to inspire some modern derivations ranging along the taste spectrum from subtle ale to abrasive varnish. Over the next seventy years brewing in New Zealand was sporadic and on a very small scale.

However, in other ways the Dusky Sound concoction was an important symbol of the British society that was just beginning to explore New Zealand and would come to dominate it from 1840. In that society, beer was simultaneously a drink, a food, a medicine, a comfort and a social lubricant. Those who decried the ravages of alcohol on the emerging colonial frontier of the 1820s and 1830s never had beer in mind. Indeed, despite its tentative beginnings so far from 'home', beer's rise to prominence in New Zealand was inevitable as an extension of a long global history and, in particular, followed an established British culture of producing and consuming ale. Something of this lengthy and complex history must unfold before returning to Cook and his successors in New Zealand.

Beer – in its simplest form a drink fermented from cereal grains – was probably the first alcoholic drink deliberately made by humans. It has been linked to the 'Neolithic revolution', which transformed nomadic hunter-gathering to stable settlement perhaps around 12,000 BCE amid refinements to various agricultural techniques for the production of grain. The first evidence of fermented beverages dates from China 7000–9000 years BCE, and almost simultaneously in Sumeria, between the Tigris and Euphrates rivers and encompassing Southern Mesopotamia.[1] Whether by accidentally fermenting wet bread or more deliberate experiments with grains, beer soon became an essential element of both diet and ritual in evolving civilisations.

Texts from 3200–3000 BCE indicate that beer was not merely an agricultural product, but part of the centralised economy of the Sumerian states.[2] The Babylonians who rose to power in Mesopotamia during the second millennium BCE brewed at least twenty distinct types of beer.

The pioneering legal code of Hammurabi, ruler of Babylon *c.* 1792–50 BCE, stipulated a daily beer ration, with amounts dependent on the social standing of the individual, along with strict and unpleasant penalties for those failing to maintain a quality beverage. Beer was also an important dietary and religious item in Ancient Egypt. It was frequently placed in tombs to accompany the dead, and 30,000 gallons a year was offered as a gift by Pharaoh Ramses II (*c.* 1200 BCE) to keep the favour of the gods.[3]

Beer reached Europe by 3600 BCE. While somewhat marginal to the Greeks and Romans who preferred wine in warmer climates,[4] beer took a firm hold among the Teutonic peoples of temperate Central Europe, and later in Britain, especially with the decline of the Roman Empire by the fifth century.[5] Traditionally a home-based activity carried out largely by women, termed 'brewsters' or 'ale-wives' in England, brewing also became an essential element of large institutions such as military garrisons and especially monasteries. Not only was beer necessary to supply pilgrims and other travellers, it was a nutritious drink for monks subject to a variety of fasting periods. It was also a versatile drink, in that it could be brewed with a wide variety of readily available materials and to different strengths for common use or ritual occasions – such as 'bride-ale' (bridal) for weddings or 'yule-tide' (ale tide) brews at Christmas.

The Assize of Bread and Ale enacted by Henry III in 1267 confirmed the centrality of ale to English culture and diet by establishing legal regulation of a maximum price per gallon according to a sliding scale of corn prices, thus guaranteeing both a profit for brewers and a fair price for consumers.[6] As Richard W. Unger depicts medieval drinking patterns in Europe: 'The prevailing presentism makes it difficult for many to comprehend a world where beer was a necessity, a part of everyday life, a drink for everyone of any age or status, and a beverage for all times of the day from breakfast to dinner and into the evening.'[7]

Until at least the late fifteenth century, the terms 'beer' (Anglo-Saxon) and 'ale' (Danish) were used interchangeably in Britain to describe the unhopped, sweet, dark drink of these times that could not be stored for long or easily transported. It was only as hops began to appear from the late fifteenth century, with their superior preservative qualities that enabled the beverage to be stored for longer, that hopped beer was distinguished from unhopped ale. By the late seventeenth century, however, with the rapid decline of the unhopped drink, 'ale' became the common term for beer – a historical quirk that defies logical explanation.[8]

From around the mid-fourteenth century, more of it was likely to be commercially brewed by men and the traditional role of women declined sharply. As Judith M. Bennett explains, the Black Death, which wiped out as much as half of the population of Britain and Europe during the late 1340s, left an environment with fewer customers drinking more per capita, and purchasing more from alehouses for on-premises consumption than from the homes of brewsters. As brewing became more specialised and falling grain prices made it more profitable, patriarchal control gradually squeezed women out of brewing and into lower-status roles as ale-sellers and servants to the industry.[9]

In 1689, the first year for which reliable production and consumption figures are available, per-capita English beer consumption peaked at about 832 pints per person or 2.3 pints per day, which represented 400–500 kilocalories or a fifth of daily energy requirements. Thereafter it dropped sharply throughout the eighteenth century, due to increasing competition from other beverages such as coffee, tea, chocolate drinks and spirits, not least gin. Yet per-capita annual consumption was still 154 litres (271 pints) per person for England and Wales during 1800–1804, and would remain in that vicinity throughout the nineteenth century.[10] Beer was essential to weddings, christenings, funerals and seasonal thanksgivings. It was a standard feature of many working environments, and especially important for labourers during harvest, haymaking and other strenuous manual tasks.

From the seventeenth century, the preservative qualities of hops further encouraged commercial brewing as beer could more easily be stored and transported, and the final transition from a female to an almost exclusively male occupation was completed. Perhaps one-third of English beer was commercially produced by 1700, and half by 1800 as the expanding industrial revolution shifted population from country to town, eroding traditions of home brewing, while rural enclosure reduced access to brewing fuel for those who remained. The twelve largest London brewers produced 42 per cent of output for the capital in 1748, and 77 per cent in 1800.[11]

The influence of these establishments went hand-in-hand with a significant change to the preference of London beer drinkers, which would also figure prominently in early New Zealand tastes. By the 1720s there were at least thirty beer styles available in London, ranging across a wide spectrum of taste and strength. Many of them were blended prior to consumption. But gradually, as it was discovered that the traditional working-class staple of strong, dark-brown, bitter beer became much better if left to mature for at least twelve months, a style known as 'porter' came to dominate – named for the thousands of porters in London who carried goods from place to place and needed a refreshing and nutritious beer to sustain them. Heavily hopped to balance the sweetness of the malt, porter was perhaps 6 per cent alcohol by volume (abv) but not fully attenuated – meaning that not all of the malt sugars were turned to alcohol during the brewing process, and some therefore remained in the beer to provide energy to drinkers. However, the time required to mature porter determined that its manufacture was largely the preserve of those with sufficient casks and cashflow to store it until ready for sale.[12]

Meanwhile, despite a reputation for the production and consumption of spirits, both Ireland and Scotland witnessed strong growth in brewing by the mid-eighteenth century, especially of porter. While the brewing traditions of lowland Scotland would later become especially significant to New Zealand, the establishment of a brewery at St James's Gate, Dublin, by Arthur Guinness in 1759 is the most enduring symbol of the Irish brewing industry. By the 1880s Guinness was the largest brewery in the world, responsible for 60 per cent of Irish output and with a strong presence in English and international markets.[13]

Given its cultural significance, provision had to be made for beer as Britain extended its commercial and political influence further afield. It became a standard English sea ration from the fourteenth century, both as a luxury to counter the hardships of life at sea, and more importantly as vital food and medicine and a widely regarded preventative, if not a cure, for scurvy – the scourge of long sea voyages that is said to have caused more losses than enemy action for the British Navy during the eighteenth century. Yet even in its hopped form, the relatively limited keeping quality and bulk of beer posed significant challenges on the long voyages of commerce and exploration that were becoming a regular feature of eighteenth-century maritime expansion. Spirits, especially brandy and later rum, were deliberately used as a substitute for beer. While these met the need of sailors for alcohol, they did not offer the perceived nutritional and medicinal benefits of beer.[14]

With the cause of scurvy remaining uncertain until the isolation of ascorbic acid (vitamin C) in 1932, the eighteenth-century medical profession was awash with theories and procedures for its cure. The British Admiralty sponsored numerous experiments to test them –

but always with the proviso that any potential anti-scorbutic for a long sea voyage should be stable for at least a year, preferably two, and not be bulky.[15]

With the almost universal faith in the medicinal qualities of beer, it quickly entered the lexicon of scurvy cures. Among the most influential were the views of Dr David Macbride, who characterised scurvy as a disease of putrefaction that could be inhibited by the production of 'fixed air' (carbon dioxide) as occurred in fermentation. Any food that fermented quickly in the digestive tract would help to counter putrefactive disease. However, as beer derived all of its fixed air from the malt of which it was made, Macbride concluded that malt rather than beer would be more effective on long voyages as it took up less room and was likely to last longer.[16] Others, drawing on North German and North American brewing traditions, advocated types of spruce beer, which had added vegetable matter such as the fresh leaves of spruce fir trees or the essence of these.[17]

Dried malt was taken on James Cook's 1768–71 *Endeavour* voyage, with a certain amount being ground each day, mixed with boiling water, allowed to stand for several hours and then issued to the crew at the rate of at least 1 quart per person per day. A prominent surgeon, Nathaniel Hulme, also apparently provided Joseph Banks with directions for brewing spruce beer, but there is no evidence that this was done.[18]

In January 1772, Henry Pelham, secretary to the commissioners of victualling, wrote to the Admiralty recommending 'inspissated juice of wort', which was made by boiling most of the water away from wort (unfermented beer), or 'inspissated juice of beer' (already fermented). This technique had emerged earlier in the century among brewers attempting to reduce excise duties by paying them on smaller quantities of beer before adding water. As beer was mostly water, a concentrate would take up less space and a strong concentrate would keep better. The Admiralty promptly ordered supplies of both inspissated juice of wort and beer for Cook's second voyage aboard *Resolution* and *Adventure*, and also trialled it on ships bound for the Falklands.[19] Cook brewed at sea in August and November 1772, but found that in hot climates and due to the movement of the ship the inspissated juice fermented uncontrollably and a considerable quantity was lost as it forced its way out of its casks. But he noted on 20 September that 'The beer made from this juce is of a very deep Colour, something of a burnt taste and without bitter and must be drinked soon after it has done fermenting otherways it turns hard and will sour.'[20]

More consistent results were enjoyed when New Zealand's first brewing began in the cooler climate of Dusky Sound, Fiordland, on 1 April 1773. Cook recorded in his journal: 'The Juce deluted in warm Water in the proportion of Twelve parts Water to one part Juce made a very good well tasted small Beer'.[21] In his official account of the voyage published in 1777, he presented the recipe in more detail.

> We at first made our beer of a decoction of the spruce [rimu, *Dacrydium cupressinum*] leaves; but, finding that this alone made it too astringent, we afterwards mixed with it an equal quantity of the tea plant [mānuka, *Leptospermum scoparium*] (a name it obtained in my former voyage, from our using it as a tea then, as we also did now), which partly destroyed the astringency of the other, and made the beer exceedingly palatable, and esteemed by every one on board. We brewed it in the same manner as spruce beer, and the process is as follows. First make a strong decoction of the small branches of the spruce and tea-plants, by boiling them three or four hours, or until the bark will strip with ease from the branches; then take them out of the copper, and put in the proper quantity of molasses, ten gallons of which is sufficient to make a ton, or two hundred and forty gallons of beer. Let this mixture just boil; then put it into casks, and to it add an equal quantity of cold water, more or less according to the strength of the decoction,

> or your taste. When the whole is milk-warm, put in a little grounds of beer, or yeast if you have it, or anything else that will cause fermentation, and in a few days the beer will be fit to drink. After the casks have been brewed in two or three times the beer will generally ferment itself, especially if the weather is warm. . . . Any one who is in the least acquainted with spruce pines will find the tree which I have distinguished by that name. There are three sorts of it: that which has the smallest leaves and deepest colour is the sort we brewed with, but doubtless all three might safely serve that purpose.[22]

On 5 April, second lieutenant Charles Clerke wrote 'got Beer on board for the People and stopt their Spirits – this Beer I think is a very palatable pleasant drink; the Major part of the People are I believe of the same opinion, for they seem to drink pretty plentifully of it'. The Swedish naturalist (and evident cocktail aficionado) Anders Sparrman added: 'After a small amount of rum or arrack has been added, with some brown sugar, and stirred into this really pleasant, refreshing, and healthy drink, it bubbled and tasted rather like champagne.'[23]

Although this beer used inspissated juice of wort (rather than beer), molasses and spruce, it seems that Cook soon concluded that he could have brewed with molasses and spruce alone, as there was no distinct advantage in inspissated juice as a fermentable, and it was the fixed air rather than the fermentable that was believed to prevent scurvy. Indeed, while Cook brewed again in New Zealand on his third voyage, at Queen Charlotte Sound, in February 1777, there is no firm evidence that the juice was used again.[24]

After the second voyage Cook reported to the Admiralty that 'wort made with malt is without doubt one of the best antiscorbutic sea medicines yet found out', and that it would be a 'Valuable and useful Article' if it could be prevented from fermenting in hot weather. However, he qualified this with the observation that 'If given in time it would, with proper attention to other things, prevent the scurvy from making any progress, but would seldom be found to cure it.'[25]

Although it may have had some limited effectiveness because germinated barley contains substantial amounts of ascorbic acid, some of which may have remained in Cook's malt, the very low rates of scurvy on his voyages were more likely due to various other substances used, such as sauerkraut, mustard and marmalade of carrots, and a strict regime of cleanliness. But so many were used simultaneously that it was impossible to determine which actually worked. Certainly, by the mid-1780s, following severe outbreaks of scurvy on a number of voyages on which malt was issued, there were growing doubts about its effectiveness as an anti-scorbutic. Moreover, while some crews were happy to accept essence of wort as a medicine, they were less enthusiastic about its use as a substitute for beer.[26]

Nevertheless, Captain George Vancouver, who had sailed with Cook on his second and third voyages, remained faithful to the remedy and brewed spruce beer when his ships *Discovery* and *Chatham* reached Dusky Sound in November 1791. As a journal of the voyage recorded:

> Thus we quitted Dusky bay, greatly indebted to its most excellent refreshments, and the salubrity of its air. The good effects of a plentiful supply of fish, and spruce beer, were evident in the appearance of every individual in our little society. The health of our convalescents was perfectly re-established, and excepting one with a chronic complaint, and two wounded by cuts in their legs, we had not a man on the surgeon's list; though, on the most trifling occasion of indisposition, no person was ever permitted to attend his duty.[27]

But in 1795, against mounting evidence, the Admiralty officially adopted lemon juice for use against scurvy, and in 1798 all issue of malt for this purpose was stopped.[28] New Zealand's first foray into beer and brewing was at an end, but it would not be absent for long.

*

As the nineteenth century opened, New Zealand became home to a motley collection of sealers, traders, whalers and escaped convicts from Australia, who generally worked hard and drank harder. From 1814 they were joined by missionaries who condemned their alcoholic excesses and despaired at the likely impact of drink on Māori. New Zealand therefore followed a pattern familiar to any number of rugged and mostly male frontiers in which drink, and spirits in particular, were ubiquitous as European exploration and commerce encircled the globe. Yet the emphasis on the demands and perils of frontier life as a cause for drinking clearly ignores the parallel influence of drinking cultures from which all of these European visitors and settlers came. There were certainly differences in New Zealand, but also much that was familiar, including a clear distinction between beer and other alcoholic beverages, not least by the most remote offshoot of the British and Foreign Temperance Society.

The drinking habits and attitudes characteristic in New Zealand were very much a reflection of both the entrenched traditions and growing resistance to particular forms of drinking within British society at the time. These would underlie developments on the new frontier. In Britain, alcohol was still a main beverage of choice through much of the nineteenth century, as there were few safe alternatives. Although technological change improved the quantity and quality of water available, and notwithstanding added urgency from the discovery of a link between cholera and waterborne sewage by John Snow in 1854, progress could hardly keep pace with the rapid growth of cities. Mains water supplies were limited even in upper-class households until the 1850s, and water storage only improved when iron piping replaced wood. Even by the 1870s there was still much unsafe water in London, and many believed that it should not be drunk unless purified with spirits.

Likewise, uncertainty about the processing and storage quality of milk remained until at least the 1870s, when the development of refrigeration and the expansion of railways enabled more efficient delivery from country to city. But per-capita consumption was still very low during the 1880s, and milk was not widely accepted as a drink until the 1930s. Other thirst-quenchers, such as soda water and cordial, were only beginning to enter common usage during the early nineteenth century, but not always with the affordability or accessibility of alcohol. Coffee consumption increased until the 1840s, when it was superseded by tea, which dropped in price with decreased taxation following the Napoleonic Wars. Ginger beer became popular around the same time, but it was also noted during the 1890s that many of those purchasing it in pubs wanted it mixed with spirits.[29] In New Zealand, the water may initially have been safer, but this did not last. Other beverage choices were shaped by the vagaries of import networks and the preferences that arrived with the settlers.

Regardless of the availability of alternatives, alcohol continued to be sought for its medicinal properties and was frequently prescribed by doctors. It assisted dentists and surgeons, quietened crying babies, and was believed to impart physical stamina to workers, act as a barrier against cold, cleanse the system, cure indigestion, and act as a painkiller, especially for the poor, who regarded pain as a disease in itself rather than a symptom of disease. While medical science began to question these ideas by the 1860s, and not least the view that fatter beer drinkers embodied superior physical strength, the use of alcohol for medicinal purposes endured along with the idealisation of a rotund John Bull, tankard in hand.[30]

Brian Harrison also observes that drinking places 'mirrored the interests and needs of their localities; broadly speaking, their two main roles were as recreation centre and as meeting place'. The inn, tavern or alehouse was often preferable to a cold, crowded and noisy home. Indeed, 'Light, heat, cooking facilities, furniture, newspapers and sociability were then obtained by the poor only at the drinking place.'[31] Paul Jennings adds that the pub was also strongly linked to conceptions of English identity and the notion that as a community institution it was unique to England.[32] In an age with a relative prevalence of solitary occupations, these establishments provided important human contact as well as serving as local news centres and venues for all manner of formal and informal meetings of trades, travellers, societies and reform campaigners. Even as organised sport and leisure began to capture the popular imagination during the nineteenth century, the public house remained a significant provider of recreation and entertainment.[33] Its role among the rudimentary settings of colonial New Zealand was to be even more pronounced.

At a psychological level, especially during a time of unsettling social change as the industrial revolution took hold, alcohol and drinking places undoubtedly eased fear in those for whom catastrophe and economic disaster were not far away. As Harrison explains: 'Such places brought temporary harmony into the disordered lives of many bored, exhausted or exploited individuals.' Yet he also stresses the positive dimension of alcohol as a social lubricant in many rituals of community interaction and celebration, not only moderating gloom but enhancing festivity.[34] To this Paul Jennings borrows from William James in positing alcohol as perhaps a variety of religious experience – 'the great exciter of the yes function in man', bringing the drinker 'from the chill periphery of things to the radiant core'.[35]

Improvements in housing, the impact of the railway on patterns of travel and meeting, changing leisure patterns and the emergence of other public buildings, such as town halls and libraries, all gradually altered the role of the public house from the mid-nineteenth century. But there were still complaints by the 1870s that England had largely failed to grasp the continental model of cafés or other alternative eating and drinking establishments. Indeed, these were to take hold only slowly in Britain amid an enduring devotion to, and nostalgia for, the communal role of the pub in both town and countryside.[36]

Nevertheless, there was always a kernel of opposition to drink, and more specifically to excessive drinking. Notwithstanding the monastic tradition, elements of organised religion had attacked drunkenness from at least Anglo-Saxon times, partly due to its association with pagan rituals and partly from a humanitarian sense of the productive and balanced life of moderation. From the early sixteenth century, England witnessed a succession of licensing legislation to preserve public order and conserve grain. While a few had always demanded total abstinence, a gradual increase in spirit drinking from the late seventeenth century, especially the 'Gin Craze' of the first half of the eighteenth, drew stronger reactions.[37] Meanwhile, the beginning of the industrial revolution, with its greater emphasis on regular work time and discipline, discouraged anything such as drunkenness that would undermine the productivity of the workforce.

The Licensing Act 1753 was the first to require alehouse licensees to provide evidence of good character. Licences also had to be renewed annually rather than being granted in perpetuity. In 1787, under the influence of politician, philanthropist and abolitionist William Wilberforce, George III issued a *Royal Proclamation for the Encouragement of Piety and Virtue, and for the Preventing and Punishing of Vice, Profaneness and Immorality*. This, in turn, empowered magistrates to enact further restrictions against the licensed trade.[38]

From the early nineteenth century, reports of growing anti-spirits and later temperance sentiment began to trickle into Britain from the United States, at the same time as a variety of other evangelical, moral reform and self-improvement movements, such as the Lord's Day Observance Society and the Society for the Prevention of Cruelty to Animals, began to gather momentum. In 1829–30, an anti-spirits movement emerged simultaneously in Ireland and Scotland. In England it was no coincidence that the movement achieved its early successes in the industrial heartlands of Lancashire and Yorkshire. The British and Foreign Temperance Society was formed in 1830, with a number of brewers among its supporters.[39]

But contradictions quickly appeared in the movement. Condemnation of spirits in predominantly spirit-drinking areas singled out the poor while leaving the wealthy free to indulge in wine. On other occasions beer drinkers got more drunk than spirit drinkers. As Harrison notes: 'So pronounced were regional variations in drinking habits that only a teetotal pledge could achieve in beer-drinking rural England and Wales what an anti-spirits pledge could achieve in spirit-drinking Ireland and Scotland.'[40]

Within this gradually evolving drink and anti-drink environment, the Beer Act of 1830 brought dramatic if perhaps short-term change. With the rise of commercial brewing through the eighteenth century came strong objections to brewer monopolies, especially to their 'tying' of licensed establishments to supplying only their beer. Some also argued that the restriction of licences by magistrates greatly increased the value of those remaining.[41] Reports of the adulteration of beer also added to perceptions of brewer corruption. In 1814, 14,000 Londoners signed a petition against adulteration and high beer prices. Many felt that the problem was a lack of competition and the failure of previous legislation to curb the power of the brewers.[42]

In 1830, the Duke of Wellington's Tory Government, ever devoted to the gospel of free trade, introduced the Beerhouse Act 'expedient for the better supplying the public with Beer in England, to give greater facilities for the sale thereof, than was then afforded by licences to keepers of Inns, Alehouses, and Victualling Houses'.[43] The Act allowed any householder assessed on the poor rate, and on payment of 2 guineas to the local excise office, to open their house between 4 a.m. and 10 p.m. to sell beer and cider only. They were free from the requirement of alehouses, inns and taverns to obtain a licence from a magistrate, but subject to a £20 penalty if they sold wine or spirits. The excise duty on beer was also abolished, leaving only that on malt and hops. The intention was to erode the brewer monopoly, encourage the drinking of beer rather than spirits, especially gin, and to bring lower prices and improved quality to the consumer through competition with existing establishments. Within six months from October 1830, more than 26,000 new beer shops opened in England, and nearly 46,000 by 1838, of which perhaps 40 per cent brewed their own beer.[44]

Yet the imagined paradise for drinkers never arrived. A temporary rise in beer consumption during the early 1830s levelled off by 1840. Many of those who did not brew their own quickly became tied to brewers for their supplies. Among those who did brew, and among the established breweries, there was increasing evidence that cost-cutting competition produced poor-quality, often adulterated, beer.[45] Real and imagined problems with social order in so many largely unregulated drinking places also drew opprobrium to the Act. As Lord Palmerston observed in 1853: 'The words "licensed to be drunk on the premises" are by the people interpreted as applicable to the customers as well as the liquor.'[46] The Act was not, however, substantially amended until 1869.

If the Beerhouse Act did not have a long-term impact on those who drank, it did galvanise the forces opposed to drink. While some moderationists remained keen to promote the consumption of beer, the 'temperance drink', at the expense of spirits, an increasingly vocal cohort of teetotallers argued for total abstinence. From the 1850s, as later chapters will explain, they also moved away from moral suasion to advocacy of prohibition by legislation to achieve their goal.[47]

However, the impact of the teetotallers on Britain should not be exaggerated. Between the 1820s and 1860s per-capita beer and wine consumption and the number of drink-selling outlets increased. The only partial victory was a slight decline in the per-capita consumption of spirits from 5 litres in 1840 to 4.4 in 1860, although it peaked again at nearly 6 litres in 1875. Certainly there was an influential and literate group of temperance 'opinion-makers' by mid-century, but the committed movement numbered well under 100,000 during the 1860s. Throughout the nineteenth century, publicans and brewers continued to enjoy wealth and respectability within communities, and the latter group secured a significant stake in local and national politics.[48] It will be seen in later chapters, however, that the cause of temperance and ultimately prohibition took a firmer hold on New Zealand and other new-world societies in which notions of a fresh start, respectability and progress through diligence and self-improvement in a new setting held greater sway.

*

As European contact with New Zealand expanded from the late eighteenth century, it brought with it a strong, if sometimes contested, tradition of alcohol consumption and of the drinking place as central to the community. Moreover, the new land was something of a blank slate, in that, with the possible exception of fermented tutu juice, no intoxicating drink was known to Māori until the arrival of Europeans. Accounts from a number of early explorers note Māori aversion to alcohol and descriptions of it as 'wai piro' (stinking water) or 'wai kaha' (strong water).[49] The prevailing European drinking culture during the early years of settlement was unlikely to affect this view.

The sealers, traders, whalers and convicts, who flirted with the coast of New Zealand from the early 1790s and began to congregate in larger numbers in the north after 1800, were hardy adventurers and perhaps more likely to have a predilection for alcohol than most of their counterparts in Britain and Europe. A transient, almost exclusively male, population generally lacks a sense of social obligation, and is subject to few restrictions around its drinking behaviour. This was accentuated by the use of alcohol as a currency in an environment where real money was meaningless. Whalers were frequently paid in food and rum, and it was traditional to indulge after the capture of a whale. The supply of alcohol in New Zealand during the early nineteenth century may also have been enhanced by the fact that no customs duty was payable in New South Wales on alcohol brought from New Zealand. There is evidence that some merchants shifted alcohol through New Zealand, and that some was landed at the Bay of Islands directly from the United States before most of it was sent on to Sydney.[50]

Nor was Sydney short of alcohol for the New Zealand market. Although the convict colony in New South Wales was almost certainly not as constantly soaked in alcohol as most accounts have suggested,[51] very quickly after their arrival in the early 1790s officers of the New

South Wales Corps exploited their position and wealth to monopolise imported rum and other spirits, and exchange these for goods and labour at very favourable rates. Despite the efforts of successive governors to control the trade, 'The Rum Corps', as it became known, ensured that alcohol remained the dominant currency of New South Wales until after 1810, when the regiment was recalled and Governor Lachlan Macquarie was finally able to establish coinage.[52]

Nevertheless, Sydney remained a vital supply base. Customs records for January to August 1830 show that New South Wales alcohol exports to New Zealand amounted to 3300 litres of beer, 11,000 litres of rum and whisky, and 7300 litres of brandy and gin – for a European population probably well under a thousand. And this takes no account of supplies secured from Hobart or other sources.[53] Notably absent from the exports to New Zealand was a significant proportion of beer; not surprisingly, as it was a relatively bulky and sometimes unstable export commodity compared to spirits. Although brewing had begun in Sydney during the early 1790s and there were fifteen breweries in New South Wales by 1833, along with another twenty-one in the more temperate climate of Tasmania, there is no record of whether any of the local production was sent across the Tasman.[54]

Of the spirits sent to New Zealand, there was no guarantee of quality at point of departure or in the hands of those responsible for distributing them. When asked his method for prolonging the rum supply, Geordie Toms, head of a whaling station in Queen Charlotte Sound during the 1830s, reportedly explained: 'When I takes out a glass of rum, I puts in a glass of water; when it gets too strong of water, I puts in turps; and when it gets too strong of the turps, I puts in bluestone.'[55]

Not surprisingly, accounts of alcoholic excess were numerous, although one always has to remember that most of them were written by those, such as missionaries, with a vested interest in emphasising depravity in the hope that it would prompt more formal British intervention in New Zealand. Samuel Marsden, New South Wales chaplain and instigator of the Church of England's missionary activity in New Zealand, observed of the Bay of Islands in the early 1830s: 'Here drunkenness, adultery, murder, etc are committed. . . . Satan maintains his dominion without molestation.'[56] Edward Markham, a visitor in 1834, labelled it 'a wicked world and full of drink'.[57]

The following year, Charles Darwin, arriving at the Bay of Islands during the historic voyage of HMS *Beagle*, lamented: 'Besides a considerable native population there are many English residents. These latter are of the most worthless character; & amongst them are many run away convicts from New South Wales. There are many spirit shops, & the whole population is addicted to drunkenness & all kinds of vice.'[58] Writing in 1837, Marsden's daughter Martha was equally disparaging: 'The publicans are under no restrictions. They can receive into their houses whomsoever they think proper, having no one to control them – hence they are totally neglectful of all order and any appearance of morality.'[59]

But a more palatable alternative to spirits was on offer to at least some in the Bay of Islands. Murdoch Riley suggests that 'beer' was made by missionaries and Māori from sow thistle juice, supplejack roots, mataī sap, kohekohe bark and bush honey, among other things.[60] But the details are unclear. With more certainty, home brewing can be inferred from the planting of hops. On 1 August 1821, Rev. John Butler noted in his journal that he had dressed a bed of hops: 'I have fourteen hills which look exceedingly fine, as the plants are very strong. I brought a single root from Port Jackson, and planted the whole fourteen hills from it last spring; and I gathered a small sprinkling of fine hops from them in the season. I hope this year they will produce a good full crop.' However, no mention is made of any production using the hops.[61]

Sometime during 1835 Joel Samuel Polack, a trader and land speculator originally from London, established the first permanent brewery in New Zealand at Kororāreka (Russell).[62] The brewery plant and a supply of hops were imported from Sydney and a brewer from Hobart. In 1840 Polack claimed that the brewery was inspired

> as a preventative to the then rapid spread of deleterious spirits that were consumed, less probably from taste, than the want of an invigorating substitute. . . . The natives after some little practice in quaffing soon became to relish it and baskets of potatoes, packages of fish, etc, soon found conveyance to Parramatta [Polack's property], in exchange for pierian drafts of New Zealand beer.[63]

No other account of this brewery has survived, and whether Polack's professed concern for the well-being of his fellow residents was greater than a desire for profit will never be known. But the brewery was evidently a success until his departure for England in 1837. He returned in 1842, but there is no evidence that he brewed again.[64]

Shortly after Polack began production, the cause of beer received another fillip from respectable elements of the Bay of Islands community. The missionaries were certainly not averse to alcohol, but they were at the forefront of growing concern about drunkenness and the consumption of 'ardent spirits' during the early 1830s, especially as they feared that Māori might be enticed away from previously abstemious habits and would then be more difficult to convert to Christianity. By 1834 most missionaries were in favour of a total prohibition against the importation and sale of alcohol.

In September 1835 a resolution against spirits was passed at Paihia, and a public meeting at Hokianga declared in favour of the 'total and immediate abolition of the importation and sale of ardent spirits'.[65] This resolution was communicated to Governor Bourke in New South Wales, who urged a public proclamation of such a law to all visiting vessels. Naturally there was resistance from some traders, but also from James Busby, the British Resident at Waitangi, who resolved not to acknowledge the measure until it was passed by a meeting of all chiefs in the north. Given the lack of formal British jurisdiction over New Zealand before 1840, there was little choice in the matter. By early 1836 the failure of a prohibition at Hokianga was evident. In April Henry Williams, the most prominent Anglican missionary, observed that a similar failure was likely at the Bay of Islands without a properly constituted authority to enforce it, especially in the context of other tensions between Christian and non-Christian Māori. Supporters of prohibition therefore turned to temperance, with persuasion rather than compulsion as their only viable option.[66]

In March 1836 a temperance meeting was held at Hokianga at which eighty Māori and five Pākehā signed a temperance declaration. Eliza White, wife of Wesleyan missionary William White, observed: 'We hope the object will be gained to prevent the drinking of this liquid fire by these poor simple natives who are so sadly led by the white people.'[67] On Wednesday, 11 May, a public meeting attended by fifty people was held at Kororāreka at which the New Zealand Temperance Society was established. Its preamble was unequivocal:

> Whereas the evils of Intemperance from the use of ardent spirits have arisen to an alarming height among many of the resident Europeans in the islands of these seas, producing among them, disease, want and, in some instances, sudden death, accompanied also by a long train of immoralities consequent upon drunkenness.[68]

James Busby was elected president, and Henry and William Williams became vice presidents. A sum of £8 10s was raised for the business of the society, which declared its intention to circulate information, influence prominent persons and request 'principal chiefs' to attend its meetings, and that they be addressed in Māori regarding the evils of ardent spirits. It was also proposed to correspond with the 'Parent society' – the British and Foreign Temperance Society in London. Members duly signed a declaration agreeing to 'abstain from distilled spirits, except for medicinal purposes, and to discountenance the causes and practice of intemperance'.[69]

There was not the slightest contradiction for members of the society and others who continued to drink beer or wine in moderation. Following the lead of Marsden, who had planted vines at the Bay of Islands in 1819, Busby was a skilled viticulturist trained in France, and later wrote a manual on the subject for Australasian use and established a vineyard at Waitangi. In December 1838, prior to the arrival of William Grant Broughton, the first Anglican bishop of Australia, Busby ensured that both wine and London porter were in good supply.[70] Undoubtedly spirits made up the great majority of alcohol pouring into New Zealand during the 1830s, but English bottled and bulk ale and porter, and mahogany beer or spirit engines (for use in dispensing from casks), were regularly offered for sale at the Bay of Islands by 1840.[71] Despite the relative ease of bringing spirits to the far end of the British Empire, beer was too important a cultural component of the British settlers who were about to descend on New Zealand for obstacles to stand in its way for long.

Under the various systematic colonisation schemes initiated by Edward Gibbon Wakefield, the New Zealand Company and its successors, and by various other less formal means of settlement, especially in northern New Zealand, the European population, including military, expanded from perhaps 2000 in 1840 to 15,852 in 1848, 26,707 in 1851, and 61,224 in 1858.[72] Obviously the amount of alcohol available for consumption expanded accordingly, although the drinking environment this produced was surely offset by a population that included the 'civilising influence' of a significantly larger proportion of women and children than were found in the troublesome Bay of Islands before 1840.

Yet all of them had ready access to alcohol from the moment of their departure for New Zealand. A 'leave taking' dinner for Canterbury Association settlers departing from Gravesend in September 1850 began thus:

> The Rev. Dr. Jackson, Bishop Designate of Lyttelton, the proposed capital of the new colony, said grace, and the party at the long tables commenced a vigorous onslaught on the viands so profusely supplied to them. Ale and porter were liberally furnished to all of the party who were not members of temperance societies, of whom, we understand, there were many.[73]

Cabin passengers bound for Taranaki in 1841 were supplied with a bottle of beer every day, a bottle of wine every third day, and a pint of spirits every fourth. Children were allocated 'a proportion of bottled beer'.[74] During an 1844 debate on the desirability of providing cheaper rates for cabin passengers to New Zealand, advocates of the scheme maintained that no more should be charged 'than was sufficient to provide the passengers with an abundance of good tea, sugar, coffee, butter, beef, pork fresh and salt, bread and biscuit, water and a liberal allowance of beer'.[75] In 1851, Canterbury Association settlers were informed: 'The Commander of the vessel is allowed to supply to the Chief and Fore-cabin passengers moderate quantities of port and sherry wine at 3s per bottle, and of ale and porter at 10d per bottle; but no spirituous liquors are permitted to be sold on board, except under the direction of the surgeon.'[76]

Upon arrival, those settlers inclined to drink were confronted with few restrictions. During 1840–41 New Zealand was under the liquor licensing control of New South Wales. There public houses could trade from 4 a.m. to 9 p.m. in summer, and 6 a.m. to 9 p.m. in winter, but had to close on Sunday, Christmas Day and Good Friday. But a lack of will and a lack of resources cast much doubt on whether these laws were enforced very seriously in New Zealand.[77] As Edward Jerningham Wakefield complained of Whanganui in 1841, 'That drunkenness exists there is not to be denied; or to be wondered at, since two licenses have already been issued for that settlement, and there is not even a constable resident there to guard against their abuse.'[78]

The first alcohol legislation under the fledgling New Zealand administration of Governor William Hobson was a Prohibition of Distillation Ordinance enacted in 1841, providing for a fine of not less than £100 or more than £500 ($16,000 to $80,000 in 2022) for any person operating a still to make spirits for other than medicinal purposes.[79] While this was perhaps a measure to protect Māori from the perceived evils of spirits, it was as much a fiscal measure designed to stifle any competition for imported spirits and the significant customs revenue derived from them. Indeed, in a colony without income or land tax this revenue and licence fees made up 2.6 per cent of all government income in the period 1841 to 1847.[80]

When a distinct Licensing Ordinance for New Zealand was enacted in 1842 following formal British annexation, it was perhaps remarkable for being more liberal than that in New South Wales. Justices of the peace (JPs) meeting in each district in April were empowered to grant retail licences to 'such persons as shall be approved of by the majority of such justices'. Every successful applicant had to give a bond plus two sureties of £50 each, which was forfeited if the licensee failed to keep the law. The licence fee was £40 in boroughs and £30 in other places. At the same time the fee for beerhouse keepers in Britain was 2 guineas (£2 2s), suggesting that the business of a publican in New Zealand was one confined to those of some means, or at least a good line of credit, and perhaps not to the rabble sometimes depicted. Hours of sale were 6 a.m. to 10 p.m. on working days, and 1 to 7 p.m. on Sunday, Christmas Day and Good Friday, with a provision to continue until midnight on any working day if a further £10 was paid. Publicans were only to accept payment in cash, and no wages were to be paid on licensed premises other than to the employees of the establishment. But all of these licensing provisions applied only to those conducting a day-to-day trade in a recognised public house. Any quantity of liquor larger than 2 gallons (9 litres) could be sold by anyone without the need for a licence. An amendment in 1844 allowed a police magistrate and one JP to issue licences, and also allowed women to hold licences.[81]

In essence, these were the measures that governed drinking in New Zealand until the early 1870s. The 1847 Sale of Spirits to Natives Ordinance will be discussed in more detail in the next chapter. It was followed by an 1851 Licensing Amendment Ordinance, which gave justices power to reduce the number of licences in a district if they felt it necessary, 'by refusing to grant renewals to those persons who shall have conducted their houses in the least orderly manner' – a provision that did not necessarily imply that they were disorderly. A further amendment in 1854 permitted the sale of alcohol in Parliament buildings by authority of the Speaker of either House, without the need for a licence. But by this time, following the establishment of a provincial government system in 1853, licensing laws were under the control of each provincial government except where they impacted on customs revenue or Māori.

The provinces generally followed the 1842 Ordinance, although with some variation in hours of sale and an increasing tendency to ban Sunday sales.[82]

Among the more idealistic and abstemious settlers there was hope that New Zealand would develop as a land without a need for public houses. A lengthy letter from 'An Inhabitant' of Auckland in March 1849, inspired by the extensive list of applicants for publicans' licences, insisted that these should be issued only on the basis of necessity.

> Unfortunately in Britain many are driven, by the want of comforts at home, to spend their leisure hours and hard earned pittances in the taproom of a pothouse, or parlour of a tavern, where all is carefully arranged to attract and seduce – but then this very practice is daily lowering the social and moral estate of the people. In this favoured settlement the condition of the community, except indeed of the most depraved, is happily the reverse. Almost every head of a family is possessed of his own home and domestic comforts, the fruits of industry and laudable well-directed enterprise. Single men have access to comfortable private boarding houses, and travellers from the interior when they occasionally arrive, as well as visitors from the adjacent colonies, if they prefer it, may meet with accommodation at the three principal hotels. It thus appears that there is no reasonable pretext for the number of applications put in; nor could the bench be justified in acceding to the disproportionate and immoderate demand for licenses.[83]

But a crude assessment of need takes no account of want. On a primitive colonial frontier with a high degree of transience, the pub was a preferred source of warmth, shelter, companionship and entertainment. Its provision of public space when little else was readily available ensured a role throughout the nineteenth century as the venue for public dinners and meetings of cultural, political and sporting purpose, amateur theatricals, auctions, inquests and all manner of business transactions. In many places the publican was equally a ferryman, a ploughman, a postman and an employment broker with boats, horses and carts for hire and provisions for sale.[84]

How much drinking went on in this minimally regulated society, and whether it was excessive by relevant mid-nineteenth century British standards, is hard to tell. By Christmas 1849, Auckland merchants Brown & Campbell were taking special precautions to offset the considerable quantities of alcohol available in the settlement. As one observer wrote: 'There will be a great amount of headaches next week in anticipation of which old Wright [Brown & Campbell's cellarman] has been erecting a soda-water machine . . . to supply the victims.'[85]

Almost without exception, New Zealand historians have argued that consumption, particularly of spirits, was excessive and that the fondness for them reflected a settler population in search of rapid oblivion to cope with the challenges and isolation of frontier New Zealand society. Grigg insists that, although economic prospects were better in New Zealand: 'For many people on the frontier alcohol offered an escape from squalor, loneliness, isolation and poverty.' It was a 'psychological prop' for those homesick for Britain and ill-prepared for the demands of colonial life. In this behaviour they were simply continuing a long British tradition of using the pub and the bottle to escape from the misery of their surroundings.[86] According to Miles Fairburn, drink was used to 'blot out the psychic pain of social isolation' and 'break down the barriers that separated people and prevented intimacy' in an undeveloped and chaotic or 'atomised' society.[87] In its submission to the 1974 Royal Commission on the Sale of Liquor in New Zealand, the Justice Department summed up the apparent problems for early New Zealand: 'The general hardness and discomfort of a pioneering society, the frequent isolation, the masculine and mobile society of bushmen, miners, shearers and swaggers, the absence of acceptable alternative activities and doubtless the poor quality of much of the liquor, are all likely to have played their part.'[88]

Stevan Eldred-Grigg claims that an average Pākehā male consumed 45 litres of commercial spirits and 14 litres of beer per annum during the 1840s – dropping to 24 litres of spirits but 167 litres of beer by the 1860s.[89] But it is simply not possible to draw this or any other conclusion from available evidence for the 1840s. Import statistics are not reliable until 1853, and there are no complete figures for New Zealand beer production until 1871.[90] Eldred-Grigg's figures may work for the 1850s if we assume that women, children and all Māori drank nothing and that European men drank all of the alcohol imported in any given year. But one cannot make any of these assumptions.

The problem with per-capita consumption is that it covers the entire population, including children who generally did not drink much, even allowing for such things as the shipboard beer allocation, and women who are assumed, by the nature of physiology if nothing else, to have drunk less than men. One possibility is to only count adult males as drinkers, on the basis that males who abstained from their portion would be offset by women who did not. Another possibility is to assume that women drank at half the rate of men. But this is pure speculation. Further, there is no reliable measure of the Māori population until 1858, and no meaningful alcohol statistics before 1881. Dramatic fluctuations in the amounts and proportions of different beverages imported each year – such as the 994,000 litres of beer imported in 1854 but only 526,000 in 1856, or the 1.05 million litres of spirits imported in 1859 but only 746,000 the following year – also cloud patterns of consumption. In 1849, Brown & Campbell claimed to have six years' supply of wine on hand, suggesting that there was stockpiling of supplies to ensure an even distribution over time.[91]

Amounts of alcohol of themselves mean little unless they are set in context. The only meaningful measure of whether drinking by the New Zealand settler population after 1840 was abnormal is by comparison with Britain, from where most of the settler population and their cultural attitudes to drink originated. The available evidence for New Zealand is sparse but revealing enough. The best measure is an estimate of pure alcohol intake across different beverages. Beer is measured at 5 per cent and 7 per cent alcohol by volume, to allow for both diversity in the types of beer available – ranging from weak harvest ale to XXXX ale and strong imported porter and ale – and uncertainty as to the proportions of each strength that were brewed. If figures of 40 per cent alcohol by volume are assigned to spirits, of which a few were stronger and others weaker, and 15 per cent to wine, it is possible to gain some sense of the impact of what was being drunk.[92] Given the prevalence of beer drinking in England and Wales but spirit drinking in Ireland and Scotland, it is worth comparing both sets of consumption statistics.

What is immediately apparent from Table 1.1 is that New Zealand consumed substantially larger quantities of spirits and wine than Britain, but substantially less beer. This pattern, which would continue until the 1870s, is hardly surprising given some of the difficulties of beer compared to spirits as an import and internal transport commodity, and the time required to establish a substantial local brewing industry. These issues will be discussed in the next chapter. Even without figures for New Zealand beer production before 1860 to augment those for imports, and the issue of home brewing and illicit distilling to be considered in subsequent chapters, the levels of pure alcohol consumption revealed here hardly present New Zealand as excessive compared to Britain during the 1840s and 1850s. No doubt New Zealand drank more,

Table 1.1: New Zealand and United Kingdom Alcohol Consumption to 1864 (litres)

Amount consumed per person annually									
	New Zealand			England and Wales			United Kingdom		
	Beer	Spirits	Wine	Beer	Spirits	Wine	Beer	Spirits	Wine
1840–44	–	–	–	138.6	3.2	–	88.6	5.0	1.0
1845–49	–	–	–	132.7	3.3	–	88.2	5.9	1.0
1850–54	–	–	–	134.1	3.4	–	95.0	5.7	1.1
1855–59	18.1	14.1	5.2	133.2	3.7	–	100.0	5.4	1.0
1858	20.2	14.9	2.7	–	–	–	102.3	4.5	1.0
1860–64	25.0	16.3	5.6	143.6	3.5	–	112.3	4.4	1.5
1861	14.5	12.4	3.8	–	–	–	108.6	3.8	1.7
1864	28.4	16.3	6.5	–	–	–	119.1	4.1	1.8

Pure alcohol*				
	With beer at 5% alcohol by volume		With beer at 7% alcohol by volume	
	New Zealand	United Kingdom	New Zealand	United Kingdom
1840–44	–	6.82	–	8.33
1845–49	–	6.58	–	8.35
1850–54	–	7.21	–	9.13
1855–59	7.35	7.27	7.68	9.27
1858	7.41	7.01	7.76	9.06
1860–64	8.59	7.57	9.09	9.86
1861	6.29	7.17	6.58	9.34
1864	8.86	7.81	9.42	10.20

* Wine consumption figures for England and Wales are not available. All New Zealand figures to 1864 are for imports only.

Source: *Statistics of New Zealand*, 1853–1900; New Zealand Census, various 1867–1901; T. R. Gourvish and R. G. Wilson, *The British Brewing Industry, 1830–1980*, Cambridge: Cambridge University Press, 1994, 30; G. B. Wilson, *Alcohol and the Nation: A Contribution to the Study of the Liquor Problem in the United Kingdom from 1800 to 1935*, London: Nicholson and Watson, 1940, 330–32; B. R. Mitchell, *Abstract of British Historical Statistics*, Cambridge: Cambridge University Press, 1962, 8–10, 408–10.

but not astronomically so. Moreover, returning to these comparative calculations in subsequent chapters will reveal a sharp decline from the late 1860s in New Zealand consumption relative to Britain and other places – a point that became particularly important to the defence of the liquor trade as the temperance and later prohibition movements gained momentum.

An equally difficult question to answer is how much of the drinking was problem drinking. Here at least there are some statistics and any number of lurid contemporary accounts, but they are not always what they seem. Under the 1842 Ordinance the penalty for drunkenness – merely the offence of being drunk rather than any requirement to be disorderly – was a fine of not less than 5 shillings or more than 20 shillings or a maximum of forty-eight hours' imprisonment in default. Any person convicted three times within the space of six months was liable for the fine and imprisonment for seven days. There was no suggestion of any penalty for the drink seller.[93] By 1854 total convictions for drunkenness were 27 per 1000 Europeans. They peaked at 41 in 1855 and fell gradually to 16 in 1861, before leaping again to 28 or 5159 convictions in 1864 as the gold rushes were in full flight – although these figures are also deceptive in that they include repeat offenders rather than the actual number of individuals who were convicted at least once. For England and Wales during the same year the rate was little more than three.[94] The question that can never be answered is how many of these drunks were waylaid by beer rather than stronger temptations.

Nevertheless, the issue was clearly one of enforcement. In March 1841 a public meeting was held at Barrett's Hotel in Wellington to consider 'the steps necessary to protect the public from the outrages of the police establishment'. The chairman, G. B. Earp, declared:

> He . . . did not stand there as the apologist of the drunkard; but this he did contend for – that the working man, after the toils of the day were passed had, if he chose, as much right to expend a portion of his earnings in a glass of spirits or ale as his wealthier or more fortunate neighbour. It was the duty of the police, if a man was proceeding quietly to his home, if he required assistance, to assist him; and not, as was the case here, if found out an hour later than usual, manacle and drag him to the lock-up-house and to maltreat him in the most brutal manner.[95]

Resolutions of the meeting were referred to the Wellington justices to ensure that the matter was raised with Governor Gipps in Sydney, although the outcome was never reported. Seven years later, in a move that has attracted a number of historians determined to portray a drunken New Zealand, it was announced that one in eight Aucklanders were convicted for drunkenness during 1847 – perhaps not unexpected with a high proportion of soldiers in a garrison town. While the editor of Auckland's *New Zealander* congratulated the settlers for 'exhibiting the most orderly character of any maritime place of equal population in her Majesty's dominions',[96] the *Nelson Examiner* scoffed at Auckland attempts to assert moral superiority by pointing out that Wellington, 'quite as important a maritime place as Auckland', and of similar population, had a conviction rate of perhaps one in forty while Nelson had one in eighty.[97] However one approaches it, Auckland was perceived as an exceptional case by contemporary standards, and there are problems in trying to draw conclusions from rates of conviction.

An argument that drinkers were seduced by an extraordinary number of drinking places in the new settlements is also debatable. In 1842 Auckland had fourteen hotels for a European population of 2895, or one for every 207 people. In 1851 there were just under thirty hotels for a European population of 9000, or about one for every 300, although one new arrival to Auckland observed that 'on the average there is one grog shop for every three of all the other trades put together'.[98] But for England and Wales, publican's licences per capita shifted only slightly from one for every 274 people in 1831 to one for 308 in 1861. If beerhouse licences are added, the figures shift from one for 168 to one for 186 people over the same three decades.[99] In sum, anyone arriving from Britain during the mid-nineteenth century should not have found anything unusual in the ratio of pubs to people. Indeed, aside from the rustic setting and the tendency to spirits rather than beer, the overall attitude to drink and drinking places was probably more familiar than unfamiliar. Surely most of those who went to the colonial pub in search of companionship, not least as a substitute for far-distant family, did so in a positive effort to expand their social horizons rather than in an air of desperation. Those who insisted otherwise were frequently idealists who imagined New Zealand as a land for new beginnings and opportunities rather than one that inherited and imitated the traditions of the old world in which alcohol consumption generally, and beer in particular, was essential and expected.

BROWN, CAMPBELL AND CO.'S DOM

BREWERY, AUCKLAND.

The Domain Brewery, Auckland, seen here around 1890, was originally established by prominent merchants William Brown and John Logan Campbell in 1850 as the Hobson's Bridge Brewery. Auckland Libraries Heritage Collections, 7_A5067

TWO

'It is a Brewer's Paradise': Beer and Settlement, 1840–1860

While anecdote abounds, the evidence for a popular notion that most of New Zealand was drunk for most of the 1840s and for many years beyond is sketchy. Rather, as David Hughes observed of early Australia, if alcohol was consumed with the psychological desperation suggested by some, then why did the settlers expend so much effort to establish a brewing industry producing a weaker beverage than easily obtainable spirits?[1] Attention must turn to the perspective of the brewer and the drinker to understand how and why beer became such a significant component of colonial life, an important feature of daily rituals and public occasions, and an indicator of prosperity and stability.

Prior to the establishment of colonial production in New Zealand, the beer drinker who could not produce their own was reliant on imports. Because statistics were not kept, there is no way of knowing the amount of beer imported into New Zealand before 1853 or its proportion in relation to spirits and wine. While much British wealth during the nineteenth century was generated through the export performance of key industries such as textiles, engineering and shipbuilding, brewing was not to the forefront. Britain exported no more than 3 per cent of its total output during the nineteenth century. Exporting was the most difficult component of the brewing trade, due to it being a bulky product and issues of durability and quality during many months at sea. If the beer was too old when exported, it was frequently undrinkable upon arrival. If it was too young, then the continuing fermentation could cause bottles and casks to explode during a rough voyage. There was also little incentive to overcome difficulties in determining price, obtaining payment from distant creditors and dealing with high shipping and insurance charges when it was all too easy to sell in local British markets that were ever expanding with the decline of domestic, home-based or publican brewing from the late eighteenth century.[2]

Nevertheless, there was a growing demand from overseas as the British Empire expanded and colonists wanted the familiar things of home. Beer was exported to North America, the East and West Indies and the Mediterranean during the eighteenth century. Although by the 1850s there were still substantial exports of porter from the likes of Whitbread in London, there was a gradual shift in taste to pale ales during the first half of the nineteenth century – especially to the pale ales produced in Burton-on-Trent.[3]

Martyn Cornell explains that 'Greed and luck were the parents of India Pale Ale' (IPA).[4] It emerged from the desire of commanders on East India Company ships to make a fortune supplying home comforts to Company employees in India during the late eighteenth century. Although the company held a monopoly, it had always allowed the commanders to conduct their own private trade, and purely by luck one of the beers they shipped was found to improve dramatically during the journey. George Hodgson's Bow Brewery, although one of the smaller operations in London after its establishment in 1752, was conveniently placed near the Company docks at Blackwell and its beer was therefore easy to get to ships. Hodgson was also willing to extend long credit to the ship commanders. As well as porter, which kept well because of its relatively high alcohol content, Hodgson sent out an October beer – a strong, well-hopped, autumn-brewed beer that was frequently kept for a year before bottling and drunk a year after that. However, it was soon found that the four- to six-month voyage to India through rough seas and slow temperature changes with regular rocking of the beer in its oak casks sped up maturation. The beer was ready to drink on arrival with the taste of a beer six times its age, and became the almost universal beer in India by the end of the eighteenth century. There is no evidence that Hodgson knew this would happen or specifically experimented with a 'new' beer.[5]

However, in the early 1820s a new generation of Hodgsons decided to cut out the middlemen and make their own shipping arrangements with India. They did their own retailing, raised the price by 20 per cent and refused to supply on any terms but cash. Inevitably, this caused considerable antagonism to the East India Company and the Madras and Calcutta merchants who were cut out. In 1822, in their determination to find other suppliers, they approached the Burton-on-Trent brewer Samuel Allsopp, who had just lost his lucrative Russian market due to an increased tariff. Enticed by the prospect of a 7500-barrel-per-annum Indian market, Allsopp attempted to replicate Hodgson's beer. Although he did not fully realise it at the time, the 'hard' water of Burton, with more calcium sulphate than London water, and better for conversion of starch into sugar when using pale malt, produced a much better pale beer than Hodgson. Allsopp's sent their first IPA to India in 1823 and were quickly followed by other Burton brewers, especially Bass. Hodgson's Bow Brewery soon faded from the market.[6]

With India secure, Allsopp and others quickly developed a much wider export network. There is ample evidence in the regularly published cargo lists from recently arrived ships that a variety of breweries, if not of beer styles, were represented in the infant New Zealand market. When it opened in 1840, Barrett's Hotel in Wellington offered 'wines, spirits, ale and porter of the best quality', while J. Howes, licensee of the rival Queen's Head Tavern, announced 'that he has always on hand of the very best quality every description of beer, ale, wine and spirits at moderate charges'.[7] Quite probably both of these establishments purchased from the cargo of the *Lady Nugent* when it arrived the following April. On offer were 1000 dozen bottles of Taylor's India Pale Ale as well as their stout and pale ale in wood, Hebblewhite's stout and ale in bottles and Tennent's bottled ale.[8] Other choices for the New Zealand consumer during the 1840s included Bass's bottled pale ale, No.3 Burton ale and porter, various Burton ales in hogsheads (54 gallons or 245 litres), Dunbar's Ale, Abbott's Ale, an unnamed 'Best London Double Brown Stout', Trueman's London Stout, Whitbread Stout, and 'Gunis's [*sic*] Dublin Stout'.[9] Generally these were the products of the larger firms that had the resources and contacts to tackle the difficult task of export to one of the remotest corners of the British Empire. Hardly surprisingly they were also the stronger and/or heavily hopped styles that travelled well.

As much of the early beer advertising simply offered 'ale', 'porter' or 'stout', there can be no certainty about preferred brands, but it is clear that ale was becoming a dominant style. Auckland merchant William Brown informed his London agent in 1849 that Bass Ale was the only variety worth sending out to New Zealand as its quality enabled it to survive in a crowded Auckland market, but the continued appearances of its competitors makes this claim doubtful.[10] In imports to New Zealand at this time there is no sign of the sweet mild ale that had become a favourite of the London working class from the 1830s, or of any other lighter styles of beer. Nor, other than regular supplies of Tooth's XXX ale from Sydney to Auckland during the late 1840s, do there appear to have been any consistent Australian imports.

Allsopp's soon developed an Australasian export network and their pale ale first reached New Zealand in the middle of 1848. Six years later its publicity also arrived, with a strong emphasis on the health-giving qualities of the product.

> Professor Liebig's opinion of this unrivalled beverage for all climates, appropriately described by the Sanitary Commissioner of the Lancet as the 'wine of Malt' is thus adduced in a letter to Mr Henry Allsopp: 'The specimens of your Pale Ale sent to me afforded me another opportunity of confirming its valuable qualities. I am myself an admirer of this beverage, and my own experience enables me to recommend it, in accordance with the opinion of the most eminent English Physicians, as a very agreeable and efficient tonic, and as a general beverage both for the invalid and the robust.'[11]

In addition to IPA, there is evidence of 'Australian ale' being brewed for export to both Australia and New Zealand. The name seems to refer to 'No.3' grade ale from Burton-on-Trent, a beer stronger, probably slightly darker and rather sweeter than a conventional IPA. As early as 1841 the Burton brewers William and Thomas Saunders advertised 'East India and Australian Beer . . . brewed by them expressly for those markets, also the Australian Strong Ale'.[12] Bass were certainly exporting to New Zealand by 1847, and by 1860 advertising for Allsopp's in New Zealand made specific reference to a special brewing for the 'Australian' colonies – orders for which could be placed with a variety of local agents.[13] From August 1859 'Marrian's No 3 Australian Ale', from the Sheffield brewer Thomas Marrian, was regularly available in Auckland and was significant as perhaps the first Burton-style ale to be brewed outside Burton-on-Trent. The merchants who imported this beer, O. R. Strickland & Co., also advertised 'Worthington & Robinson's Celebrated Burton Ale', Bass's Ale and Barclay, Perkins Stout. Ind Coope and Salt & Co. were other Burton options available in New Zealand by the early 1860s.[14] Indeed, so apparently ubiquitous were these strong ale imports in some localities that one Wellington correspondent, 'Freetrader', complained in May 1858 that it was difficult to secure affordable weaker beer such as dinner ale for household use. 'I find a difficulty in obtaining Colonial Ale as the Publicans appear from some cause or other reluctant to serve you with it and press upon you English Ale which, in my opinion, is only the precursor to spirits drinking.'[15]

Variety notwithstanding, imported beer was relatively expensive and one can perhaps understand the greater consumption of spirits. A dozen bottles of ale, perhaps 9 litres at most and likely less depending on the size of the bottles, cost 21 shillings in Wellington in April 1840, before falling gradually to 12 or 13 shillings, where it generally remained until at least 1845. By contrast, a gallon (4.5 litres) of rum never cost more than 10 shillings 9 pence and fell as low as 4 shillings for most of the period 1842–45; a gallon of brandy, aside from a few spikes to 15 shillings, was generally 6 or 7 shillings; and gin was generally 7 to 10 shillings a bottle, although as low as 2 shillings in 1842. Only wine, which reached 32 shillings per dozen bottles during 1842–43, was consistently more expensive than imported beer.[16] Prices from the 1850s also reveal significant variations in different parts of New Zealand. In 1855 a hogshead of imported beer cost £10 in Auckland, £9 in Wellington and Canterbury, but only £5 in New Plymouth – although it is not certain whether the latter was 'imported' from Britain or Nelson. Auckland and Canterbury were generally the most expensive places to buy beer, New Plymouth consistently the cheapest. However, despite the expense and periodic inconsistency of supply, there was clearly a market in which the consumer was willing to pay for what they preferred rather than making do with what they could get.

The sharp variations in quantities of alcohol imported (Table 2.1) also suggest some challenges for the dedicated beer drinker. While the actual liquid quantity of beer imported kept pace with and eventually exceeded that of spirits, by far the greater portion of the imported pure alcohol volume was spirits. Further, whereas spirits and wine could be stored and distributed at a fairly even rate of supply to meet demand, the comparatively limited keeping qualities of imported beer, remembering that it may already have spent six months or more maturing on its journey to New Zealand, made stockpiling less likely. In short, the beer drinker desirous of a familiar pint or two from home almost certainly encountered a more volatile market in terms of price and availability than the confirmed spirit drinker.

One way to bypass obstacles of price and supply was to brew your own. Yet the skill of domestic brewing, already noted as being in decline during the eighteenth century, was in wholesale retreat by the time settlers embarked for New Zealand during the 1840s. The collapsing economic position of the rural working class and the departure of large numbers for the expanding industrial cities ensured that the cost of brewing equipment, materials and fuel were increasingly beyond the means of many, and that the chain of brewing expertise handed from generation to generation was broken. It is estimated that private brewing constituted two-thirds of all beer in England in 1700, half in 1800, one-fifth in 1830, one-eighth by 1851–55, and one-fortieth by 1866–70. These developments paralleled a rapid concentration in the commercial brewing industry, where the large firms of London and Burton in particular gradually squeezed out small firms and publican brewers to such an extent that by 1914 the ten largest brewers produced a quarter of all beer in England.[17]

Table 2.1: Alcohol Imports to New Zealand, 1853–1860 (litres)

	Beer	%	Spirits	%	Wine	%	Total
1853	466,619	38.4	586,202	48.2	163,715	13.4	1,215,536
1854	994,574	50.2	787,462	39.8	197,573	10.0	1,979,609
1855	749,685	43.9	549,320	32.2	407,694	23.9	1,706,699
1856	526,408	39.7	588,129	44.4	210,207	15.9	1,324,744
1857	1,106,387	51.2	738,084	34.2	314,155	14.6	2,158,626
1858	1,238,812	53.4	912,795	39.4	166,651	7.2	2,318,258
1859	1,318,796	49.0	1,052,371	39.1	320,015	11.9	2,691,182
1860	1,419,597	57.5	746,821	30.2	303,595	12.3	2,470,013

Source: *Statistics of New Zealand*, 1853–1860.

No doubt some colonists made it their business to be equipped with all of the skills for life in the new pre-industrial New Zealand society, but one should not overestimate the number who had some skill in malting their own barley or maintaining a reliable yeast for fermentation. Indeed, the evidence for their activities is scarce. In July 1848 a Nelson newspaper sought to encourage the home brewer by publishing 'Directions for the management of barley intended for malting', supplied by a brewery in a 'neighbouring colony'.[18] Three years later Thomas Smith of the Bridge Street Brewery in Nelson seemed to sabotage his own business prospects by offering malt and hops for anyone interested in brewing their own, while in September 1852 Charles Homeyer of New Plymouth repeatedly advertised two kinds of 'home-brewed farmhouse ale' – a mild dinner ale and a good pale bitter ale, described as 'a good stomachic', in 2-gallon kegs at 1 shilling 6 pence per keg. Alternatively, farm produce or firewood would be taken in exchange.[19] In October 1853 Vicesimus Lush, vicar of Howick, Auckland, recorded a near calamity with his family beer supply.

> Blanche busy brewing: when the large tub was almost full of warm liquor, Allie [his son] must needs go peeping over and suddenly lost his balance and went plump, head foremost, into the beer. Fortunately . . . he was out again in an instant but got a good ducking and a terrible fright. How thankful we should be that the beer was not scalding hot.[20]

Some others who fell into their beer during the next decade were not so lucky.

The case of *Head v. Brown*, heard in the Christchurch Resident Magistrates' Court in November 1857, also offers an interesting glimpse of home production – and consumption. During a long dispute over the terms of an employment agreement as farm servants and whether or not the plaintiff and his wife had done the work required of them, the defendant noted that there was constant friction over the amount of beer available. Although the property was brewing 180 litres a fortnight and all of it was being drunk, the plaintiff repeatedly complained about a shortage of beer. His wife apparently avoided this problem, in the opinion of the defendant at least, by retiring to bed every afternoon with a bottle of gin. Eventually the resident magistrate adjourned the case and announced that it was going to the Supreme Court. But there is no sign that it went any further.[21]

If the import supply was problematic and home brewing limited, conditions were therefore ideal for the establishment of local breweries. Following Polack's brief endeavour at Kororāreka, it is not entirely certain when and where New Zealand's second commercial brewery was established. In later years a number of prominent firms were inclined to embellish their history to ensure they were present at or near the beginning. Typical is John Logan Campbell, who claimed to have opened a brewery in Auckland early in 1842, where evidence clearly shows that his first Hobson's Bridge Brewery did not begin operations until December 1850.[22] The most likely successor to Polack was William Smithson in Auckland, who advertised his Norfolk Brewery from January 1842:

> Unadulterated Beer, from English Malt. W. Smithson begs to inform the inhabitants of Auckland that he has commenced brewing Ale and Table Beer from Imported English Malt, at his Brewery, adjoining the New Court House, and can supply the former at 2s. 6d. per gallon, and the latter at 1s. 6d. per gallon, Cash. Also, Yeast at 3s. per gallon, suited for Bakers and private Families.[23]

Next after Smithson was probably Wellington's Te Aro Brewery very late in 1842, followed in quick succession by the rival Wellington Brewery, two in Nelson and probably one each in

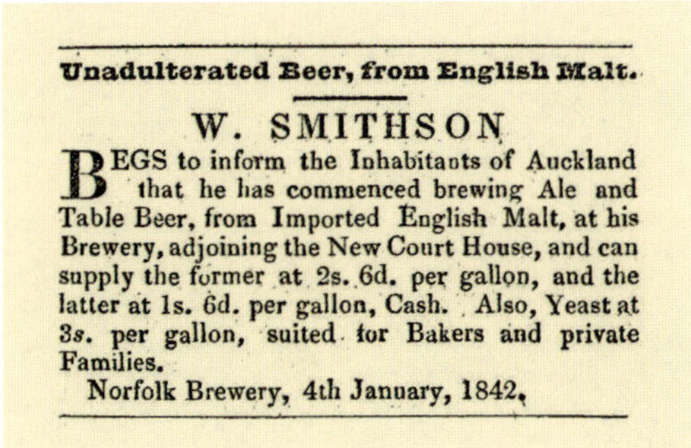

Unadulterated Beer, from English Malt.

W. SMITHSON

BEGS to inform the Inhabitants of Auckland that he has commenced brewing Ale and Table Beer, from Imported English Malt, at his Brewery, adjoining the New Court House, and can supply the former at 2s. 6d. per gallon, and the latter at 1s. 6d. per gallon, Cash. Also, Yeast at 3*s*. per gallon, suited for Bakers and private Families.

Norfolk Brewery, 4th January, 1842.

This 1842 advertisement for W. Smithson's Norfolk Brewery is almost certainly the first advertisement for a local brewery to be published in a New Zealand newspaper. *New Zealand Herald & Auckland Gazette*, 15 January 1842, 1

Auckland and New Plymouth. Wellington could have had its first brewery even earlier. In March 1841 a brewery plant was advertised for sale by one Frank Losack, who was leaving the settlement due to 'domestic calamity', his wife having died of dysentery the previous month leaving five young children, and 'thereby resigning his sanguine expectations for success as a brewer here'. He was happy to lend expert advice to any purchaser as he had been for many years connected with the 'famed Windsor Brewery' in Berkshire.[24]

The following year it was reported that 'enterprising gentlemen' were establishing a brewery in Wellington on the basis that if it was worthwhile to import barrels of beer, it was more worthwhile to import a few pounds of ingredients to add to the local water.[25] But there is no sign of anything tangible until January 1843 when the Te Aro Brewery appeared in a subscription list for the Wellington anniversary fête. Two months later the apparently much larger Wellington Brewery began advertising 'ale and table ale in draught or bottle in any quantities from 2 gallons and upwards'.[26] In early April the *New Zealand Colonist* was most impressed with the premises.

> Since our last publication, we have been gratified by a visit to the Wellington Brewery, erected by Messrs Drake and Northwood, a building of no mean pretensions, both as to the size and the arrangements, and we cannot help congratulating our fellow colonists on this great addition to our self-production and comforts. The outlay must have been very considerable, but we hope the proprietors will reap the benefit of their spirited undertaking, which we have no doubt they will. Mr Masters has also a small brewery, and another one is about to be erected on Te Aro Flat.[27]

Four days later a rival publication added that 'Real Port Nicholson [Wellington] ale and beer can now be purchased, and we think the colonists will encourage that which is produced in their own place.' It was also suggested that there were good prospects for the increased cultivation of barley around New Plymouth as a good quantity of malt would be needed for the Wellington Brewery.[28] Whereas Smithson in Auckland had started with imported malt, it seems that the transition to local malting was swift as growing conditions were good and the supply would be cheaper and more reliable than importing. The Te Aro Brewery seems to have produced little more than its advertising, but the Wellington Brewery was a flourishing concern throughout the 1840s. In January 1847 it advertised for 1000 to 1500 bushels of Chevalier barley for malting, and in April simply for 'good malting barley' without specifying a variety. In July 1849, a new owner announced that ale would be 'brewed under the experienced management of Frederick Abbott, Esq., late Manager of Abbott's Bow Brewery, London, (famous for India Ale), and warranted pure malt and hops'.[29] But there is no sign of the brewery continuing into the 1850s.

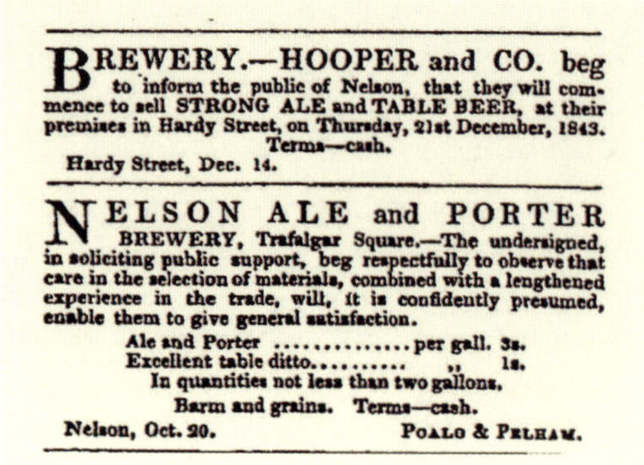

BREWERY.—HOOPER and CO. beg to inform the public of Nelson, that they will commence to sell STRONG ALE and TABLE BEER, at their premises in Hardy Street, on Thursday, 21st December, 1843. Terms—cash.

Hardy Street, Dec. 14.

NELSON ALE and PORTER BREWERY, Trafalgar Square.—The undersigned, in soliciting public support, beg respectfully to observe that care in the selection of materials, combined with a lengthened experience in the trade, will, it is confidently presumed, enable them to give general satisfaction.

Ale and Porter .... per gall. 3s.
Excellent table ditto.......... „ 1s.
In quantities not less than two gallons.
Barm and grains. Terms—cash.

Nelson, Oct. 20. POALO & PELHAM.

Advertisements for the two breweries operating in Nelson in late 1843. While Poalo and Pelham's Nelson Brewery closed before this advertisement was published, or very soon after, Hooper marked the beginning of a six-generation Nelson brewing dynasty. *Nelson Examiner and New Zealand Chronicle*, 6 January 1844, 1

Meanwhile a more enduring brewing tradition had taken root in Nelson. In October 1843 Poalo Portello and Charles Pelham, trading as the Nelson Brewery, rather diffidently announced that 'The undersigned, in soliciting public support, beg respectfully to observe that care in the selection of materials, combined with a lengthened experience in the trade will, it is confidently presumed, enable them to give general satisfaction.' They offered ale and porter at 3 shillings a gallon and table ale and table porter at 1 shilling.[30] But the partnership was dissolved in October 1844 – perhaps in the face of a more concerted and talented effort from George Hooper.[31]

A baker by trade, Hooper arrived in Nelson from Gloucestershire in October 1842. With capital provided by Thomas Renwick, an Edinburgh-trained doctor who acted as surgeon on the same ship, Hooper & Co. began to advertise strong ale and table beer from December 1843. As well as a brewery in Nelson they already had a maltings in Waimea.[32] By February 1844 there was repeated praise for the 'spirit and activity with which business is conducted at this brewery. It is an evidence that the public are not slow to patronise the "Nelson brewen" in preference to imported malt liquors.'[33] Although they were briefly obliged to import barley from Sydney, suggesting a limited local supply, the *Nelson Examiner* declared in June 1844 that 'The superiority of Hooper's over any colonial ale brought into this settlement is admitted by all who have had an opportunity of judging. This has induced a respectable Wellington publican to send here for a supply which was shipped a few days since.'[34] Here the reference to 'colonial' is presumably to beer brewed in the colony rather than for export to it. The following year Hooper & Co. chartered a cutter to take their beer to Auckland, where 'The superior quality of this beverage will be sure to command a ready sale'.[35] By the end of the decade their beer was regularly being dispatched to Wellington, New Plymouth, Auckland and the Bay of Islands.[36]

Early in 1846 Hooper's brewery underwent considerable expansion, as much to keep pace with demand as to ensure a good stock of ale could be kept in order for it to age 'which it so frequently stands in need of'.[37] This expansion may have been behind Renwick's efforts to bring out a second brewer from Scotland. But this proved a considerable challenge to his business associate there, who informed him in October 1846 that brewers were in short supply and subsequently received very high remuneration. A 'gentleman' who could manage the entire brewing process could earn anything from £150 to £1400 per annum, while a 'working man' who understood a little about the brewing process could expect £80. Renwick's correspondent wanted to know whether he and his partner were 'a couple of thickheaded fellows who have been learning nothing from your present brewer', in which case they would need to go for 'the scientific man – the master', but 'if the quality of the brew is not necessarily required to be first rate' then they could go for the working man.[38] It appears that no brewer was found until one arrived from London in the mid-1850s, but the quality of the beer evidently remained first-rate. During 1845 the brewery produced 624 hogsheads, and in 1846 it malted 5000 bushels of barley and produced 800 hogsheads. Several Nelson farmers had entered agreements to sell their whole barley crop to the brewery at a stipulated price.[39] By the early 1860s, now trading as Hooper & Dodson, they informed the public that their porter, available in draft or in bottle, was as good as any imported product. To confirm the point they presented the opinions of several 'leading medical men'. S. A. Cusack declared that 'I consider it to be most wholesome and can recommend it for the use of invalids requiring a stimulus, combined with nourishment and a tonic'. F. W. Levine added that 'Having used messrs Hooper & Co's bottled porter for several years, I have pleasure in recommending it as an agreeable, wholesome, and unadulterated article.'[40] On these sound principles Hooper & Co. and its subsequent business manifestations were to remain part of the Nelson market until taken over by Dominion Breweries in 1969.

It seems that the only other commercial brewer operating during the early 1840s was Richard Secombe, who apparently established his first brewery in New Plymouth before the end of 1843. But as Table 2.2 reveals, there was a minor rush during the second half of the decade and a more dramatic one from 1857 to 1860.

While eleven of the fifteen breweries established up to the end of 1850 were in the North Island, reflecting the distribution of population, twenty-four of the thirty-two established during the next decade were in the South Island. The vast majority were in the six main centres of European settlement prior to 1860 – Auckland, New Plymouth, Wellington, Nelson, Christchurch and Dunedin. As testimony to the difficulties of land transport and communication, only one establishment, Henry Dodson's Wairau Brewery in Blenheim, was at any significant distance from a port. Nelson, with its advantages of climate for the growth of barley and hops, had the most breweries, but the timing and distribution elsewhere was not even. Whereas Auckland, Wellington and Nelson had breweries within two years of settlement and New Plymouth and Christchurch within three, Dunedin had to wait eight years.

Most breweries were small operations managed by the brewer and perhaps one or two employees. It is impossible to determine how many brewers had practical or commercial brewing experience prior to arriving in New Zealand. A few breweries, such as the Hobson's Bridge Brewery in Auckland and the Packers' Albion Brewery in Christchurch, were business ventures in which the owners did not manage day-to-day brewing. Several, such as Wood & Rogers in Auckland or Travers & Oldham in Christchurch, closed or were taken over by their competitors within two years of opening. The majority lasted less than a decade – a reflection of the high rates of transience during the colonial period or the multitude of challenges to brewers that will be explained in more detail in the next chapter. Where two or more breweries existed in the same market for any length of time, as was generally the case in Auckland and Nelson, and in Christchurch from the late 1850s, survival was probably due to the limited capacity of any one brewery to supply a thirsty public.

But a number of these breweries did endure. Indeed, antecedents of seven of the ten establishments that merged to form New Zealand Breweries in 1923 can be seen by the end of 1860, followed in the next decades by James Speight & Co.'s City Brewery (1876) and George McGavin & Co.'s Union Brewery (1882), both in Dunedin, and David Barry's Gisborne Brewery (1897). The other seven – Lion and Captain Cook in Auckland; Staples in Motueka and later

Table 2.2: Breweries Established in New Zealand, 1835–1860

Year	Brewery	Location	Proprietor
1835	Polack's Brewery	Kororāreka	Joel Samuel Polack
1842	Norfolk Brewery	Auckland	William Smithson
	Te Aro Brewery	Wellington	Joseph Masters
1843	Hooper & Co.	Nelson	George Hooper and Thomas Renwick
	Nelson Brewery	Nelson	Poalo Portello and Charles Pelham
	Taranaki Brewery	New Plymouth	Richard Secombe
	Wellington Brewery	Wellington	Thomas Drake and Thomas Norwood
1846	Auckland Brewery	Auckland	S. A. Wood and S. H. Rogers
	Bridge Street Brewery	Nelson	Thomas Smith
1847	Albert Brewery	Auckland	George Partington
	Unknown	Wellington	William May
1848	New Plymouth	New Plymouth	William George
	Waimea Brewery	Waimea	Henry Redwood
1850	Hobson's Bridge Brewery	Auckland	William Brown and John Logan Campbell
	Kai Warra Brewery	Wellington	Charles Cameron
1851	Smithson & Scholes	Auckland	William Smithson and John Scholes
	Victoria Brewery	Wellington	James Gerard
1853	Avon Brewery	Christchurch	Taylor, Crawford and Jaggar
	Harley's Brewery	Nelson	Charles Harley
	Stoke Brewery	Stoke	Tom Tidd
1854	Albion Brewery	New Plymouth	Charles Homeyer
1855	Phoenix Brewery	Christchurch	Richard Taylor
	Raglan Brewery	Nelson	John Smith
1856	Dunedin Brewery	Dunedin	H. J. Mackinnon and Frederick Wickens
	Motueka Brewery	Motueka	John Staples
1857	Collingwood Street Brewery	Nelson	T. Field
	Crown Brewery	Christchurch	William May
	Sawyers Bay Brewery	Dunedin	William Strachan
	Wanganui Brewery	Whanganui	Foreman and Sole
1858	Kaiapoi Brewery	Kaiapoi	Henry Jaggar
	Suffolk Brewery	Christchurch	William Manning
	Wairau Brewery	Blenheim	Henry Dodson
	Hancock & Scholes Brewery	Auckland	Thomas Hancock and John Scholes
1859	Coad's Brewery	New Plymouth	E. Coad
	Great Northern Brewery	Auckland	Richard Secombe
	Otago Brewery	Dunedin	G. Burke and H. J. Mackinnon
	Soho Brewery	Nelson	John Smith
	Victoria Brewery	Christchurch	John May
1860	Albion Brewery	Christchurch	Henry and Richard Packer
	Anderson's Brewery	Dunedin	George Anderson
	Banks Brewery	Nelson	Archibald Banks
	Christchurch Brewery	Christchurch	Samuel Manning
	Reliance Brewery	Christchurch	Travers and Oldham
	Dunedin Brewery	Dunedin	Anderson and Ireland
	St George's Bay Brewery	Auckland	Jupp and Co.
	Ward's Brewery	Christchurch	Archer C. Croft and H. C. Ward
	Windsor Brewery	Nelson	Mark Blythe

Wellington; Crown, Manning & Co. and Ward's in Christchurch; and Strachan's in Dunedin – are worth considering briefly given their importance to much that follows.

In Auckland John Logan Campbell began to exercise his significant long-term influence over the New Zealand liquor industry from 1840. After medical training in Edinburgh, he departed for Australia to pursue sheep farming, but abandoned that plan after witnessing the effect of a New South Wales drought. He reached New Zealand in March 1840, and shortly afterwards, in partnership with William Brown, a Scottish lawyer, established Auckland's first merchant firm, Brown & Campbell. Among other business as general merchants, wine and spirit merchants, shipping and commission agents, land speculators and auctioneers, they established the Hobson's Bridge Brewery in 1850, later known as the Domain Brewery. Over time Campbell became a central and wealthy player in the Auckland financial elite, buying Brown out of the partnership in the 1870s.[41] In 1880, after leasing his brewery to several others and briefly mothballing it, Campbell returned to his brewing interests with renewed vigour. In 1897 his firm amalgamated with that of the German-born Louis Ehrenfried. In turn Campbell & Ehrenfried amalgamated with Secombe's Great Northern Brewery in 1914 to form the Lion Brewery.

Richard Secombe had been brewing in New Plymouth since 1843, before shifting to Auckland in 1856 and leasing the Albert Brewery from George Partington. In December his first brew attracted immediate praise. 'Although quite a new article, it possesses a flavour and body that indicate what would, with age, be a near approximation to many imported samples.'[42] In 1859 he acquired a 6-acre section in Khyber Pass Road and began constructing the Great Northern Brewery. Meanwhile, early in 1858 Thomas Hancock had entered into partnership with John Scholes, who had previously been in partnership with the pioneering Auckland brewer William Smithson. In 1859 Hancock became licensee of the Captain Cook Hotel in Khyber Pass Road and shifted the brewing operation there. About 1862 the partnership with Scholes was dissolved, and in 1868 Hancock entered a partnership with his son-in-law Samuel Jagger under the name of the Captain Cook Brewery.[43]

The Christchurch arm of the New Zealand Breweries merger began with the establishment of William May's small Crown Brewery in 1857, followed by the partnership of Archer C. Croft and Hamilton Ward in June 1860. While May brewed on his own account, it seems that Croft, a doctor, and Ward, from a wealthy County Down family, were businessmen rather than practical brewers.[44] Indeed, Ward's determination to prosper from his New Zealand career did not entirely impress Charlotte Godley, wife of the leader of the Canterbury Association settlement, who wrote in May 1852 that he had become 'too colonial' and that if his father had sent for him to come home 'he might have learned that there are things in the world to care for besides "stock" and "money"'.[45] When the partnership with Croft was dissolved in April 1862, Ward continued on his own until 1867, when the business was sold to Thomas Douglas and Henry Lee, who continued it under the Ward's name.

At the same time Samuel Manning & Co., established probably in 1860 and expanded on a new site in 1865, was beginning its rise to prominence in Christchurch. Manning had served his brewing apprenticeship at his father's successful Suffolk Brewery, established in Christchurch in 1858.[46] Meanwhile, in Dunedin, the Aberdeen-born William Strachan established a small brewery at Sawyers Bay, near Port Chalmers, in April 1857. He moved the brewery to Dunedin in 1861, and managed it as the Victoria Brewery until his death in 1888. The brewery continued as a company until it was acquired by Speight's in 1908.[47] The final pre-1860 connection with the events of 1923 is that of John Staples and his two sons, who closed their ten-year-old Motueka Brewery in 1866 and relocated to Wellington, where they were soon to dominate. As with others who endured in a colonial market where many came and went, they evidently possessed business acumen and beer that people wanted to drink. Both were essential.[48]

The styles of beer produced by these breweries, and the preferences of their consumers, are not easy to determine. Many breweries simply advertised 'ale' and 'porter'. Others advertised their weakest offerings, for regular family consumption, as 'table beer' or 'table ale' and occasionally 'table porter' and 'small beer'. At the other end of the scale were 'strong beer', 'strong ale' and 'stout'. Alternatively, especially from the 1850s, brewers used an X designation for strength – with X as the weakest and XXXX the strongest.[49] Although porter was a staple of most breweries, its gradual decline in Britain and the rise of stout can also be seen in New Zealand. The term 'stout', as originally applied to beer, meant strength rather than style or colour. During the eighteenth century one could have 'pale stout' or 'stout porter' – a stronger version of porter. During the nineteenth century, however, stout emerged as a strong, dark beer in its own right, and a number of breweries included this in their repertoire as well as porter.[50]

This does not suggest that each distinct beer was brewed separately, other than that the malt for darker and lighter beers required a different amount of roasting. While the larger nineteenth-century British brewers were beginning to use sparging, where extra water is sprayed onto the mashed grains to extract as much of the fermentable sugars as possible, smaller brewers were more likely to drain off the first and strongest wort as perhaps their XXXX beer, before adding hot water and mashing the same grains a second and possibly a third time for successively weaker beers. It was also common to blend stronger and weaker beers to produce a mid strength.[51] No doubt particular New Zealand brewers established a local reputation for one or other of their ales, porter or stout, but such details have not survived. Nor can anything be determined about the different proportions of beer styles consumed until the early twentieth century, and even then the distinction was simply between ale and stout.

*

With a brewing industry becoming established, it is evident that beer was a staple of many public and festive occasions during the early years of settlement, and viewed as a more suitable drink than spirits. Having noted the substantial dinner held to mark the first Auckland cattle market in December 1850, the *New Zealander* added that 'This was very well for a sort of inauguration, but we are glad to learn, that, under a judicious presiding influence, the refreshments provided for Market days are now limited to a substantial luncheon of cold meat and a glass of good ale.'[52] Five years later two thousand Nelsonians enjoyed liberal quantities of sandwiches, Nelson ale, London porter and Devon cider while celebrating the fall of Sebastopol during the Crimean War.[53] At the end of the decade, the presentation of new colours to the 58th Regiment in Auckland concluded when 'the non-commissioned officers and men sat down to a plentiful dinner of roast beef and plum pudding, with ale and tobacco *ad libitum*'.[54] The following year, Rev. Henry Harper reported a discussion about wedding preparations with a Canterbury station hand who was engaged to a housemaid. The station hand asked if he should go to Christchurch to order some 'hard stuff' to give the guests something to drink. '"No," I said. "It won't do, take my advice, order enough bottled beer to go round, which will do no harm, but no grog."'[55]

Brewery openings were popular civic occasions of their own and clear testimony to the status of brewers. A lengthy account in January 1862 of Secombe's new Great Northern Brewery grandly concluded that:

> as an example, among many, of the future commercial and manufacturing greatness of Auckland, this concern is deserving of notice. It is not ephemeral; it is a substantial and solid fact, and one which our English readers will accept as a proof that there are among the settlers of New Zealand . . . men who are not only worthy of the parent stock, but who are eminently qualified to reproduce in the South Pacific, a counterpart of the British islands.[56]

The opening of a large new malthouse in Christchurch in March 1870, erected in partnership between Augustus Moore & Co. and Ward & Co., featured a plentiful supper and toasts to the prosperity of the business, the province and the 'agricultural interest', in the new building, which was decorated with flags and a large image of John Barleycorn, the mythical personification of alcohol, holding a foaming pot of ale.[57] The following December, the opening of Robert Whitson's new Albert Brewery was attended by the Auckland provincial superintendent, the mayor and other 'principal citizens'. Loyal and patriotic toasts, speeches and a luncheon were only the beginning, as nearly 150 people reconvened in the evening for a ball, which 'gave additional eclat to the inauguration of the brewery, at which, till an early hour this morning, the assembled guests continued to "chase the hours with flying feet"'.[58]

Beyond festivity, high hopes were held that the brewing industry would invigorate the colonial economy and produce a commodity worthy of export. George Hooper's first brew in December 1843 prompted the *Nelson Examiner* to observe that 'As this climate is particularly favourable to brewing, we may hope at an early day to export malt liquor to a large extent.'[59] In September 1845 prominent settler Danforth Greenwood wrote that:

> brewing and the supply of provisions were the only two ways to make a substantial return on investment in the settlement. An experienced brewer said to me the other day, 'Why, sir, you may brew here all the year round, and eight months out of the twelve equal to October. They cannot brew at all in Sydney, and but indifferently in Van Dieman's Land; and you have the Indian and China markets within six weeks' sail, and ships sailing there from here in ballast; and you have neither malt-tax nor hop-duty. It is a brewer's Paradise.'[60]

This would not be the last slight on Australian beer to come from New Zealand. In response to the excellent quality and quantity of barley being produced throughout the colony, Auckland's *New Zealander* predicted that 'if breweries or distilleries were established, so as to furnish a constant market to the farmer, the cultivation of such grain would be highly remunerative', and there was also potential for a quality ale to be exported.[61] While beer exports did not eventuate on any scale until the twentieth century, 1000 bushels of barley were sent from Nelson to a large Sydney brewery during 1848.[62] All of these themes, and an incipient trans-Tasman rivalry, are evident in the response to Dunedin's first brewery advertisement in January 1857.

> The climate of Otago is one that will admit of brewing being carried on to perfection and we hope Messrs Mackinnon and Co endeavour to reach that point. Were a really good article produced, besides supplying the home demand, the trade might become a very extensive export trade. The Australian colonies are large consumers of beer, and produce a very inferior article. We do not understand the reason why beer of a first-rate character has not been produced in New Zealand. Under any circumstances the supply of beer at a cheap rate to our labouring population will tend to reduce the consumption of spirits – a change beneficial both on sanitary and economical grounds.[63]

Similarly, the *Lyttelton Times* hoped that the establishment of William May's Crown Brewery in April 1857 would assist the local economy by encouraging the growth of barley on the Canterbury Plains. 'Times are bad among the farmers if we may judge from one circumstance. We hear that, at a meeting of the inhabitants of the Lincoln Road, it was agreed not to subscribe towards the building of a church nor towards a clergyman's stipend. The reason is alleged to be the hardness of the times from the low price of grain.'[64] The implication that investment in beer was a higher priority than a minister surely caused apoplexy to an emerging temperance movement.

Hop cultivation was also encouraged for local and export purposes. Although a 2-acre experiment in New Plymouth during the early 1840s produced inferior hops, Charles Hursthouse claimed that the land and climate were suitable and 'The native women and children would make capital hop pickers, and poles are cheap and plentiful.'[65] Hops were being grown in the Auckland Government Garden from 1841, and by July 1844 spare cuttings had been given to William Smithson, who was preparing 2 acres for a hop garden.[66] In the same month a Wellington correspondent emphasised the suitability of the climate and export potential, while suggesting that a lack of suitable hops was preventing full advantage being taken of the barley supply for brewing good beer. He had hop seed to give away to any who would plant it.[67] In a letter to the *New Zealander* on 30 March 1857, shortly after his arrival in Auckland from New Plymouth, Richard Secombe addressed the matter in more detail. While he was glad to see successful hop production in Auckland after extensive cultivation in Nelson and Taranaki, he insisted that the home market needed to be supplied before any thoughts of export. 'We will take ten tons per annum . . . and should be glad to know there was a prospect of next year's supply being produced in the settlement. Hops are one of those articles that are much better produced here than when imported, for they lose all or nearly so of the essential aroma by shipping however carefully packed, and the best quality of pale hops are never exported.' Secombe promised that friends in New Plymouth would provide 2000 plants to a grower in Auckland, and he was willing to arrange cuttings for nothing more than the cost of transit to anyone else interested in growing them.[68]

Breweries sustained the colonial economy in another less obvious way by providing a regular supply of yeast and grains for bread making. How yeast was nurtured on the long journey to New Zealand or by individual brewers once here is not recorded. The perils of supplying yeast are revealed in the unfortunate case of Robert Low, an employee of the Wairau Brewery, who drowned with two others in September 1858 while crossing a river on the way back to Blenheim after collecting yeast left at Waitohi (present-day Picton) by a passing ship.[69] The majority of early brewers followed their advertisements for various types of ale and porter with a note stating that yeast was always available.[70] As David Veart points out, colonial bakers faced constant difficulties in producing leaven from scratch in a setting where commercial yeast was not available. A number of imported cookbooks, including Isabella Beeton's *The Book of Household Management*, had instructions on how to wash brewer's yeast for bread making to remove the bitterness that it generally possessed.[71]

Beer also assisted colonial growth by providing workplace sustenance, as the numerous advertisements for harvest ale or harvest beer attest. Many of them addressed farmers specifically, leaving little doubt that British traditions of providing rural workers with a refreshing incentive to productivity were carried to New Zealand.[72] In response to a resolution of the Dunedin Total Abstinence Society in January 1864, one correspondent insisted that

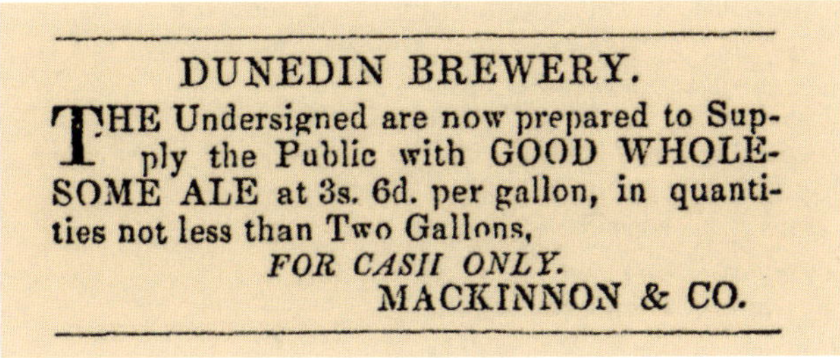

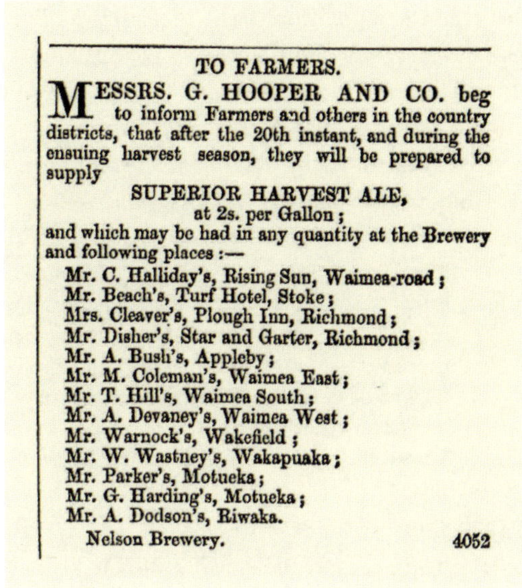

ABOVE: An 1857 advertisement for Dunedin's first brewery, established in late 1856. *Otago Witness*, 10 January 1857, 2

RIGHT: Hooper & Co. of Nelson was typical of many breweries that offered a lighter harvest ale to farm workers during the summer. *Nelson Examiner & New Zealand Chronicle*, 26 November 1859, 1

'No labourer obliged to use a pick or shovel or an axe could labour as hard as he is obliged to do, did he abstain from his usual quantum of beer.'[73]

The lengthiest discussion of drinking during the harvest is more enlightening for what it reveals about the relationship between Māori and beer. Under a Sale of Spirits to Natives Ordinance of 1847 it was made illegal to sell or supply spirits to Māori or to give wine or beer in any quantity that would 'cause intoxication'. The penalty was a fine of £10, and the only exemption was for medicinal spirits. The governor had the power to prevent the sale of alcohol in any area so as to keep it from Māori, although this provision was not used until the wars of the 1860s when there were clear signs of increased Māori consumption.[74] It is evident that Pākehā politicians saw the 1847 Ordinance as a paternalistic measure to protect Māori, although at a Queen's Birthday dinner for the 'natives' of Auckland in May 1849, 'Cold roast beef, plum pudding, bread and ale were the substantial viands ranged along the tables at which some 500 natives took their seats in the most orderly manner.'[75] That many Māori were also happy with the prohibition is revealed in various petitions to Parliament during the 1850s, such as that from a group of Waikato and Waipā chiefs in 1856:

> This is our word to you – do not let any intoxicating drinks come to this land to Waipa, but let it be kept there [Auckland] at the inns. Let the Maori law, which has been asserted to by all the chiefs of this place and Waikato, be sacred, namely that the Natives be not allowed to drink spirits . . . This is the worst thing hitherto brought to New Zealand . . . we have determined, therefore, not to allow it to be used.[76]

But some Māori objected to specific legislation for them that did not also apply to Pākehā. A large public meeting of Māori at Motueka on 22 November 1859 to consider a recent proclamation from the resident magistrate regarding the sale of spirits and other liquor to Māori declared its opposition to the measure. '[T]hey did not thus like to be singled out and treated like slaves.' They had adapted to Pākehā habits, abided by laws, were British subjects under the same Queen and entitled to 'any and all of the white man's privileges'. Europeans had first introduced alcohol to Māori, and, although some certainly got drunk, this was no reason to condemn all for the problems of a few. A purported translation of the resolutions of the meeting was duly supplied to the *Nelson Examiner*.

> Now, listen! This is what we all of this meeting say: – Let us have beer. The beer we drink is not stolen by us. If we get drunk we go to the lock-up, and we are fined; and if we do not pay the fine we are sent to prison. If we go to work for the Pakeha and there is beer given to the white men, are the Maories to drink water – that would be wrong. One of our people has been working at the brewery for six months, and if the Governor does not choose to allow the Maories to have beer, we will build a brewery and make our own. One word more – Beer was made to be drunk.[77]

The *Examiner* conceded that Māori had won the argument, and noted that the law had seldom been enforced in Motueka, thus allowing them to develop a taste for beer which was suddenly forbidden to them. 'They ask whether beer is good for the Pakeha, and not for the Maori; "if it is good, it is good for us as well as for you", say they; "if it is bad, why do you let it be sold at all, and take money for allowing people to sell it?" '[78] A Māori brewery was never established and no specific mention can be found of Māori in the nineteenth-century brewing industry, other than 'Charlie' and 'Hemi', who worked for Richard Secombe in Auckland during the late 1850s and came to the attention of the court for stealing and killing a goose.[79]

*

For all that brewing was heralded as making a positive contribution to the colony, already by the 1850s there were signs of the challenges that were to stalk the industry for much of its subsequent history – temperance advocates, tied-house controversies and complaints about prices.

While the New Zealand Temperance Society of 1836 opposed only spirits and not beer or wine, its successors during the 1840s reflected the growing division in Britain by declaring themselves in favour of total abstinence from all alcohol. The Auckland Total Abstinence Society, formed in 1842 and in frequent correspondence with the British and Foreign Temperance Society in London, held regular tea parties and urged the public, generally unsuccessfully, to take a formal pledge not to drink. The society also established a library to provide alternative recreation away from the public house. A Nelson Temperance Society was established by August 1843, and a branch of the Independent Order of Rechabites, a secret and avowedly prohibitionist fraternal society, also operated briefly in Nelson during the same year, followed by another in Whanganui during the early 1850s.[80] In February 1844 the Port Nicholson Total Abstinence Society in Wellington sent a memorial signed by sixty members and others to Governor Robert FitzRoy:

> We, whose names are immediately undersigned, being total abstainers from all intoxicating drinks . . . are deeply concerned at the widely extended and increasing prevalence of intemperance, and believing it to be one cause of the distress which now exists, and that it is also displeasing to God, feel it our duty to use our utmost endeavours for its suppression.

In response to a request from the Society to grant land for the erection of a Temperance Hall, FitzRoy replied that he was happy to comply with this 'very proper and reasonable request', and added, 'There can be no doubt that total abstinence from intoxicating liquors has been beneficial, and that although not necessary for all temperaments and constitutions, it has been highly advantageous to thousands.' While this stopped short of a complete endorsement, land for the hall was quickly selected.[81]

A gathering of nearly two hundred people at a meeting of the Nelson Total Abstinence Society in August 1850 heard perhaps the first specific attack on beer from the New Zealand temperance lobby. 'A barley pudding was made for the occasion, and a comparison was drawn between this and a gallon of ale – a gallon of ale, at a cost of 2s, leaving the appetite of a family unsatisfied, while a barley pudding, agreeable and nutritious, at a cost of only 1s (one-half the price of the ale) would dine a family of five persons.'[82] But as Jim McAloon points out, a quarter of those listed in the Total Abstinence Society pledge book for the 1840s returned to drinking, including one who subsequently became a publican. Membership of temperance societies was as erratic as many other enthusiastic endeavours in the early years of settlement and ought not to be exaggerated.[83]

During the 1850s, opponents of alcohol became more organised and vociferous, and their cause began its gradual shift from an emphasis on the desirability of individual temperance or abstinence to one advocating total prohibition by legislative means. Both Auckland and Otago established branches of the Maine Law League, inspired by the US state of Maine, which in 1851 had enacted a law prohibiting the sale of all alcoholic beverages except for 'medicinal, mechanical or manufacturing purposes'. By 1855 a further twelve states had followed Maine to total prohibition. It would not be long before legislative incursions on the habits of the drinker and the conduct of the brewer reached New Zealand.[84]

Meanwhile, although supporters of temperance generally wanted to dispense with public houses altogether, at the very least they sought to oppose any brewery monopoly over them. As commercial brewing expanded in Britain during the eighteenth century, larger brewers sought to 'tie' public houses to guarantee outlets for their beer and found easy pickings among publicans who became indebted to brewers for their beer supply. Usually, ties involved the brewery supporting the development of a business by lending money to a licensee at a reduced rate of interest in return for an agreement to take only the beer of that brewery. Alternatively, a brewery could buy an establishment and install its own licensee or offer considerably reduced rent in return for exclusivity. In Britain, a House of Commons select committee claimed in 1817 that about 30 per cent of public houses were tied to brewers by ownership or loans, and 50 per cent in London. Various legislative attempts to break these monopolies failed until the initial success of the Beerhouse Act in 1830, but the number of tied houses soon increased.[85]

From the late 1840s there was evidently enough competition in Nelson and Auckland to sustain ties. When Thomas Smith offered his Bridge Street Brewery for sale in 1853, he stated that there were 'several public houses in good situations with the business'.[86] In the same year a convoluted letter criticising the tied-house arrangements of both Nelson breweries suggested that publicans or 'brewer tenants' ought not to be given licences if they were a 'front' for others and were not at liberty to supply any beer the customer demanded.[87] Later in the year, Nelson magistrates apparently indicated that they would refuse licences to those publicans taking beer from only one brewery. Naturally Dr Renwick, the financier of George Hooper's Nelson brewery, sternly resisted any proposal along these lines. At a meeting of the Nelson Provincial Council to discuss the licensing ordinance in December 1853, he stated that it was the usual practice of breweries to restrict their tenants to their beer only, and that, as publicans and brewers had a vested interest in running an orderly business, there was no need for any intervention by magistrates in licensing matters.[88] The magistrates retained their control in this first of many jousts.

For 'Anti-Monopoly', writing from Auckland in April 1852, the problem was not so much the tying of public houses as the lack of them that allowed the inflation of beer prices.

> [A]t Parnell (where I reside) there is but one public house, the consequence is that the legalised monopoliser charges Her Majesty's lieges one shilling and four pence for a quart of porter, something like 25 per cent extra profit on the town price. At the same time, if the trade was thrown open to a greater extent, Colonial ale would be sold at 6d per quart. At Hobartown, good ale is sold at 6d, common (and that even better than that generally sold in Auckland) at 4d per quart, and porter at less than 1s, which would not only decrease the consumption of spirits, but encourage the manufacture of a better colonial ale, which I consider would have a less injurious tendency than the rum at present consumed by the working classes.[89]

The following year 'Anti-Humbug' of Nelson wanted to know why the brewers had added 50 per cent to the cost of beer without reference to the cost of barley which had fallen significantly three weeks earlier.[90] No reply was forthcoming.

Due to variations between settlements and between styles of beer, conclusions about price during this period are difficult. As a rule it appears that colonial beer during the 1840s was generally £4 to £5 per hogshead and 1 to 2 shillings per gallon for smaller quantities – or 40 to 60 per cent less than the price of imported beer.[91] During the next decade the price remained stable with only isolated increases. In August 1854 Richard Secombe informed his 'numerous friends and the public' that, due to increases in the cost of barley, firewood and labour, he was obliged to raise the price of his ale from 2 shillings to 2 shillings 6 pence per gallon.[92] In February 1858 the new Dunedin Brewery, without the restraint of a competitive market, offered ale in barrels at 3 shillings a gallon, but 3 shillings 6 pence a gallon for lesser quantities.[93] But Hooper & Co.'s price list for November 1858, from a Nelson market with at least four breweries, offered a much lower range.

Table 2.3: Hooper & Co. Price List, 1858

	Gal.	Hhd.
Best XXX ale	2s 6d	£6
Superior XX ale	1s 6d	£6
Table X ale	1s	£6
Best porter	2s	£5

Prime bottled ale: 8s per dozen bottles returned.
Gal. = gallon; Hhd. = hogshead

Source: *Nelson Examiner & New Zealand Chronicle*, 1 December 1858, 1

These prices suggest different markets for different styles of beer. That the XXX ale and porter were 10–15 per cent cheaper by the hogshead indicates that they were intended for sale to publicans. That the XX and X table ale were cheaper in smaller quantities is a sign that they were most likely purchased from the brewery or its agents for home consumption. Alternatively, the brewery may have delivered. In 1846, the Epsom Brewery regularly advertised that it sent a dray to town twice-weekly to deliver and collect home orders. Similarly, Dunedin's Marshall & Copeland announced in July 1863 that 'orders from the city accompanied with a remittance or reference in town will be promptly tended to'.[94]

Perhaps early transport difficulties and therefore fluctuations in supplies of barley and hops to different parts of New Zealand did create some volatility in the price of beer during the 1840s and 1850s, but for the most part New Zealand mirrored Britain in keeping its beer prices remarkably stable, although slightly more expensive, over a long period. From the 1830s to 1914 British brewers generally followed a price scale ranging from 10 pence to 2 shillings a gallon depending on the style and strength of the beer – with 1 shilling 6 pence to 2 shillings the most common. Duties were relatively unchanged, and various fluctuations in the price of malt and hops were not sufficient to be reflected in prices.[95] By the mid-1860s the brewers of Canterbury had agreed on standard prices to publicans – ordinary colonial ale and porter at £5 15s per hogshead, or 2 shillings 6 pence per gallon for lesser quantities. When dealing with one brewer only for not less than three months, publicans would receive a discount of 10 shillings per hogshead or 6 pence per gallon for lesser amounts – but no discount if an account was not settled on or before the due date. Likewise, in 1869 George Field's Alton Brewery in Timaru was selling ale at 2 shillings 6 pence per gallon, while ten years later Francis Franks of the Eclipse Brewery, Temuka, and Lorenz Riesterer of Wellington's Junction Brewery were among many still selling at the same price.[96] But as Chapter Four will explain, the imposition of a beer duty in 1880, and the determination of brewers to pass it on to the consumer, would start to push prices upwards.

In April 1861 the *Lyttelton Times*, in the midst of a lengthy and remarkably verbose editorial on the virtues of good beer, observed that:

> Of all the aids to temperance that have at various times been proposed, there is none so efficacious as cheap beer of good quality. The comparison of a glass of good beer sold at threepence, with one of bad spirits for sixpence, will be found a greater assistance to the temperance cause than any amount of pledges or restrictive laws. The latter seek to impose fetters on the natural appetites; while the supply of wholesome liquor would provide them with a harmless and agreeable employment.[97]

In answer to the call, there were now perhaps thirty active breweries in New Zealand, most of them less than five years old and some in the midst of a fairly short existence. Hooper & Dodson in Nelson was easily the longest established, while Richard Secombe in New Plymouth and Auckland, John Scholes in Auckland, and William May of Wellington and Christchurch were others who had endured for more than a decade. They, even more than the publicans, faced few legal or moral constraints on their business, and beer and breweries were frequently praised as a sign of New Zealand progress. While there was certainly an active temperance movement in each of the main settlements, this was still more of an anti-spirits movement. But within a year the relative tranquillity of colonial society and its drinking habits would alter dramatically and all things connected with beer willingly followed suit.

CAPTAI
IMPERIAL
HANCOCK & Co MALTSTERS

Established by Thomas Hancock and John Scholes during the 1850s, Auckland's Captain Cook Brewery became one of the most prominent and innovative breweries by the late nineteenth century. Lion Technical Library

THREE

Golden Ale? The Rush to Brew, 1860–1870

In 1871, when the first reasonably reliable census was taken, New Zealand had sixty-nine breweries producing at least 20 million litres of beer. With the addition of imports this enabled a consumption of 81 litres per annum for each Pākehā man, woman and child in the colony. Beer was now the dominant beverage in terms of pure alcohol intake. As the *West Coast Times* observed of Hokitika in December 1876, 'As a community we are more beer drinking than spirit drinking, and we get through a fair share of the sparkling liquor.'[1] Such dramatic expansion from the thirty or so breweries operating in 1860 was hardly surprising given the frenetic demographic and economic boom of the 1860s New Zealand gold rushes, especially in the South Island, and the gradual assertion of dominance over Māori in the north. With a growing population and significant expansion beyond the initial areas of settlement came both demand for alcohol and infrastructure to support an expanding brewing industry and beer market. But if gold provided impetus, demand did not always ensure success. Although gold seekers and soldiers provided a market for as many as eighty new breweries during the 1860s, few of these breweries survived to make a significant long-term contribution to the industry. While the drinking habits of miners and soldiers were heavy and sometimes spectacular, the consumption patterns of the 1860s were the high tide. From the end of the decade per-capita consumption in New Zealand plunged dramatically, to perhaps only half of that for the United Kingdom by 1880, and to less than a third by the end of the century. There were technical and financial challenges aplenty for brewers and an array of potential disasters close at hand. Only the best and the luckiest prospered in a rapidly changing market.

Well before Gabriel Read discovered gold at Tuapeka, Otago, in May 1861, triggering the most substantial gold rush to New Zealand, general patterns of migration had doubled the European population from 26,707 to 59,413 between 1851 and 1858. Although the various systematic colonisation schemes of the 1840s had not flourished, the escalation of the Highland Clearances during the late 1840s and the severity of the Irish famine at the same time were proof enough to many of the need to pursue opportunities elsewhere. But as in California and eastern Australia, gold transformed New Zealand's population, infrastructure and culture on a much greater scale. The population of Central Otago soared from perhaps 300 in May 1861 to 24,000 by the end of 1863. Although it plummeted to 7000 by April 1865 following new gold discoveries in Nelson, on the

West Coast and later at Thames, the legacy was a network of new towns, roads, railways and other amenities. Gold made Otago the most prosperous province in New Zealand. By 1870 it contained one-quarter of the entire European population and provided one-third of all exports.[2] In the North Island, growth was also driven by military infrastructure as campaigns against Māori escalated and eventually led to substantial land confiscations. But as gold prosperity tapered off by the end of the decade, Colonial Treasurer Julius Vogel initiated an extensive programme of overseas borrowing by central government – £20 million during the 1870s – encouraging renewed immigration with assisted passages for more than 100,000 settlers and financing new settlement and extensive public works such as roads, bridges, railways and telegraphs.[3]

As the total European population climbed beyond 250,000 in 1871 and to 490,000 by 1881, the previous isolation of the various settlements slowly gave way to more regular coastal shipping, the establishment of telegraph services and eventually railways.[4] But such expansion was only relative. Beyond the four small major cities, there were few provincial towns of substance and road and rail communications within the rural hinterland remained difficult.[5] In simple terms the number of potential drinkers and drink suppliers increased almost eightfold in the two decades after 1860. It was a considerable undertaking to supply beer to such a rapidly expanding population. Once again, the initial impetus came from imports. When the rush to Otago began in mid-1861, Dunedin had only three small breweries, with another soon to be established. When the rush shifted to the West Coast in late 1864 and to Thames in 1868, there was virtually no existing European population let alone a brewery.

Yet despite the difficulties of exporting beer and of moving it to isolated inland settlements once it arrived, the early 1860s witnessed a remarkable triumph of will. During 1861, 1.5 million litres reached New Zealand. Imports for the following year leapt to 3.5 million litres (almost half of it through Dunedin), and 1863 brought 5.1 million, followed by a peak of 5.2 million litres in 1864 – a 239 per cent increase in three years, against a 73 per cent increase for

Figure 3.1: Alcohol Imports, 1853–1900 (litres)

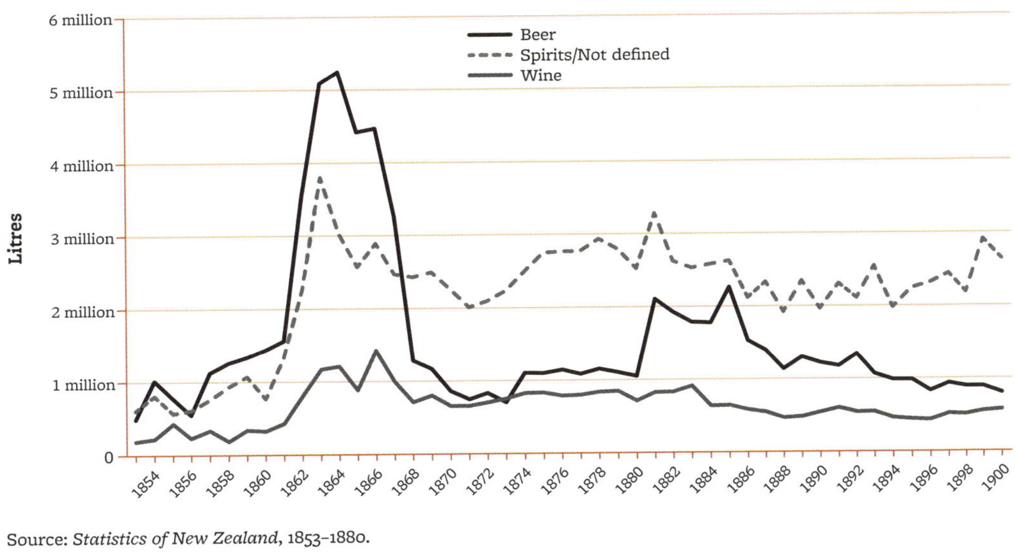

Source: *Statistics of New Zealand*, 1853–1880.

the total population and a 78 per cent increase in males aged over fifteen. Put another way, there were 14.5 litres of imported beer for each person in the colony in 1861, and 28.4 in 1864. But from 1867 (3.2 million litres) to 1868 (1.2 million) the market collapsed, never to return on anything like its former scale. Why this happened is not clear. Perhaps it was a realisation of expanding colonial production as the major gold rushes began to taper off, although the suddenness of the drop makes this doubtful. Perhaps it was linked to the departure of imperial troops during 1866–67 following the securing of relative peace in the North Island. Certainly, the demand for rum, the standard military spirit ration, also fell away sharply at this time. But there is no evidence to indicate that troops were routinely supplied with imported beer rather than local product. Whatever the explanation, the ratio of imports to local production never recovered. During the 1870s the amount of imported beer never made up more than 5.3 per cent of the total available for consumption. Although it spiked at 8.6 per cent for a single year in 1881, it settled at 3–5 per cent for the remainder of the nineteenth century.

Two features remained unchanged from the import patterns of the 1840s and 1850s. More than 90 per cent of the beer came from the United Kingdom, with a small quantity from Victoria and a tiny amount from New South Wales. Secondly, there was still an abundance of Burton pale ale of one sort or another and of London porter and stout. But there was also an increasing amount of scotch ale, the presence of which heralded a significant Scottish influence on the taste and style of New Zealand beer during the later years of the nineteenth century. Indeed, the Scots made a disproportionate contribution to the export trade. At its peak Scotland produced no more than 6 per cent of the total British beer output during the nineteenth century, but provided one-third of exported beer by the 1890s. Moreover, Scottish beer was also re-exported from English ports after export from Scotland – thereby obscuring its origins and meaning that the total figure was likely to be much higher. The relevant import statistics for New Zealand specify only 'United Kingdom' as the point of origin.[6]

The Scots began with exports to the West Indies and the Australian colonies from the 1830s, with the export volume increasing markedly after 1850 and especially after 1875. Three factors contributed to the success of this trade. Firstly, notwithstanding the taste for pale ale, strong Scottish ales also travelled relatively well compared to many other styles. Secondly, to maintain quality and avoid tampering along the way, beer was most effectively exported in bottles, and the Scots were pioneering and skilled bottlers. Thirdly, and above all other considerations, the leading Scottish brewers were skilled merchants, prepared to make considerable effort with the packaging of beer and to secure necessary business networks, especially with other Scots.[7] Similar expertise and determination was not evident with English brewers, a point made by the Scot Robert Whitson at the opening of his new Albert Brewery in Auckland on 1 December 1870.

> He could truly state that the English people thought anything would do for the colonial market. He had tried to get the good stuff from home, but had been rarely successful. . . . When he first came to the country, he was astonished to find the rubbish which was being drunk here. He had brought out some Scottish ale with him, and this he must say was very much appreciated.[8]

Whitson's search for the beer of his homeland was not entirely fruitless. As early as November 1840 D. S. Durie & Co. announced 'superior bottled Scotch ale' for sale in Wellington, and both bottled and barrelled ale was regularly imported to Auckland.[9] Ales from Younger of Edinburgh and Younger of Alloa were advertised in Christchurch by 1859, and were readily available in Dunedin by the early 1860s. The Royal Oak Hotel, in the Octagon, was one of a number of establishments advertising bottled Alloa and Edinburgh ales by 1863. At the height of the West Coast gold rush it was possible to obtain Jeffrey's Edinburgh Ale in stone bottles or Younger's Edinburgh Ale in wood. By 1864 the products of William McEwan's Fountain Brewery, Edinburgh, were also available in New Zealand.[10] McEwan won renown as one of the most important exporters from the 1850s. His ability to establish reliable connections with colonial merchants was essential, not only in terms of guaranteeing regular payments over distance and time, but also in terms of gaining information about the local market, especially the eventual shift away from heavier beers in warmer colonial climates. Among McEwan's New Zealand agents were the Scottish merchants W. & G. Turnbull of Wellington, John and Robert Gilfillan of Auckland and John Barr & Co. of Dunedin.[11] There is more than pure coincidence in these networks of Scots dealing with Scots.

Other publicans and merchants secured their own market niche. In September 1863 J. McCubbin of the Otago Hotel in Dunedin announced that he had obtained the sole right of sale for Tyson's Carlisle Sparkling Bitter Ale. 'For clearness of colour or delicacy of flavour, this ale is unsurpassed, being at the same time a splendid tonic.'[12] Guinness, or 'Guiness' and even 'Guineass', first reached New Zealand in April 1851 and was relatively common by the early 1860s.[13] By late 1865, Parker & Garside, brewers of Hokitika, were supplementing their own production by importing directly from the brewery of Robert Manders & Co., Dublin, for bottling in New Zealand, and by late 1868 Strike and Blackmore of Greymouth were repeatedly advertising for 100,000 dozen porter bottles to enable bottling of imported beer.[14] But as the next chapter will show, the bottling of imports also raised the spectre of adulteration and other forms of tampering.

*

Even with the comparative flood of imports during the early 1860s, these were the years when the local industry secured its prominent position. Growth was spectacular but also predictable, as breweries followed the initial pattern of gold discoveries and population increase in the South

Table 3.1: New Breweries, 1861–1870

	North Island	South Island
1861	1	2
1862	1	8
1863	–	7
1864	–	5
1865	4	13
1866	2	5
1867	1	3
1868	3	5
1869	6	8
1870	3	5
Total	21	61

In a few cases the establishment date of a brewery is not certain. These are assigned to the first year in which they were known to be operating.

Island before eventually expanding further afield in the north in the wake of military incursions against Māori and the opening of the Thames goldfield.

Having lagged behind its counterparts for breweries during the 1850s, Dunedin as the main entry point to the Otago goldfields acquired ten new ones between May 1861 and the end of 1864. The most enduring were: William Strachan's Victoria (1861); the Well Park (1862), named after Tennent's long-established Wellpark Brewery in Glasgow; Marshall & Copeland's Water of Leith (1862); and the Red Lion (1862), which eventually prospered under Maurice Joel. The first of the goldfields establishments were Hunter & Co's Creswick Brewery at Clyde and A. D. Duncan's Milton Brewery, both opened in 1862. The following year Queenstown acquired three, Balclutha one and Milton another. As the rush moved to the West Coast, Hokitika opened six breweries during 1865. Two years later there were at least fifteen breweries on the West Coast, at Hokitika (seven), Greymouth (two), Brighton, Charleston, Kumara, Moonlight's Gully, Ōkārito and Westport. Three more appeared the following year, including the Monteith family's Phoenix Brewery at Reefton. Meanwhile, following the establishment of James Paul's long-lived Egmont Brewery in New Plymouth in 1864, the North Island received two more at Ngāruawāhia in 1865 to serve the military and new settlers as the Waikato war came to an end. Several others

Marshall & Copeland's Water of Leith Brewery during the late 1860s. Established in 1862, this was the most prominent brewery in Dunedin before being surpassed by Speight's during the late 1870s. It closed in 1887. Hocken Collections, University of Otago, 509-262b

James Paul's Egmont Brewery, New Plymouth, established in 1864, was taken over by DB in 1969. Department of Labour Collection, Alexander Turnbull Library, PA1-0-367-48

followed around Auckland and then Thames in 1868–69 after further gold discoveries. Among the soon-to-be significant provincial towns, Invercargill's first two breweries opened in 1864, followed by breweries in Ōamaru in 1865, Timaru in 1866, Napier in 1869, Hamilton in 1873 and Gisborne in 1874. In the older established brewing centres where population growth was more gradual, there was no need for such rapid expansion. Auckland had no new brewery from 1861 to 1869, Christchurch had none from 1862 to 1873, Nelson only one during the 1860s and Wellington only two from 1857 to 1876.

At least a third of the breweries established during the 1860s failed to last beyond the turn of the decade, and a number of others closed during the early 1870s. Some were victims of shifting population as gold declined and its seekers and followers moved elsewhere. Others, as discussed shortly, fell victim to bad luck, bad management and a multitude of other calamities. It seems that only three of the twelve Otago goldfields breweries of the 1860s lasted more than five years, while the number in Westland fell from perhaps sixteen to six between 1868 and 1871. However, there were also at least thirty-five new breweries during the 1870s as a new wave of migrants arrived. The distribution of this growing population was largely rural and engaged

Table 3.2: Growth of the Brewing Industry, 1867–1881

	Number of breweries	Beer produced per year (L)	Average production per brewery per year (L)	Employees per brewery	Population per brewery
1867	51	12,498,340	245,065	5.47	4,316
1869	79	16,258,459	205,803	–	–
1871	69	20,346,509	294,876	5.30	3,716
1874	82	19,308,689	235,471	5.24	3,653
1878	91	25,831,144	283,858	5.63	4,554
1881	99	20,930,511	211,419	5.31	4,949

Table 3.3: Beer Production by Province, 1871–1881

	1871		1874		1881	
	Number of breweries	Average production per brewery per year (L)	Number of breweries	Average production per brewery per year (L)	Number of breweries	Average production per brewery per year (L)
Auckland	10	348,206	12	316,387	15	288,916
Taranaki	2	76,146	2	61,825	4	94,429
Wellington	9	130,742	7	178,566	11	228,204
Hawke's Bay	2	190,295	4	97,720	5	137,802
Nelson	7	216,516	11	137,761	14	80,596
Marlborough	3	130,620	5	115,613	5	84,000
Canterbury	13	146,181	17	222,285	19	281,747
Westland	6	492,604	6	245,070	7	66,789
Otago	17	489,381	18	355,815	19	371,041
New Zealand	**69**	**294,876**	**82**	**235,471**	**99**	**211,419**

Source: New Zealand Census; *Evening Star*, 17 September 1869, 2. The 1867 figure for beer production was acknowledged as a significant underestimate due to incomplete returns.

in primary occupations. The majority lived on farms or in towns of fewer than a thousand people. Consequently, there was scope for a network of local breweries to serve an increasingly widespread population. Indeed, although the total population increased by 91 per cent during the 1870s and the number of breweries increased by 43 per cent, output increased by only 9 per cent. On one hand this reveals a settled population with a more even gender ratio that was drinking considerably less per capita than it had during the peak gold years. On the other hand, it confirms that most of the new breweries were small local establishments, each catering for a specific region isolated from other suppliers. Their brewing capacity was considerably lower than some of their longer-established main-centre counterparts.

The South Island continued to dominate both quantity and capacity (Tables 3.2 and 3.3), producing 74 per cent of all beer in 1871 and 65 per cent in 1881. Otago was especially dominant, with a quarter of all breweries and 41 per cent of all beer production at the beginning of the decade. But its breweries were rivalled for efficiency by the six remaining in Westland, which had the same average annual output of about 490,000 litres in 1871. Those of the Auckland province, including the Thames goldfields, were next, with an average of nearly 350,000 litres, but the thirteen in Canterbury averaged only 146,000. Ten years later Otago still accounted for 32 per cent of all beer, but Canterbury was rising fast with 24 per cent. Nelson, which had been so influential during the 1840s and 1850s, produced only 4 per cent, reflecting the general inability of the province to gain ground in the scramble for people and infrastructure.

*

Little is known about the majority of those who contributed to the brewing industry during the 1860s and 1870s. Like many things in this frenetic period of movement and expansion, they and their breweries came and went in a few years or less without leaving any detailed record. It is clear that the majority were English; a significant and influential minority were Scots; some, such as Hamilton Ward in Christchurch and William Crawford of the Gisborne Brewery, were Irish; Thomas Surman of the Wakatip Brewery, Queenstown, was American; and George Goodger of the Swan Brewery, Cromwell, was Canadian. Canute Peterson Beck of the Manuherikia Brewery, Alexandra, was a Dane. Among a sprinkling of Germans were William Nicholas Scheib of the Frankton Brewery, Queenstown, W. C. J. Kortegast of the Crown Brewery, Hokitika, Louis Ehrenfried of the Phoenix Brewery, Thames, and Joseph Kühtze who began working in Otago around 1870. The last two names will reappear regularly in much that follows.

There was no shortage of expertise among the new brewing fraternity in New Zealand. In Nelson, George Hooper apparently engaged a brewer from London during the mid-1850s.[15] Thomas Field, born near Tunbridge Wells, Kent, had fourteen years' brewing experience before arriving in Nelson in November 1854 and establishing his Collingwood Street Brewery in 1857.[16] In 1858 the Somerset-born brewer Rogers Deacon arrived to work in Christchurch, having previously made an arrangement with Richard Packer, who was about to establish the Albion Brewery.[17] At almost the same time, William Manning, previously a maltster, started his Suffolk

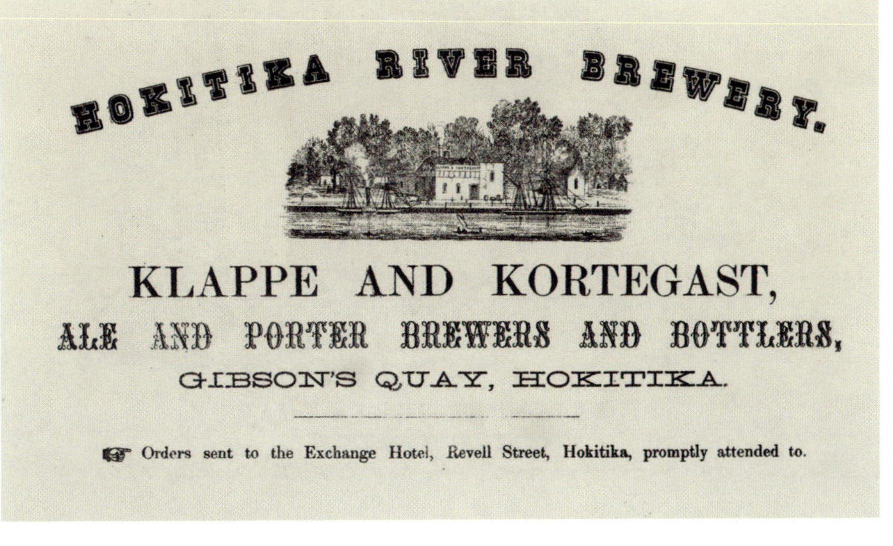

An 1866 advertisement for Klappe and Kortegast's Hokitika River (later Crown) Brewery, established the previous year. Alexander Turnbull Library, PUBL-0145-adv-001

Brewery in Christchurch.[18] James Forward, who opened a brewery in Temuka in 1865, had previously brewed in Salisbury.[19] A Glaswegian, Robert Stewart, worked in breweries in Brisbane, Sydney and Christchurch from 1862, before shifting to Greymouth in 1880 and establishing Stewart's Brewery, which by the mid-twentieth century was the last small brewery operating outside the control of the big two players – New Zealand Breweries and Dominion Breweries.[20]

Others took longer to return to their true calling. Robert Whitson, whose views on the quality of imported English ale have been noted already, was a brewer in Leith and then London, but a farmer in New Zealand before eventually taking over both the Hobson's Bridge and Albert breweries in Auckland during the early 1860s and assisting his son-in-law William Crawford to establish the Gisborne Brewery in 1874.[21] Henry Jaggar, described as a brewer when he arrived in Canterbury in August 1851, worked as a cooper making churns and dairy utensils, and seems to have worked briefly for the Avon Brewery before establishing his Kaiapoi Brewery in November 1858. In turn, this brewery was transferred to carpenter George Harrison Davis and his wife Emma, who had migrated from the same Oxford parish of St Thomas. Jaggar's sister Susannah married the South Canterbury brewer, and former school teacher, Francis Franks. There were likely to have been family and pre-migration origins linking other brewers.[22]

Arguably the most talented brewer attracted to New Zealand was William Dawson. Born in Aberdeen and learning brewing under the tuition of his father in Durham, he worked briefly in Burton-on-Trent and then for William Younger & Co. in Edinburgh before reaching Dunedin in July 1873 and beginning work at the Well Park Brewery. Three years later with fellow Well Park employees James Speight, a traveller for the brewery, and Charles Greenslade, a miller and maltster, he left to establish the City Brewery, more commonly known as Speight's. William Dawson senior joined them as a brewer in 1879. Despite the trading name of the firm, which was

Ehrenfried brothers' Phoenix Brewery, Thames, *c.* 1868.
Louis Ehrenfried moved to Auckland in the mid-1880s and, in conjunction with his nephew Arthur Myers, established an enduring brewing dynasty. Auckland Libraries Heritage Collections, 7_A8611

entrenched as one of New Zealand's leading breweries by the late 1880s, Dawson, as head brewer, in many ways contributed most to the character and reputation of the beer.[23]

In the nature of a colonial setting where rapidly expanding demand often outweighed skilled providers, many others turned to practical brewing without any previous experience. But unlike the modern generation of craft brewers, a good number of them were more likely attracted by the prospect of profit and progress rather than the ideals of the craft. Dunedin was especially encouraging to these mainly veterans of the goldfields. Indeed, when Samuel Marks sought to lease his White Star Brewery in 1866 while he returned to England, he offered instruction to anyone without brewing experience who wished to take it over.[24] William Strachan, who had worked as a brassfounder in Aberdeen before acquiring brewing experience in Melbourne, opened the Victoria Brewery early in 1857. Maurice Joel was originally an engraver attracted to the Victorian goldfields in 1853, before taking over the Red Lion Brewery in July 1864 to relieve his financially imperilled brother-in-law Joseph Woolf, who subsequently took his own life. James Wilson, a partner in the Well Park Brewery with his brother-in-law George Duncan, was originally a professional photographer in Scotland and Dunedin. George McGavin, originally from Forfarshire, spent time on the Victorian and Otago goldfields, and worked variously as a carrier, publican and contractor before establishing the Union Brewery in 1882.[25] In Christchurch, Augustus Moore, possessed of incomplete medical training but still willing to perform medical services when required, also turned his hand to farming and liquor retailing during the 1850s before acquiring his first brewery in 1865. His counterpart in the Christchurch trade, Richard Taylor, started in Lyttelton in 1853 as licensee of the Robin Hood Tavern, described as a 'low grog shop' and gambling den operating in a 'shameless way'. He opened the Phoenix Brewery in 1855, was refused renewal of his publican's licence in 1859, took over Travers & Oldham's small brewery in 1861, and finally retreated to a quieter life of cattle farming in 1866.[26]

Two of the most significant North Island breweries were founded on little if any formal training. Thomas Hancock, after trying various jobs with little success in New Zealand during the 1840s, went to Australia at the start of the Victorian gold rushes and with various Auckland associates did business as storekeepers, gold buyers, cartage contractors and brewers of hop beer. When the business failed to survive the move to new diggings, Hancock returned to Auckland almost broke, leased the Royal George Hotel in Newmarket and then purchased the Captain Cook Hotel with a substantial mortgage. Of his move from publican to founder of the Captain Cook Brewery, he recalled in the early 1880s: 'As to my knowledge also of the business, it was picked up in a haphazard sort of way. I never served an apprenticeship to it.' It seems that Hancock acquired most of his brewing knowledge from his business partner John Scholes.[27] At around the same time Bernhard and Louis Ehrenfried came from Hamburg via Ballarat to Dunedin in 1862. They were storekeepers at Waikouaiti and Hokitika, lost most of their money as guarantors for a failed business associate in Otago, and finally turned to brewing in Thames from 1868. After Bernhard's death the following year, Louis secured the business with the acquisition of numerous tied hotels before relocating to Auckland in the mid-1880s, where he acquired the late Robert Whitson's Albert Brewery. Shortly before Ehrenfried's death in 1897, his firm combined with that of John Logan Campbell.[28]

If there is one certainty about the nineteenth-century brewing industry it is that women were rare. The 1869 list of those holding brewing licences includes only one woman, Emma Davis of Kaiapoi, who secured the licence on the death of her husband in 1868. She was awarded second prize for ale at the Canterbury A&P Show in 1871 and continued the brewery until being declared bankrupt early in 1873.[29] Only four licences were issued to women in Otago during the nineteenth century, all for provincial breweries where a widow needed to continue the family business. Similarly, following the death of James Speight in 1887, his wife Mary Jane became a shareholder in the business with William Dawson and Charles Greenslade, until her own death in 1896.[30] Meanwhile, in 1889 Mary Innes had taken over management of both the Te Awamutu and Waikato breweries following the second bankruptcy of her husband, Charles. Following his death in 1899, she arranged loans to clear his debts and became the sole owner of Waikato Brewery before arranging a deed of partnership with her eldest son, Charles Lewis Innes.[31] Aside from Davis, it is not known whether any of these women had a practical involvement in brewing.

*

A thirsty colony was no guarantee of profits and long-term prosperity to brewers, trained or not. The attrition rate for breweries was high, and even among the best there were challenges that needed to be negotiated on an almost daily basis. Before examining these, it is important to look further into just how thirsty the colony was.

Most accounts suggest that gold poured out of the ground at a rate only exceeded by that at which grog was poured down the throats of the miners. An apparently vast number of pubs crowded shoulder to shoulder in Hokitika, Thames and many smaller gold towns, offering all sorts of enticements to part miners from their hard-earned precious metal almost as soon as they got it. Indeed, there is a common impression that most of those who became prosperous on the goldfields did so from supplying rather than mining. As one correspondent noted of Ōkārito on the West Coast in December 1865, 'the storekeeping interest is overdone. Everybody who does not dig sells grog and everybody who digs drinks copiously, much of the liquor having never passed through Her Majesty's Customs.'[32] Alcohol, whether imported or locally produced, found a market where it could hardly keep pace with demand. Yet there is a tendency to get carried away with the alcohol rush both as a business venture and in terms of consumption. Lloyd Carpenter explains that the profitability of goldfields merchants has been oversimplified and exaggerated.[33] This was certainly true of the brewers.

As with the early settlement period, it is no easy task to determine the kind of alcohol culture that prevailed and the specific place of beer within it. Philip Ross May, the most skilled chronicler of a New Zealand gold rush, observed of the West Coast in 1865 that 'Legend notwithstanding, drunkenness was rife in the goldfields community.'[34] Indeed, from the opening of the primitive calico-covered 'wooden skeleton' at Collingwood in 1857 that grandiosely called itself the Hotel Commercial, to the eighty-nine licensed establishments for a Dunedin population of around 22,000 in 1864, to the £2669 paid in brandy duty at Hokitika in April 1866 compared with only £18 6s on soap, images of gold and drink go hand in hand in most minds. Very quickly after gold discoveries there were apparently numerous licit and illicit drinking establishments, private stills and an especially high import of spirits.[35] It is also doubtful that the diggers held their liquor any better than most of the population. The medical superintendent of Hokitika hospital claimed that insanity on the goldfields was largely the result of delirium tremens, and eventually a special room was built at the hospital for these cases. While Hokitika was the sixth-largest town in New Zealand in 1867, it had the third-highest number of convictions for drunkenness.[36]

It has also been suggested that the high rate of spirits consumption on the goldfields and the New Zealand frontier more generally until the 1870s had a more deleterious effect than the consumption of beer in Britain. Due to their greater strength, spirits were a much faster road to oblivion. As Jock Phillips put it: 'On vile liquor men's behaviour quickly deteriorated.'[37] Drinking by miners and rural workers such as bushmen and station hands was also apparently sporadic, with abstinence while engaged in hard frontier work being periodically traded for a chaotic 'spree' or 'binge' in which men came to town determined to make up for lost drinking time. This was certainly the case for Samuel Lisle, who was convicted in October 1863 for stealing a can containing a gallon of ale from Whitson's Albert Brewery in Auckland. In mitigation, he said he had been drinking since he came in from the bush and did not know what he was doing. He was sentenced to three months' imprisonment with hard labour.[38]

Accounts of alcohol-induced chaos are undoubtedly right for some of the people some of the time, but they need to be kept in perspective. Too many contemporary observations came from people who appear to have had little experience of or empathy with the normal nineteenth-century drinking culture of the working class. Others are no doubt derived from the visible extremes of behaviour, and therefore obscure much drinking that was mundane and uneventful. Nor is it clear why contemporary writers and later historians assume that spirits were consumed rapidly and destructively merely because they were spirits. Most drinkers, especially Scots and Irish settlers with a long-established tradition of spirit drinking, would be unlikely to imbibe in the same physical quantities as for beer. As for the spree, Rev. Henry W. Harper's observation in July 1861 is a good indication of the range of possibilities:

> [S]o far as I have seen, people here are temperate, but there is a curious custom among many of the station hands; for many months they stick to work, never showing any craving for drink; then comes their annual holiday; they draw a considerable amount for wages, and travelling to some shanty of a public house, or to Christchurch, proceed to 'knock down their cheque', giving it to the landlord, and bidding him to treat all-comers as long as it lasts. Needless to say that all they get for their hard-earned money is a sore head and empty pocket. You may argue with them, and they gravely plead that to 'have a burst' is necessary for health after the long monotony of station life and fare, and that it beats any medicine. Few are habitual drunkards, at least in the country districts. Of course, there are many who save money, and in a few years' time are in a position to start for themselves, and not a few who have overcome temptation and are thoroughly temperate.[39]

On the goldfields Frank Harbottle was perhaps not unusual in his recollection that while he 'went on the spree and got jolly well tight' with other miners it was 'the only time I was ever senseless in my life'.[40] A correspondent for the *West Coast Times*, covering the diggings in Ross in September 1865, suggested that drinking was a welcome sign of prosperity. On his first night at the Bendigo Hotel he described a crowded dance room where 'a tolerable amount of money was spending and several gentlemen were already in a state of beer and excitement'. At 2 a.m. a drunken fight started and ended in a general melee, but 'These occurrences are considered as favourable indications of the richness of a new goldfield, as the more drinking there is the more gold is supposed to be getting.'[41] When Harper moved to Hokitika in the late 1860s he organised a Parish Festival with games and entertainment for adults and children. In reply to a comment that the event was just like those of England he replied, 'I dare say you noticed a difference today; no beer, no broken heads, no scrimmaging, and no drunkenness.'[42] A perusal of biographies of those who became prosperous from the goldfields and related activity also

W. Nahr's Westport Brewery during the late nineteenth century.
Alexander Turnbull Library, PUBL-0104-001

suggests that drink was not so debilitating as to get in the way of a great deal of hard work, personal progress and long-term economic development.[43] In short, some indulged furiously, while others moderated their habits over time or stayed away altogether.

The nature and extent of military drinking also needs to be considered here, although by definition it was largely confined to distinct periods and locations during the mid-1840s and 1860s when troop numbers grew, as tensions with, and then campaigns against, Māori increased. There was certainly a strong drinking tradition in the British army, sanctioned in part due to beliefs about the medicinal value of alcohol, and a rum ration of half a pint a day remained standard during the nineteenth century along with provision of beer when available.[44]

Charlotte Macdonald also notes that 'Preparations for war, training and drill, parade and sentry and picquet duties, the routine mundanity of life in uniform, on a naval ship or on barrack ground, are supremely tedious most of the time (hence the rum, the booze and the brothels, the cards and the gambling).'[45] Various reports towards the end of 1863 suggested that drunkenness among the troops in Auckland was a common source of complaint. But as one correspondent described the anomaly of the situation: 'In one corner of a Camp — A Canteen, to assist the men to drink, and consequently to drunkenness. In another corner — The Triangles, where the man is flogged for being drunk. In another corner — A Hospital, to assist in recovering, as quickly as possible, the man who has been flogged.'[46] Yet it is evident, not least in the case of the 70th

Regiment at the Miranda Redoubt on the Firth of Thames around Christmas 1863, that officers also took an active rather than merely reactive approach to controlling drinking.

> Captain Tighe, our commandant at this camp, took a strong measure on Thursday morning. He seized all the beer, and had it removed from the canteens up to the front of the redoubt where it yet remains piled in its barrels and guarded by a posted sentry. This has caused a good deal of dissatisfaction, not only by the canteen proprietors but also by some of the sober men of the different corps; they say even their dinner beer was disallowed. I hear this morning that the said beer is to be sent up to Auckland forthwith, and no more allowed to be sold or landed at Pukorokoro.[47]

While officers of the 40th Regiment reportedly treated their men to a substantial dinner and a liberal supply of ale without incident at Christmas 1866, the departure of the 18th Regiment from Whanganui towards the end of 1867 suggests that they were appreciated more as consumers than defenders.[48] As the *Wanganui Chronicle* remarked: 'In a pecuniary point of view, the departure of the 18th will be a loss to our storekeepers, but one which they will speedily get over. The beer trade is the only one that will be materially affected, and suffer most.'[49] Before departing, however, the men had one final outing to the Whanganui races.

> The Grand Stand was thoroughly tested today by about 160 soldiers, who moved about in it in such a way that if it had got a weakness in any part of it, it would have been discovered. After regaling themselves on colonial beer, every muscle was exerted to give a momentum sufficiently great to bring the Grand Stand to the ground, a consummation that would evidently have greatly delighted the men of the gallant 18th; but it stood like a rock. This was a precaution wisely taken by the Jockey Club, who have manifested a regard for the public safety.[50]

These and other displays, and no doubt the emergence of fearsome military threats from Tītokowaru and Te Kooti during 1868, prompted something of a moral panic about military drunkenness. In September, the House of Representatives passed New Zealand's first central government licensing legislation, the Public Houses Act 1868, which empowered the governor to suspend licences and the import of alcohol to any district where the military was engaged.[51] William Fox, of whom more in the next chapter, insisted in the House that drunkenness was due to the moral flaws of the troops. Others attributed it to a lack of firm military discipline. According to the *Evening Post*: 'The disorganisation at the Front, both in matters civil and military, is unparalleled: drunkenness and insubordination almost unchecked, and very few officers found willing to encounter the daily humiliation they have to submit to from the want of management existing.'[52] However, by the time the last British regular forces departed New Zealand in 1870, there was no lingering sense that their drinking exploits were any more or less pronounced than in any other part of the British Empire.

Evidence from per-capita alcohol consumption statistics, sketchy though it is until the 1870s, also fails to paint frontier New Zealand as a paradise of spirited inebriation. Moreover, the available figures are for Europeans only. If Māori were added, although there are no statistics for this purpose until 1881, the per-capita rates would fall even further.

Table 3.4 confirms the general impression established from the early 1840s that New Zealand consumed substantially larger quantities of spirits and wine than England and Wales in particular and the United Kingdom in general during the 1860s, but substantially less

Table 3.4: New Zealand and United Kingdom Alcohol Consumption, 1864–1878 (litres)

Annual consumption per person

	New Zealand			England and Wales			United Kingdom		
	Beer	Spirits	Wine	Beer	Spirits	Wine	Beer	Spirits	Wine
1864	28.4	16.3	6.5	–	–	–	119.1	4.1	1.8
1865–69	–	–	–	163.2	3.4	–	130.9	4.4	2.0
1867	71.6	11.0	4.4	–	–	–	130.9	4.5	1.1
1870–74	–	–	–	173.6	3.8	–	141.4	5.0	2.4
1871	82.2	9.6	2.1	–	–	–	134.6	4.8	2.3
1874	68.0	8.2	2.7	–	–	–	151.4	5.7	2.4
1875–79	–	–	–	184.1	4.3	–	150.9	5.7	2.3
1878	67.0	6.9	2.0	–	–	–	152.7	5.3	2.1

Annual per-capita consumption of pure alcohol*

	With beer at 5% alcohol by volume		With beer at 7% alcohol by volume	
	New Zealand	United Kingdom	New Zealand	United Kingdom
1864	8.83	7.81	9.30	10.20
1867	8.64	8.27	10.08	11.21
1871	8.30	8.92	9.95	11.62
1874	7.12	10.15	8.48	13.19
1878	6.47	10.01	7.81	13.07

* Spirits are assumed at 40% and wine 15% abv. New Zealand beer figures for 1864 are for imports only. The production for the fifty-one breweries recorded in the 1867 census was acknowledged as an underestimate due to incomplete returns. Figures in the table are for Pākehā only.

Source: *Statistics of New Zealand*, 1853–1900; New Zealand Census, various editions, 1867–1901; T. R. Gourvish and R. G. Wilson, *The British Brewing Industry, 1830–1980*, Cambridge: Cambridge University Press, 1994, 30; G. B. Wilson, *Alcohol and the Nation: A Contribution to the Study of the Liquor Problem in the United Kingdom from 1800 to 1935*, London: Nicholson and Watson, 1940, 330–32; B. R. Mitchell, *Abstract of British Historical Statistics*, Cambridge: Cambridge University Press, 1962, 8–10, 408–10.

beer. This was hardly surprising given the scale of the local brewing industry and the relative difficulty noted earlier of beer compared to spirits and wine as an import item. But relative differences in patterns of consumption do not justify the historical emphasis on spirits over beer beyond the 1860s. In fact by the 1871 census, and presumably sometime earlier, beer exceeded spirits as the dominant source of pure alcohol, and the rate of spirits consumption was in sharp decline. At no time, even during the 1860s, was overall consumption unusual compared to Britain. Without question, some worked particularly hard to sustain the per-capita consumption figures, but there is no way of knowing how many of them did this via a healthy and habitual pint after work and a whisky at bedtime, or how many saved their entire quota for a periodic binge. There are simply too many unanswered questions to enable a clear picture of how much drinking was going on in an ordinary environment on a day-to-day basis in the goldfields or anywhere else.

Of course these data represent licit consumption only. Stevan Eldred-Grigg claimed that there was both a substantial amount of smuggling and illicit production during the nineteenth century. But by its very nature, the true extent cannot be determined. While some claimed that both were rampant, others were doubtful.[53] Convictions for 'sly grog', which were common enough during the nineteenth century, refer merely to selling without a licence, and reveal little beyond anecdote about whether the alcohol was self-produced or purchased for resale. As noted already, the skills of domestic brewing, not least being yeast maintenance in an isolated setting, were in sharp decline. Likewise, official encouragement of legal commercial whisky production in Scotland from the early 1820s, and more determined efforts to pursue excise evaders, had sharply reduced traditions of illicit production.[54] In both cases, bulky equipment was also required to produce in any significant quantity. In assessing whether New Zealand alcohol consumption was excessive by contemporary standards, any level of illicit production needs to be set against some continuation of domestic brewing in England and Wales, and some smuggling and illicit distillation in Scotland and Ireland.[55] In short, aside from issues of the erratic supply of licit alcohol that undoubtedly affected isolated areas at particular times, there is no compelling reason to assume that those in New Zealand had a peculiar propensity to smuggling or skills for illicit production.

As noted in Chapter One, judgements based on rates of conviction for drunkenness should also be treated cautiously. Working from the highly unlikely assumption that each adult male convicted for drunkenness was only convicted once, the rate of 5.8 per cent for 1858 dropped to 3.1 per cent in 1861, and peaked again at 5.3 per cent at the height of the gold rushes in 1864, but fell to 4.5 per cent by 1874 and 2.6 per cent by 1881. Expressed in these terms, one may wonder whether public drunks were especially ubiquitous among the total population. By definition the periodic 'spree' also suggests that convictions occurred in clumps in very particular areas, and that there were long periods when most people were not exposed to the troublesome drinker.

Figure 3.2: Convictions for Drunkenness, 1854–1880

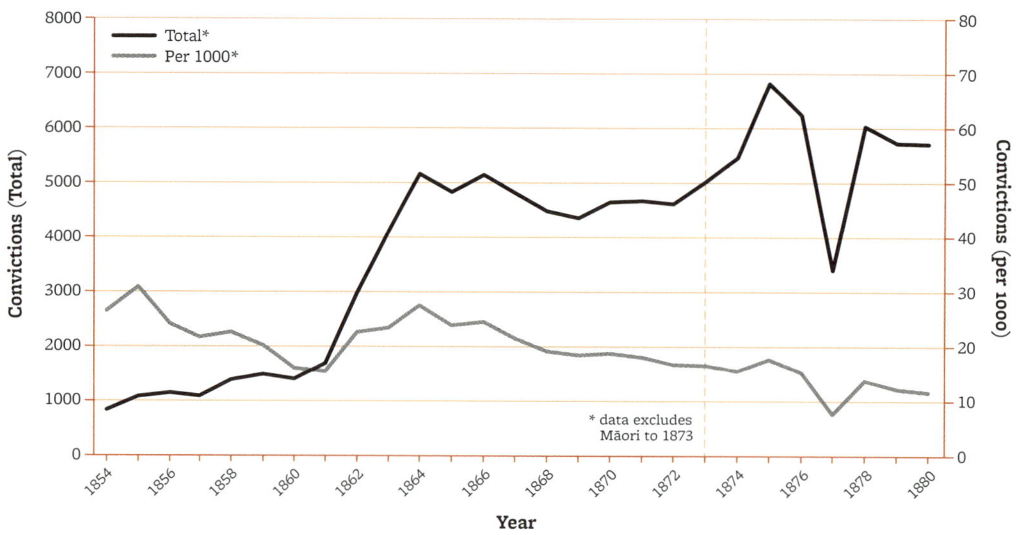

Source: *Statistics of New Zealand*, 1854–1880.

Nor do pub numbers tell us anything reliable about consumption. Although they were sites for drinking, as previously noted they had many other roles to play in a developing society. While initial gold-rush and other settlements featured many temporary and insalubrious shanties cobbled together to capture the demand of a shifting frontier population, many of these businesses quickly became permanent multi-storeyed establishments. As well as single bedrooms, bathrooms and a diverse menu, the better ones offered dancing to orchestral accompaniment, hosted regular public dinners and balls, and supported a wide range of sport and games.[56] Over time, competition in a crowded market undoubtedly raised standards. Many publicans also had important responsibilities to their locality. When the first publican's bush licence was issued to George Simpson at Lyell Creek on the West Coast in March 1864, he was required, in lieu of an annual fee, to provide a boat 'at a charge not exceeding one shilling for each person' crossing the river. A second licensee without these obligations was required to pay £20 a year.[57] In 1869 Joseph Giles, the goldfield's warden and resident magistrate at Westport, although generally opposed to the proliferation of licensed establishments, nevertheless endeavoured to ease travel problems up the Buller and Īnangahua rivers by granting a free publican's licence to those willing to keep ferry boats available at all times.[58]

*

While the colony was not wallowing in alcohol in the manner suggested by many contemporary and later accounts, there was certainly a level of consumption that offered substantial rewards to brewers with the good management and good luck to satisfy it. Above all, long-term survival depended on access to sufficient capital – initially to acquire suitable premises, good brewing equipment and a regular supply of raw materials, later to expand these and develop efficient supply networks if the product proved popular. Successful breweries, as noted already, were able to tie hotels to ensure some guaranteed market. Some also began to employ travellers who found business further afield. Undercapitalised breweries, especially those in rural localities with a small local market, could not do any of these things or even fully insure their premises against the all too frequent perils of fire and flood.[59]

No doubt many small concerns, especially those without significant capital or experienced brewers, began life in the sort of primitive, hastily assembled wooden structures that characterised rapidly expanding gold towns such as Queenstown and Hokitika. But most fell in the wake of prosperous establishments that were able to procure good equipment and build multi-storeyed landmarks in mainly single-storey surroundings. The ideal, to avoid the need for pumping between one brewing stage and the next, was for a brewing tower in which raw materials and water were taken to the top floor, from where gravity moved them through mashing, boiling and fermentation to the bottom, where the beer was racked into casks for conditioning. Initially much brewing equipment was imported from Melbourne or from England, but by the 1870s Dunedin firms such as A. & T. Burt and Sparrow & Co. were manufacturing boilers, vats, copper piping and other equipment for the New Zealand market.[60] Steam power and refrigeration also came relatively early to New Zealand breweries. When Richard Taylor advertised the sale of his Phoenix Brewery in Christchurch in April 1863, the plant had a 120-hogshead capacity and included a steam engine, water tanks able to hold 270,000 litres, grinding equipment and pumps to take water to any part of the brewery. The sale included a brewhouse, malthouse and a dwelling house of six rooms.[61] At the end of 1863, a report on Marshall & Copeland's Water of Leith Brewery emphasised the sophistication of

Staff of J. Mandl & Co., Hokitika, during the 1890s. Joseph Mandl is possibly at far left.
Barry Thompson Collection, Alexander Turnbull Library, PAColl-0605-1

the establishment, especially a water filtering bed and bottling and corking machines. Of the facilities at the nearby Well Park Brewery the reporter observed, 'it becomes at once evident that they have been planned by a person well acquainted with mechanical science'.[62] By early 1866 Pizzey & Arkell's Phoenix Brewery, Hokitika, featured a custom-built 4500-litre vat constructed from white pine (kahikatea), with two more in production. A lengthy description of the brewery in December 1868 included cellarage for 300 hogsheads, a cooperage, a cask washing room, an engine house and 12-horsepower engine, cooling fans and a mash tub 'fitted up with Steele's patent mashing machine, combined with a very ingenious arrangement for directing a flow of hot liquor onto the malt, so as to extract almost every particle of saccharine matter'.[63]

But technology of itself did not ensure survival. For their Swan Brewery and cordial factory established at Cromwell in 1874, George Goodger and Joseph Kühtze imported a steam engine, a bottling machine and a soda-water machine from Melbourne. They also built a new granary and malt kiln. 'With such excellent establishments as we possess at our very doors, and in a neighbouring township,' opined the *Cromwell Argus*, 'we can not see why a single gallon of Dunedin manufacture should be allowed to come into the district.'[64] But by April the following year, for reasons unknown, the partnership was dissolved and Kühtze departed for Dunedin, where he experienced two major brewery fires and a bankruptcy during the next five years. Meanwhile Goodger failed to make a success of the brewery, which was passed in at auction in 1877, and his dire financial circumstances arising from failed investments in water-race developments for mining at Bannockburn likely contributed to his death in 1883.[65]

Weather also caused its share of havoc for brewers. In August 1867 Henry Edmeades' Te Aro Brewery lost a significant amount of stock in a Wellington flood, and the following February the Temuka Brewery was swept away altogether. Until the Leith was strengthened with a stone wall around 1870, nearby Dunedin breweries suffered frequent damage from floods. Marshall & Copeland lost 14 feet of their yard in one, and a significant number of casks and malt tanks in another.[66] During the early 1860s Christchurch's Albion Brewery managed to embrace both extremes of the catastrophe spectrum. On 23 May 1861, it was destroyed along with five other buildings in a fire that caused £10,000 damage. Soon after being rebuilt, it sank into mud during heavy rain and collapsed.[67]

The long list of other breweries, obscure and prominent, damaged or destroyed by fire included Taranaki's Bell Block Brewery in March 1862, Francis Franks' Pierian Springs Brewery at Waikouaiti in 1872, and George Hally's Cambridge Brewery in 1881.[68] It was perhaps a blessing in disguise that the total destruction of the wooden Well Park Brewery in 1872 enabled its replacement with an imposing stone structure. Less lucky was John McNeil, who, after resorting to the courts to evict the insolvent Michael Houlihan from Dunedin's Cosmopolitan Brewery and then establishing a new partnership to run it, saw the premises burn to the ground almost immediately in March 1883.[69] While most of these disasters stemmed from

A bottle of Extra Stout from Simpson & Hart's Black Horse Brewery at Wetherstons, Central Otago. *Brewnews*, Lion Technical Library

fires not being sufficiently extinguished in the malt kiln, Taranaki experienced different fire challenges. In January 1875 Trigger's Brewery in New Plymouth, which had already sustained damage from a whirlwind eighteen months earlier, was the target of an unknown arsonist who placed an incendiary bundle against an outside wall. Extensive damage was limited by falling rain.[70] The arsonist of the Hawera Brewery in September 1883 was more easily identified when the owner, John Burton, was arrested for insurance fraud.[71] Likewise, when the Albion Hotel in Reefton, owned by Stuart Monteith, brewer, was destroyed by fire in May 1895, there was reason to believe the blaze was no accident as the insurance value of £300 was considered to be twice the value of the aged building.[72] Much luckier was the owner of Alexandra's Criterion Hotel. When fire took hold on a cold winter night in July 1903, most of the nearby water was frozen. Buckets of beer were then pumped from barrels in the cellar 'and the fire was finally extinguished by a liberal application of Simpson and Hart's XXXX before any serious damage had been done'.[73]

Richard Taylor of the Phoenix Brewery, Christchurch, was especially cursed. His brewery was destroyed by fire in December 1858 when the kiln furnaces were left too hot overnight. With a public subscription it was rebuilt, only to be destroyed again in early 1865. In between, Taylor's brother George died in February 1860 two days after scalding injuries received when he fell into a boiling brewery copper.[74] A similar accident to a brewery employee in Taranaki in June 1863 apparently caused no danger to life despite '[the] excruciating agonies of the poor man, from whose legs the skin hung in shreds'.[75] The following year it was reported that Richard Secombe's son was also recovering slowly from scalding after falling into a boiling vat in Auckland.[76]

Brewing science, by which is meant an understanding of how to turn raw materials into beer with consistent results, was another major challenge. Even beyond the rather rudimentary conditions of colonial New Zealand, brewing science was decidedly haphazard until the later nineteenth century, and the number of aspiring brewers who were thwarted by problems inexplicable to them can only be imagined. Although the three basic stages of brewing beer – mashing, boiling and fermentation – had common scientific underpinnings, there were numerous variations in the way British and European brewers carried them out. Moreover, brewers carefully guarded their recipes and secrets to the point where scientific development was far less than in other industries such as dyeing, calico printing or glass making. At mid-century, many British brewers, especially in rural areas, were contemptuous of the role of chemistry in brewing, and some were completely ignorant of the use of a thermometer or saccharometer. There was more scientific practice among the larger city brewers from the 1830s and 1840s, especially as lighter-coloured ales surpassed heavier winter-brewed beers in popularity. Following the lead of the Burton brewers in particular, more of them used chemical analysis and treatment of their water, such as filtering it through layers of gypsum, to minimise variations and assist with the high hopping rates necessary for pale ales, while soft London water, better for bringing out more of the saccharine matter from malts, was preferred for full-flavoured stouts. From the second half of the nineteenth century, doubts about the validity of brewing science were overtaken by a general acceptance of its credibility.[77]

Perhaps the greatest uncertainty centred on yeast for fermentation. While brewers had long been aware that it was needed, none understood how it caused fermentation from brew to brew. Many added dregs from one brew to start the next. Others dried yeast on twigs and added these. During the 1830s chemists in France and Germany proved that yeast was a living organism, but many scientists still viewed fermentation as a chemical reaction. During the early 1870s

Louis Pasteur defined the anaerobic life of yeast in beer and wine when it produces alcohol from sugar. His book *Études sur la Bière* was published in 1876 and translated into English in 1879. His work was added to by Emil Hansen at the Carlsberg Brewery in Copenhagen, who demonstrated the nature of yeast infections, different species of yeast and how to isolate pure types of yeast. However, this work was not translated into English until 1896.[78]

How quickly up-to-date brewing knowledge reached New Zealand is unknown. At the very least, all of the brewers found a climate similar enough to Britain and Western Europe and more conducive to brewing than that of their trans-Tasman neighbours. As the *Otago Daily Times* noted in September 1872, 'Some of our brewers who have had experience in Australia, inform us that great difficulty attends the business of brewing there, owing to the usually high state of the temperature and the sudden changes of weather. The expenses of production are therefore increased and the quality of the article markedly affected.'[79] Undoubtedly some of those with previous brewing experience had obtained it, like William Dawson, in modern and progressive establishments. Others certainly made a point of acquainting themselves with the latest ideas and information, such as George Duncan did on a return visit to Scotland before opening the Well Park Brewery in 1862.[80] A visitor to Ward's Brewery, Christchurch, in mid-1881 declared:

> Time was when people would have laughed heartily at the idea of any approach to scientific knowledge being requisite or even desirable in connection with brewing. . . . As a matter of course the chemical changes involved were as complicated then as now, but the rapid strides in scientific and technical research have enabled men to appreciate those changes and to direct and modify them in a variety of ways so as to secure beverages of uniform flavour and strength.[81]

But not everyone was so impressed with scientific brewing. When C. F. Todhunter, a partner in William Vincent's City Brewery, Christchurch, returned from England after brewing studies, 'he thought he knew all there was to know about it and found fault with everything'. Vincent's existing brewer left in protest.[82]

Pure water was the essence of good beer, but not easily or cheaply had even in the new settlements of New Zealand. Spring or well water was preferred, whereas stream or creek water needed to be filtered, often through a bed of charcoal, sand and chalk.[83] When Richard Secombe leased his first Auckland brewery from George Partington in 1856, the proviso was that if his well did not yield sufficient quality water to make 50 hogsheads every seven days Secombe could spend up to £50 on a new well with the amount to be deducted from his rent. Secombe embarked on further expensive and difficult well-sinking when he established the Great Northern Brewery in 1859.[84] A decade later, Horace Simpson Brown was finally able to prosper at the Timaru Brewery when he secured a regular water supply. As the *Timaru Herald* remarked:

> The water was for a long time deemed almost useless for brewing purposes on account of the large quantity of vegetable matter contained in it, but since wells have been blasted through the rock, a different quality of water has been obtained, and the most successful experiments have been made in brewing. . . . We trust the brewery will keep the character it has obtained for its beer, and in a small degree assist to check the importation of commodities which should be locally supplied.[85]

But the Captain Cook Brewery was less fortunate during the 1870s. Just as Thomas Hancock was considering expanding his plant and network of tied hotels, complaints were made that his beer tasted bad and eventually he conceded that 'it stank'. Despite reading, taking advice and cleaning the brewery, a succession of brews valued at £800 had to be thrown out – again providing a counter to the notion that colonial drinkers consumed anything in search of oblivion. Eventually Louis Ehrenfried, who claimed to have been all but ruined by contamination during his early brewing days at Thames, suggested a problem in the water supply. Finally, seepage was found from an old drain. With a new well, but expenses and losses amounting to £2000 ($430,000 in 2022), good beer flowed again. Notwithstanding the cooperation that seemed to prevail between apparent brewing rivals – although not of course to the extent of sharing recipes – others were surely not so lucky.[86]

Even before Dunedin's dramatic gold-rush expansion, there were difficulties in securing a good water supply for the town. Most streams were used as drains and polluted with dead animals, sewage and other rotting refuse. Many soon became stagnant pools of mud, prompting future Dunedin luminary Alexander Bathgate to declare that he usually drank water but had now been forced to substitute colonial ale.[87] If this was Dunedin ale, it very soon came from wells as the Leith was not considered safe for drinking by the late 1860s. Nor were some country brewers much better off. Theyers and Beck lost the initially good water supply for their Manuherikia Brewery to mining debris.[88]

The final entry in this catalogue of challenges, the quest for suitable malting barley, also demonstrates elements of a cooperative spirit within the brewing industry. Although much malt was imported to meet the rapidly expanding South Island demand during the 1860s, its relative expense at 1 shilling 6 pence per bushel encouraged many to pursue local options. In Christchurch during 1866 the Canterbury Brewers' Association, a body that only seemed to exist during that year, offered prizes for the best samples of malting barley grown during the next season – £20 to the grower of not less than 300 bushels and £10 to the grower of not less than 100 bushels. At the same time Williams & Whitmore of the Oamaru Brewery offered a silver cup for the best 100 bushels of barley produced in North Otago, while Marshall & Copeland imported 1000 bushels of seed barley from England for distribution throughout Otago. In 1872 the combined Dunedin brewers gave the Otago A&P Society a prize of £30 and a gold medal for the best 500 bushels of Otago barley.[89] By 1871 there were 23,000 acres in barley throughout the country, with 50 per cent of it in Canterbury, a proportion that rose to 60 per cent by 1878. By this time, a maltings had been established in Heathcote Valley near Christchurch which would endure to dominate the New Zealand brewing industry. In the late 1880s a maltster was recruited from England, and by 1889 the business had secured contracts to supply the large Toohey & Co. and Tooth & Co. breweries in Sydney. By the turn of the century there were 48,000 acres in barley, with around 40 per cent in Canterbury but significant pockets in Nelson, Marlborough and Otago; 90 per cent of all barley was grown in the South Island.[90]

Despite an expanding crop, there were frequent complaints that farmers were not growing or preparing suitable barley and that too much of what they sent to market contained oats, wheat and other seeds.[91] In 1871 Thomas Hancock lamented that no farmer in the Auckland province would 'bother himself with growing barley, while he could rear sheep for the boiling down establishment'. He was apparently compelled to spend £3000 on Canterbury barley when the money ought to have been retained in the Auckland economy.[92] For their part, farmers often complained that brewers were unreliable purchasers who frequently did not deliver on promises to buy that which had been grown specifically for them.[93]

There is no better indication of the peaks and troughs of the incipient nineteenth-century brewing industry than the career of Augustus Moore. Having been a farmer and liquor retailer during his first decade in Canterbury, he acquired Richard Taylor's Phoenix Brewery site after a fire in early 1865. He had trouble with his brewer, Edward Watson, and lost a civil case against him, although the nature of it is not recorded. Moore won an award for ale at a Melbourne Exhibition in 1866 and another for porter at the Canterbury A&P Show in 1868. By 1869 the brewery had expanded significantly and produced 590,000 litres, with its porter described as a 'specialite de la maison'. But by July 1870 Moore was bankrupt with liabilities of £3000–4000. He made an offer to his creditors and was in business again by September 1872, when he took over the much smaller Windmill Street Brewery. But he was bankrupt again by August 1874 with a deficiency of £617, which he attributed to 'bad malt'. After a further discharge from bankruptcy, he briefly managed a small brewery for investors and won prizes for ale at the Canterbury A&P Show in 1875. Two years later Moore bought a brewery in Timaru, substantially improved it and renamed it Phoenix. But this venture did not last long, and he was later a traveller for Ward & Co. and 'financially turned his attention to manufacturing chemical compounds' – most likely for soft drinks.[94]

Moore's career confirms that the ability to make good beer was no guarantee of success, but that it was also relatively easy in a network of constantly changing local breweries to find another opportunity. But the winds of change were already starting to blow. As Chapter Five will detail, economic depression during the 1880s, the rise of prohibition and improvements to transport networks would all enable larger, more efficient breweries to encroach on local markets. Most of those who failed in the later years of the nineteenth century never returned.

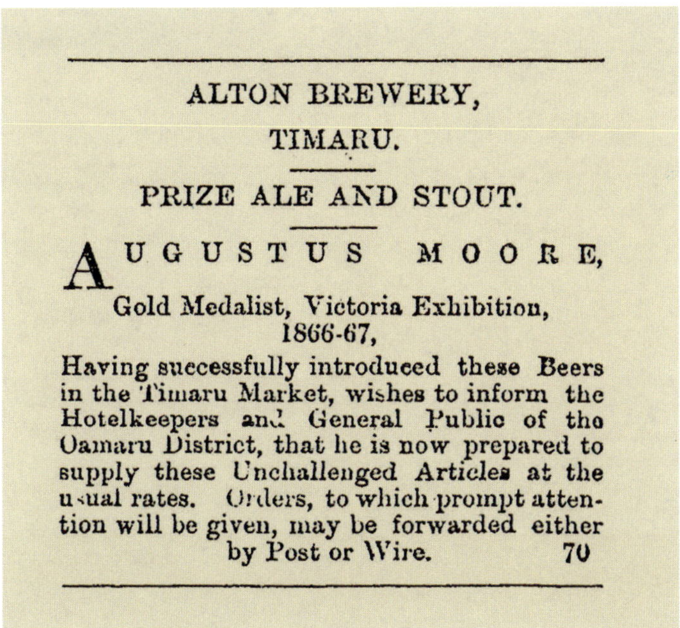
ALTON BREWERY,
TIMARU.

PRIZE ALE AND STOUT.

AUGUSTUS MOORE,
Gold Medalist, Victoria Exhibition, 1866-67,
Having successfully introduced these Beers in the Timaru Market, wishes to inform the Hotelkeepers and General Public of the Oamaru District, that he is now prepared to supply these Unchallenged Articles at the usual rates. Orders, to which prompt attention will be given, may be forwarded either by Post or Wire. 70

Augustus Moore's brewing career regularly veered between prizes and bankruptcy. This 1877 advertisement represents probably the last brewery he managed. *Oamaru Mail*, 26 July 1877, 1

The yard of the Black Horse Brewery, Wetherstons, Central Otago, during the 1890s. This was one of the most significant breweries outside the main centres until its closure in 1923. Hocken Collections, University of Otago, 509-262d

FOUR

'Trifling with the Business of the Country': Regulating Beer, 1868–1893

From 1 January 1869, life in the New Zealand brewing industry was never quite the same. Before that date brewers had been entirely unregulated aside from the requirement that, as they were not required to have a retail licence, they should not sell their product in quantities less than 2 gallons. Breweries situated within a proclaimed goldfield were also required to obtain the annual business licence that applied to all goldfield businesses. But slowly at first and then in a seemingly never-ending wave, brewers and all associated with the industry were enmeshed in a web of bureaucracy that would eventually shape every aspect of their operations, from the way beer was brewed, its distribution, its price, to where and how it was consumed.

Some measures were very specific to beer and brewing; at other times the industry was collateral to reforms and regulations with other objectives. Many of these things will be at the core of subsequent chapters. For the moment it is important to consider the fortunes of the national alcoholic beverage, as it assuredly now was, from the Distillation Act 1868 to the Beer Duty Act 1880, the Licensing Act 1881 and the rise of prohibition. But at the same time as politicians and lobbyists began their earnest and interminable debates about the role of beer, and alcohol more generally, in New Zealand society, this period witnessed the full flowering of a discussion perhaps more fundamental to the confirmed beer drinker – was the colonial product any good? Did New Zealand possess the climate and raw materials to produce decent beer, and could it hope to match the British standard with which almost everything in the colony was compared? The answers to these questions, as with those facing the legislators, were many and varied.

*

The licensing and other regulation of New Zealand brewers under the Distillation Act 1868 had something to do with myth and perception, a lot to do with spirits, but little to do with beer. In fact, only 10 of the 159 clauses of the Act specifically addressed the brewing industry. In part the New Zealand approach was influenced by sceptical consumers and the growing number of critics of the licensed trade in Britain who had long accused brewers and publicans of adulterating beer to make it last or to make it go further in the interests of profit. At the end of the eighteenth century, as higher grain prices during the French Revolution and subsequent Napoleonic Wars tempted brewers and publicans to dilute beer, there was certainly evidence of a more sophisticated approach to adulteration. The trend continued as small brewers, publicans and, after 1830, new beerhouse owners struggled to compete in a market increasingly dominated by a few large breweries. Among a range of drugs and plants apparently used to preserve or restore strength to beer were *Cocculus indicus*, copperas, quassia, capsicum, and hartshorn shavings. Sulphuric acid was also used to 'harden' new beer so that it did not need to be stored as long for aging. However, more commonly water and small beer were used to 'stretch' a barrel. Between 1813 and 1819 there were nearly a hundred convictions in England for defrauding the revenue by this means, although there appears to have been little concern with health aspects of adulteration until the passing of the Adulteration of Foods Act in 1860 and its successor the Adulteration of Foods, Drink and Drugs Act in 1872. Even then, much of the evidence for adulteration in Britain appears circumstantial rather than actual. In 1867 *The Lancet* conceded that there was no hard evidence for the use of *Cocculus indicus* in beer, but, given that it was imported into Britain and not used for medical purposes, it could be speculated that it was probably being used surreptitiously by brewers. The appointment of public analysts during the 1870s greatly reduced the adulteration of beer with narcotics, but the use of water and small beer was still very common.[1] In New Zealand the Adulteration of Food Act 1866 made it an offence to sell or manufacture adulterated products irrespective of whether they were deleterious to health, and allowed any suspecting purchaser to have items analysed by the colonial analyst on payment of a fee. Any manufacturer found guilty faced a £10 fine or imprisonment.[2]

No evidence has survived that New Zealand brewers adulterated their beer with nefarious substances and there were no prosecutions under the 1866 Act. However, it is apparent that bottlers and publicans were less scrupulous, not least because of the practice of importing beer in bulk and bottling it upon arrival. In 1863 it was reported that men in Dunedin were asking for empty beer bottles of imported varieties in good condition and with the labels intact. These were sold to bottlers or publicans who filled them with colonial beer for sale at the higher prices prevailing for imports. Forged labels were also available from Melbourne. Imported ale that was found to be bad on arrival was also sold cheaply at auction to publicans who then blended it with good colonial ale and made a profit by selling the resulting concoction. The Dunedin *Daily Telegraph* reported on 15 February 1864 that a number of respectable brewers such as Marshall & Copeland were being consistently undercut by a sophisticated fraud.

> We now have bottlers in the city who offer ale and porter of any well-known home brand that may be asked for, all drawn from the same tap. The facsimile labels of Byass, Bass, Jeffery, Allsopp and other established names are lithographed and pasted on to bottles in compliance with special orders, or in accordance with the demand of the market, and so expert have these nefarious dealers in their trade become that it is now very difficult for any but the most experienced to detect externally the imposition. Even the verdigris which collects on the wire placed over the corks on the voyage out, which used to be a means of detection, is imitated, the names of the English houses are branded on the corks, and the freshness of the tinfoil and capsules is made to look old by means we do not care to learn. . . . A process of doctoring follows by means of carbonate of soda, sugar and yeast, and the public swallow the trash under the belief that they are indulging in the genuine article.[3]

In 1870 a report of the parliamentary committee on weights and measures and adulteration of food concluded that while colonial-made beers were all found to be pure, a sample of imported beer, 'bearing on the bottle the brand of a large and well-known English firm, was found to contain poisonous drugs, namely nux vomica and cocculus indicus'. The committee believed that the most popular English brands were being forged not in New Zealand but in Australia, and resolved to pass on the result of their analysis to brewers in England.[4]

In July 1873 Bass, which already had a history of protecting its trademark and would shortly receive the first formal registration under the Trade Mark Registration Act 1875, fought back.[5] Its agent in Auckland went to the courts in pursuit of at least nine publicans and merchants who had sold ale with counterfeit Bass labels. Several publicans were required to publish statements of apology, although all claimed innocence of being involved in a fraud. But at least one who refused to give information as to where he obtained his counterfeit Bass was charged with a breach of the Trade Marks Act on the basis that a refusal to comply with the demand to give full information was prima facie evidence that he knew that the trade mark was a counterfeit. Having been advised by his lawyer to plead guilty, the defendant agreed to give the required information. Further prosecutions of this kind were pursued in Thames a fortnight later, with the same eventual willingness to cooperate. But whether Bass representatives found their way to the original counterfeiter was not reported.[6]

Yet it was the ambiguous content of spirits, especially whisky, rather than beer that attracted most attention. Under an ordinance of 1841 all distillation of spirits in New Zealand was prohibited, not least as a deterrent to those who may have been tempted to supply to Māori.[7] However, by the mid-1860s there was growing concern with illicit distillation – real and imagined. Aside from the extremely dubious, if not dangerous, quality of the local product, some of which was apparently also used by unscrupulous merchants and publicans to dilute imported scotch, the government was also concerned at the loss of revenue. When duty on imported scotch was raised from 9 to 12 shillings a gallon in December 1864, demand declined sharply from 84,581 gallons in 1862 to 56,151 in 1866. This, notwithstanding the suggestion in the previous chapter that the extent of illicit production has been exaggerated, does suggest that during the gold rushes at least some were satisfying their demand locally.[8] Indeed, at least fifteen illicit stills were seized during 1866–67, with William Seed, secretary of customs, noting that 'many of these were of the rudest possible description, and the spirits distilled in them must have been of too inferior a quality for general consumption, but yet no doubt such as would readily be disposed of in the outlying districts and among the Natives'.[9] The perception of a surfeit of mean spirit during the 1860s had also emerged in one of the many favourable reviews of Robert Whitson's Hobson's Bridge Brewery, Auckland.

> Had such an ale as the one now to be procured been supplied some time back at the same cheap rate, we will venture to say that one half of the drunkenness which results from the use of fiery spirits and heady inferior English ales which are commonly exported would have been prevented. If people must drink something other than water or tea, it is certainly desirable that they should be supplied with that which is neither hurtful to the constitution nor which owes its chief qualification to the fact 'it burns all the way it goes down'.[10]

In similar vein, William Theyers of Alexandra won praise for ales and porters that 'to a very great degree form a substitute for the present pernicious system of drinking so much raw spirit'.[11]

Seed and a number of politicians concluded that as the government could not effectively suppress illicit production, it was better to legalise domestic distillation and undermine the illicit industry by offering consumers a safe and legal alternative. At the same time, there was agitation from farmers, especially in Otago and Southland, to legalise distilling as another outlet for local grain, and from those who felt that expenditure on a local rather than imported commodity was preferable. The alternative of attempting to curtail illicit distillation by significantly reducing the duty on imported spirits failed to address the other prevailing concern of the government – that excise on spirits should continue to provide a significant source of revenue.[12] Accordingly, and with very little public or parliamentary debate, the Distillation Act 1868 legalised distilling with an excise duty of 6 shillings a gallon – half that on imported spirits. Penalties for illicit production and sale were increased, with provision for a fine of up to £500 ($65,000 in 2022) or six months' imprisonment for distillers, and a £50 fine and five years' suspension of licence for publicans selling the illegal product.

As much for the sake of completeness as to close off any remaining avenue to illicit distillers or beverage adulterers, beer made a belated appearance in the Act at clause 125. This required all breweries to register annually with the Customs Department and provide a full description of their premises, including the nature of the building and cellars, the design and capacity of the brewing plant, average monthly production and raw materials used. A customs officer could inspect the brewery at any time; brewing without a licence was punishable by a fine of at least £10 but a maximum of £100 ($13,000); a brewer could be fined up to £20 ($2600) for failure to paint his name 'in letters not less than two inches in length' upon every dray, cart or other vehicle used by the brewery; no brewery could simultaneously be used for wholesale or retail dealing in spirits, and it had to be at least 100 yards from such premises (a £50 fine); and no more than 6 gallons of spirits could be kept at a brewery (a 40-shilling fine per extra gallon). Clause 131 set itself determinedly against adulteration, although not against the seemingly more common sins of dilution or passing one beer off as another.

> Any public brewer who shall use or cause or permit to be used in the brewing of any beer ale or porter or shall put into or mix therewith or the worts thereof respectively any cocculus indicus, nux vomica, tobacco, tobacco-juice, opium, aloes, faba amasa, grains of paradise, cannabis indicus or any extract decoction infusion or preparation thereof respectively or any other deleterious or poisonous matter or thing shall be liable on conviction to a penalty of two hundred pounds for the first offence and for the second or any subsequent offence shall be guilty of misdemeanor and upon conviction shall be liable to a fine not exceeding five hundred pounds and to be imprisoned for any period not exceeding two years with or without hard labour irrespective of any other penalties imposed by this Act and all such beer ale porter and worts shall be forfeited and may be seized by any Inspector of Distilleries or other authorised officer.

It was also an offence punishable by a £50 fine to possess these items or to sell or deliver them to any brewer or retailer of beer 'otherwise than for some medicinal or domestic purpose'.[13]

Only the *Otago Witness* addressed the Distillation Act in detail. In a lengthy editorial it supported the distilling of whisky in New Zealand but regretted the imposition of an excise duty as this was a very expensive means of collecting revenue and an excise establishment would lead to 'the same absurd network of restriction as exists in Great Britain'. It also railed against the excessive controls on breweries merely to ensure their separation from distilleries and spirit dealing. 'And all this in order that New Zealand grown barley may have a chance of being made into whisky in the colony. If no better way can be devised of encouraging native industry than this, we had surely better be without it.' The Act, according to the *Witness*, was 'trifling with the business of the country'.[14]

In the end, the *Witness* had little to worry about, as the career of New Zealand distilling was brief and inglorious. The New Zealand Distillery, with William Larnach among its leading investors, began operations in Dunedin in October 1869. Auckland's Crown Distillery followed in 1870.[15] The Dunedin company wasted no time in bringing Duncan McGregor from the Glenury Royal distillery near Aberdeen to New Zealand to assist with the design and layout of its distillery and the production and marketing of the whisky. It then widely advertised that it was using Scottish expertise.[16] While there was a good supply of barley, water and coal in Otago, peat had to be imported from Islay at £10, a ton along with suitable bottles. At its peak the Dunedin operation employed thirty-five people, and production increased from 95,000 to 250,000 litres between 1869 and 1873 and profits from £775 to £41,000 ($6.25 million in 2022). The efforts in Dunedin won praise from the *New Zealand Herald* in November 1870 as the first produce from the Auckland distillery was awaited.

> [W]e will be anxious to see what kind of stuff is turned out, believing as we do that it will be considerably more wholesome than a great deal of the imported rubbish. We have tasted a sample of the whisky manufactured by Howden & Co at their distillery at Dunedin, and barring its new and raw condition, which we take to be the effects of lack of age, it seems to be spirit of really sound quality.[17]

Rawness, as a later select committee on distillation freely conceded, was a consequence of selling too quickly in order to make an initial profit.[18]

Yet the long-term future for companies that had been founded on a political whim was always precarious. In 1871 it was reported that revenue from imports of all spirits had been reduced by half, and that the government would have to consider increasing the excise on local products to cover the shortfall. Julius Vogel, as colonial treasurer, apparently set aside earlier concerns about illicit production, arguing that the local industry had been established only to benefit agriculture, but that the use of 42,000 bushels of barley a year did not offset the loss of revenue on imports. Consequently, the duty on New Zealand-manufactured whisky was increased from 6 to 9 shillings in 1873. By this point it was widely accepted, despite initial optimism, that the local product could not compete on quality with Scotland and would therefore not survive if its price advantage was removed.[19] As one Dunedin distillery employee informed a select committee, 'We have never been able to make our whisky like Scotch.'[20] The committee then determined that closing of the preferential tariff gap for local distilleries would not be an act of bad faith provided the distillers were compensated. When the New Zealand Distillery Company stopped production on 30 June 1873, it received £20,000 compensation ($3.1 million in 2022) and was converted to the Albion Brewing Company. The Crown operation in Auckland received £7500 ($1.1 million).[21]

Although the provisions of the Distillation Act remained in force, successive governments declined to issue any more licences. Conrad Bollinger in *Grog's Own Country*, his entertaining history of New Zealand licensing laws and not a few conspiracy theories, claims that the closure of the New Zealand distilleries was due to the desire of Vogel to appease powerful Scottish distillery interests concerned at declining imports into New Zealand at a time when the New Zealand government was also pursuing loans on the British money market.[22] Yet, aside from a lack of evidence for this claim, and the probability that local production only replaced illicit

Scandinavian men enjoying beer during a picnic at Lowry Bay, Wellington, December 1896. Petersen Collection, Alexander Turnbull Library, 1/2-52226-F

production, the volume of imports was already increasing during the early 1870s, before the New Zealand distilleries closed.

With the closure of the distilleries and the New Zealand wine industry scarcely an embryo, the brewing industry now had the market for locally produced alcohol almost entirely to itself. Moreover, the provisions of the Distillation Act covering brewers were not rigorously enforced. The first inspection of Otago breweries in July 1869 revealed that most were ignorant of the Act – or claimed to be.[23] When William Crawford applied for a brewer's licence in Gisborne in 1874, the local court had no idea under what Act licences were to be issued or what the fee should be. Crawford also said he had no idea, but that the fee was £10 in Auckland.[24] No such laxity would operate during the next decade and beyond.

*

As the New Zealand brewing industry expanded during the 1860s, predictions of the imminent demise of a local market for imports were frequent. In late 1863 *The Press* noted that 'The large increase in our local brew, induced by the climate and water being both so favourable for the manufacture, is likely in a short time to drive out of the market all English ales with the exception of the best brands. Consignors at home would do well to note this.'[25] In March 1865 the report of the jury judging beer at the New Zealand Exhibition in Dunedin declared that:

> [I]t is desirable that it should be stated that the Otago brews are not the mere wash which is so generally known elsewhere as 'colonial'; and there can be no doubt that, the Province being excellently suited for the production of sound and fine malt liquor, whether as regards temperature or the possession of soil admirable for the growth of barley, Otago beer will soon become an important item in the export trade.[26]

Samuel Marks of the White Star Brewery, Caversham, attempted to occupy the middle ground by advertising in 1866 that his No.4 Pale Ale was brewed in competition with English ales. 'This ale must not be compared with the common Colonial ales, but is intended to compete successfully with the best English brands, being eminently suited for family requirements or hotel business.'[27] On the other hand, a lengthy account of the Dunedin brewing industry published in September 1872 identified a broader context underpinning debates about style and quality in New Zealand. It concluded that while the prejudice against colonial ale had largely

A display by Manning & Co. at the New Zealand International Exhibition, Christchurch, 1906–07. Samuel Manning established his Christchurch Brewery in 1860. One of the three leading breweries in the city, it was part of the New Zealand Breweries merger in 1923. Alexander Turnbull Library, 10x8-1170-G disappeared, 'a considerable quantity of English ale is consumed more from custom than from any actual difference in the quality, which is clearly proved by persons drinking the best brands of Dunedin ale for English'.[28] Perhaps unscrupulous substitutions by publicans had performed a useful service after all. Equally, this is early confirmation of the triumph of brand and image over taste, which would become so much a part of the rise of brewing empires and the struggles of their smaller competitors during the twentieth century. During the mid-1880s the Petone Brewery emphasised that the water for its products was 'equal to the celebrated Burton water' and they consequently possessed 'many of the characteristic qualities of the Burton Ales'.[29] Manning's in Christchurch still felt it necessary during the 1890s to address the market for imports.

> BOTTLED ALE AND STOUT: EQUAL TO ENGLISH AND AT HALF THE PRICE. . . . THESE ALES are Acknowledged to EXCEL IN QUALITY all others bottled in New Zealand, and are Rapidly Superseding the various brands of English Imported Ale. The Extra Stout is Brewed from Malt Specially Imported from the Maltsters who supply MESSRS GUINNESS AND CO., DUBLIN.[30]

By 1910 they were enticing the North Island market with a slightly more cautionary approach. 'Beneficent as well as pleasant, equal to the best English Ale. Beware of cheaper Colonial imitations, chemically-matured, and buy only MANNING'S Winter brewed Pale Ale, purest and best.'[31]

In 1870 the *Auckland Weekly News* claimed that breweries were increasing in number and prosperity where some other industries struggled to compete with imported products in terms of both quality and price. Domestic beer's success could be traced to the advantage it received in relation to customs duty on imported beer.

> Our colonial ale will bear very favourable comparison, both for flavour and freedom from adulteration, with the general class of English-brewed ale; and we feel certain, from a thorough acquaintance with the nature of many local industries which have been attempted and have failed, that, had they received a little fostering care by means of a slight protection from an excessive amount of foreign competition, they might yet have been adding their quota towards the general prosperity of the community.[32]

Meanwhile, in a report to the Joint Committee on Colonial Industries in August 1870, Hooper & Dodson claimed that local ales had superseded imports in Nelson from the early 1860s and that imports had fallen from 19,000 to 4500 gallons during the previous seven years. Local hops were also being used in preference to imported varieties. The only impediments to the local industry were prohibitive duties in the Australian colonies which stifled a potential export trade, the lack of a New Zealand manufacturer of bottles and the need for growth in the New Zealand coal industry to reduce fuel costs.[33]

The confidence of the brewing industry can also be seen in the plethora of competitions and prizes that enveloped it from the mid-1860s. Following the lead of Dunedin's New Zealand Exhibition in 1865, the regular agricultural and pastoral (A&P) shows throughout the country began to include categories for local ale, porter and stout among those for livestock, vegetables and other produce. Indeed, competition was sufficiently intense to raise the spectre of skulduggery. While the report of the 1869 Canterbury A&P Show noted that 'They were all remarkably good samples, and compared with a year or two ago, spoke volumes for the enterprising spirit of our brewers', there was some suspicion regarding the winning entry by Manning & Co. '[T]here was a general impression both amongst judges, and others, that it was not purely a colonial manufacture, as required by the conditions of the show, but in fact contained a large proportion of English beer.'[34] But it was success beyond the colony that mattered most. As early as 1851, Hooper & Co. won a medal for malt exhibited at the Great Exhibition in London. They won further medals with bulk and bottled ale and porter at the Melbourne Intercolonial Exhibition of 1866–67, as did Marshall & Copeland of Dunedin, Moore & Co. of Christchurch, Harley & Sons of Nelson and Whitson & Co. of Auckland. Shortly afterwards Marshall & Copeland began an intensive advertising campaign throughout New Zealand for their 'exhibition beer'.[35] But there was little prospect of this being enjoyed by anyone beyond the colony. Although a small amount of Dunedin beer from Wilson & Birch was shipped to California in 1871, where it was declared to be 'light, sparking and brisk like the best English bottled beer' and predicted to become popular, there is no evidence that the consignment was repeated.[36] Marshall & Copeland also exported small amounts to Fiji, Honolulu and Australia during the 1870s. However, restrictive Australian, especially Victorian, trade tariffs presented a barrier to further expansion.[37]

By the end of the 1870s every brewery of significance, and some others besides, was adorning its advertisements with references to prizes of one sort or another, and hotels were similarly keen to stress to patrons that they had prize-winning beer on tap. In 1879 Speight's Strong Ale was highly commended at the Sydney International Exhibition, and gold medals were won in Melbourne the following year. By the mid-1880s Speight's was winning praise much further afield. In a letter from Glasgow on 9 November 1885, Mr Moore of Robb, Moore & Co., ship store merchants, explained that he had sent a sample to the excise chemist.

> [H]e has pronounced [Speight's] one of the finest beers he ever analysed, it having the high specific gravity of 10.80, the greatest amount of spirit and the least quantity of acid he ever came across. When I tell you that every shipment of beer going out of the Clyde must be sampled by the Excise Officer and then analysed by their chemists, I consider this opinion is better than anything else I could say on it.[38]

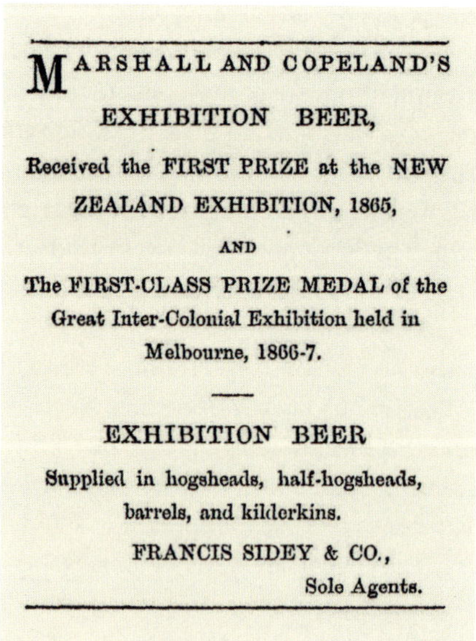

Marshall & Copeland's Water of Leith Brewery was one of the most successful in Dunedin from 1862 to 1887, and one of many to advertise the prize-winning quality of its beer.
Evening Post, 8 May 1868, 3

Unfortunately for Speight's it would be many years before their or any other New Zealand beer was exported to such discerning consumers in significant quantities.

Others were less easily swayed towards the products of colonial breweries, and there is a long history of scathing critiques of 'colonial', as the optimism of new settlements gave way to some scepticism as to whether they were developing to British standards. In September 1859 the new owner of Te Aro Brewery, Joseph Dixon, directly addressed his critic.

> A STATEMENT, calculated to do me serious injury, having been made in the Provincial Council, by one of the honourable members, that 'Colonial Ale was not fit to drink', I beg to state that whatever may have been the quality of the Colonial Ale in former times, since I have taken the Brewery, I have brewed as good and more wholesome ale than any ever brewed in the Colony before, and respectfully invite the Public to come and taste, and judge for themselves. JOSEPH DIXON, Te Aro Brewery.[39]

The following month the Canterbury Resident Magistrates' Court heard from a plaintiff who had purchased two barrels of beer at 40 shillings each from the surplus stores of a recently arrived ship. After tasting a sample and finding that the beer was of colonial manufacture and not English as he had believed, the plaintiff returned it to the auctioneer and claimed a refund of the cost, 1 shilling expenses and £4 damages due to loss of profit. The defendant argued that the plaintiff had purchased the beer 'with his eyes open, at a price cheaper than imported beer' and that the beer was sound and drinkable. Judgment was for the plaintiff in the amount of £4 1s as the resident magistrate concluded that while there was no loss of profit, and the beer was sound, the issue was about the nature of goods included in a sale without proper notice being given.[40]

In January 1870 'Egometipse', a long-time resident of Burton-on-Trent, informed *The Press* that 'after many years trial [I] have come to the conclusion that it will never pay to brew beer in New Zealand worth keeping and consequently worth drinking; but it must contain certain deleterious ingredients, the name of which are best known to those who use them'. He regarded New Zealand water as particularly unsuitable for brewing with pure malt and hops, and scoffed at 'those who imagine that the article can be exported to the neighbouring colonies'.[41]

The use of sugar was certainly a point of contention in some parts of the brewing industry at this time. It had been variously used in British brewing for centuries as a cheaper alternative to all-malt brewing, to increase alcoholic strength, to add colour and to alter flavours. But its use was officially banned in Britain until 1847. Thereafter, sugar could only be used if duty was paid, and consequently it tended to be used only sparingly until a further legislative change in 1880. However, as brewers sometimes struggled to produce good-quality raw materials in new colonial settings, the use of cane sugar was more common outside Britain. In Australia from at least the 1830s brewers used it to combat problems with nitrogenous barley and other issues with malt quality. Even then, the *Australian Brewers' Journal* argued in 1889 that sugar was used only to a very small degree and primarily 'to please the palate of some consumers'.[42]

In New Zealand, with a climate more akin to Britain, there appears to have been very little comment about the use of sugar beyond occasional proposals that it should be banned altogether to protect the market for barley. At its monthly meeting in late May 1880 the Canterbury Corn Exchange and Farmers' Club passed a resolution urging the government to enact legislation prohibiting the use of added sugar in the manufacture of beer so as to guarantee that only locally grown grain would be used. One speaker insisted that 'instead of its being a thirst quencher [colonial beer] was generally believed to be an inveterate thirst-maker and

A late 1880s advertisement for Speight's.
Lion Technical Library

stomach-deranger. [And asserted] that it possessed all the essential ingredients for the making of bad vinegar, and they were all aware of the injurious effects of such.'[43] The following year, a correspondent under the heading 'Colonial Poison' complained to Dunedin's *Evening Star*: 'I wish to call the attention . . . to the draft colonial beer sold in South Dunedin before the public get poisoned. It is bad enough to get bread made from potatoes, but when a man is confined to his bed two days through drinking one glass of colonial beer it is quite time the authorities looked after the public health. I am, etc., SUFFERER.'[44]

The *Otago Daily Times* trenchantly addressed the sugar theme in October 1882 as it embarked on the most sustained condemnation of local beer. A lengthy editorial noted that while Dunedin led New Zealand brewing and produced perhaps the best beer in the southern hemisphere, 'It will not, of course, bear comparison with the English article, but compared with Cape beer or even Auckland or Christchurch beer, it may be pronounced first rate. That is to say it is excellent good Colonial. But here comes the question – is Colonial a positively or only a comparatively good thing?' Setting aside the views of teetotallers who condemned all beer as 'a bitter and bad thing in every conceivable respect' there was, according to the writer, a widespread opinion that most beer, wine and spirits consumed in the colonies was not of good quality – 'being beyond doubt or question vile adulterate stuff'. The particular problem was the amount of sugar used in colonial breweries which gave the beer 'its peculiar flavour and character; and as long as the brewers continue this practice, for whatever reason, they will fail to produce the best possible beer that could be brewed in New Zealand'. While the very best New Zealand beer was not injurious to health, it was certainly not as wholesome or palatable as English beer: Given the very good quality of barley and hops that could be grown in New Zealand, the *Times* wondered: 'Why should not our beer be entirely a colonial product, free from all admixture of imported sugar, and in all respects equal to the light ales of the old country?'[45] When the indignant partners of James Speight & Co. attempted to refute these accusations and pointed to the various awards they had won, their letter was rejected by the editor as 'too cheap an advertisement'.[46] Perhaps the most comprehensive contribution to the debate came from a correspondent to the *Evening Post* in February 1897 who employed census data to conclude that proportions of malt and hops were declining and sugar increasing as the volume of beer production increased. 'In 1890 there were 848,134 bushels of malt and 441,364 lb of hops used, and 6,022,732 gallons of beer were produced. In 1895 only 328,059 bushels of malt and 424,838 lb of hops were used, but 6,249,278 gallons of beer were produced. The use of sugar had, however, increased from 1,345,985 lb in 1890 to 1,607,144 lb in 1895! How can New Zealand beer be called malt liquor after this?'[47]

Just as it had been observed that some drank English beer out of habit, some perhaps criticised the colonial product on the same basis. Beyond the subjectivity of personal taste, it is impossible to know how much of the perception of beer quality in nineteenth-century New Zealand was also entwined with broader debates about the culture and identity of the colony. Some were anxious to adapt, to prosper and to see all that was positive in the new settler society – beer included. Others remained determinedly British in outlook and judged all developments through that comparative lens. There were many positions in between.

Certainly, local brewers did not always have total control over the condition in which their product reached consumers. While dangerous adulteration by publicans seems not to have been a factor, a raft of court cases involving 'bad beer' highlight the difficulty in transporting beer from brewery to pub and storing it after arrival. Typical is the claim by Vincent & Co.

of Christchurch against Dennis Hoare in June 1880. The brewery sued for non-payment on 4 hogsheads of XXX colonial ale delivered to Hoare's Timaru hotel the previous January. Hoare claimed that the beer was 'hard' (contaminated) in April, but other witnesses claimed that the beer was good two months after delivery. The brewery stated that it was well known that its XXX colonial ale should not be kept for longer than two months during summer and was duly awarded the case with costs.[48] Early the following year John Bennie, an experienced brewer and maltster in Whanganui, sued another brewer, William Lingard, for £33 10s for 4 hogsheads that he had purchased from Lingard for a publican but which had gone bad. In defence Lingard insisted that Bennie had tasted the beer at his brewery and found it 'very good', but 'Mr Bennie did not treat the beer fairly as he allowed it to be shaken about in removal'. Even Bennie's own brewer, William Trimble, was of the opinion that 'the less handling beer got in the summer time the better'. Lingard won. Trimble's future employment is not recorded.[49]

Assessments of quality are further complicated by an inability to determine how beers of the past actually tasted. Even with surviving recipes, changes to barley, hops and yeast varieties all mean that it is difficult to replicate the product from a century or more ago. After touring

A vat of beer being worked, probably in the Nelson district, around 1910. F. N. Jones Collection, Alexander Turnbull Library, G-9347-1/1

numerous breweries in 1889, Australian brewer Alfred Lawrence claimed that not only were New Zealand beers considerably superior to Australian because they used significantly more malt, and all-malt in the best class of beer, they were also generally brewed to a higher alcoholic strength.[50] Beyond this there is simply not enough known about the origins and training of brewers and the exact specifications of their products to draw worthwhile conclusions, with the exception of some parallels between the Scots and their beer.

It is apparent that from the outset Speight's, with William Dawson to the forefront, produced a sweeter, stronger beer than many on offer in New Zealand, one more akin to the heavier styles of Scottish ale that have sometimes been characterised as 'alcoholic porridge' in comparison to their lighter English counterparts. An examination of the Speight's brewing book from 1879 to 1891 reveals production of nothing less than XXX ale and increasing amounts of XXXX ale, strong ale, stout and various 'special' beers as the decade progressed.[51] Dawson was constantly experimenting with new techniques and ingredients, and was a particularly enthusiastic brewer of stout. One recipe included gentian root, juniper berries and treacle. He also used liquorice and wheatmeal.[52] Journalist and idiosyncratic beer historian Pat Lawlor, in his *Froth-Blowers' Manual*, makes it clear that early-twentieth-century New Zealand was firmly divided into two camps based around liking or loathing of Speight's. Of his first encounter sometime before 1910 with what he described as a 'stern and solid' beer and a 'beer of integrity', he noted that his initial grimace and dislike of the high alcoholic content prompted the barmaid to suggest that in future he have it as a shandy.[53] There is no reason to doubt that the origins and training of other Scots brewers also had an impact on the style of beer produced. This is certainly evident in Whitson's expressed opinions noted earlier, while Robert Stewart in Greymouth was another to brew a maltier ale than his local competitors.[54] Equally there is no reason to doubt that brewers of Kentish, Irish, German or other extraction imparted local traditions and training to their work.

*

The irony of the rise in popularity of New Zealand beer from the 1860s and the consequent decline in imports was that it led to the introduction of an excise duty which was soon passed to the consumer as a price increase. Again, this was a development that owed much to British precedent. From the thirteenth century, supposed 'infringements' of the English Assize of Bread and Ale had allowed local authorities to collect a de facto licence fee from commercial brewers. During the sixteenth and early seventeenth centuries the government used 'requests' for 'loans' to obtain revenue from common, or commercial, brewers. In 1620 James I tried to ban all 'victuallers' from brewing in an effort to force them to secure beer only from common brewers who could be more easily taxed. While this and a similar attempt by Charles I in the late 1630s were fought off, excise duty on commercially brewed ale and beer became firmly established during the English Civil War. It remained in force until the Beerhouse Act of 1830 replaced excise with licence fees for all brewers and retailers, although a malt tax remained until 1880.[55]

Thus far New Zealand brewers had paid taxes only in the form of tariffs on imported sugar, malt and hops. This amounted to about 2 shillings 8 pence per hogshead – or 2.5 per cent if a hogshead was sold at about £5. With the halving of the sugar duty to come into force in 1880, this would reduce to 2 per cent – compared with 18 to 23 per cent of the price in Britain. Higher wages in New Zealand also meant that local drinkers were better off even if the tax rates were reversed. As Colonial Treasurer John Ballance observed in 1878, the consumer 'almost escapes taxation on the article'.[56] Moreover, with the expansion of New Zealand malting and hop cultivation the revenue from imports was in sharp decline – that on imported malt falling from £2654 in 1876 to £1599 the following year. The duty on imported beer had plunged from £52,268 in 1866 at the height of the West Coast gold rush to £29,266 in 1878. In presenting the Grey Government's financial statement in early August 1878, Ballance proposed a beer duty of 1½ pence per gallon (6 shillings 9 pence per hogshead) to take effect retrospectively from 7 August, before the Act had been passed by Parliament. Although not a temperance advocate, Ballance did argue that beer was not altogether a necessity and it was not a bad thing if a tax reduced consumption. However, his more pragmatic view was that a beer tax was a way of avoiding income tax.

> [W]e think the beer drinker will concur in the great moderation of our proposal if we ask him to contribute a beer-tax of 1½d a gallon. . . . He may certainly save his tax by drinking so much less beer; but, as the habit of indulgence is stronger than self-denial, there is no ground to anticipate in that respect an appreciable falling off.[57]

The total tax would be 8.5 per cent – the existing 2 per cent and a new 6.5 per cent from duty – which would yield revenue of £31,250 based on an annual consumption of 5 million gallons.[58]

The response was swift and unequivocal. The *New Zealand Herald* insisted that this would be a tax on drinkers rather than brewers or publicans, as both would simply pass on a price increase. Indeed, Auckland publicans had indicated an immediate price rise of 2 shillings a gallon – much higher than the 1½ pence duty. The *Herald* further questioned the appropriateness of a tax that would fall mainly on those least able to afford it.

> Colonial beer is not by any means a luxury of the rich. It is consumed almost exclusively by the poor. There can be but slight argument against it on the ground that it causes drunkenness, for the increasing consumption of beer must lessen the demand for spirits. There is another argument against the tax too, in the fact that it is a local production, and that the material is now in great part of Colonial production.[59]

The *Auckland Evening Star* responded in verse with 'Ballance's Beer Bill'.

> Twas said by Ballance in his speech,
> 'The beer-tax none will grudge it,'
> And spite of brewers not one foot from
> His budget will he budge it.
>
> 'Three half-pence on a gallon is
> not much to any man;
> The publicans can stand it if
> The Auckland public can.'
>
> The brewers, however, view
> This new tax as an ill,
> And 25 per cent stick on,
> To balance Ballance's Bill.
>
> Should publicans their price advance,
> The working men will fear
> That Ballance's beer tax has brought
> Their barrel to its bier.[60]

A reaction to the unsuccessful attempt by the Grey Government to introduce a beer tax in 1878. The measure succeeded two years later as the brewing industry became increasingly regulated. Auckland Libraries Heritage Collections, 661_151

Following antagonism in Christchurch on 7 August, when a customs officer accompanied by a constable visited each brewery to take stock in anticipation of the passing of the duty,[61] *The Press* raised similar objections to a tax on the working man. 'Beer, to a large number of the people of the colony, is as much a necessity of life as tea and sugar. . . . Colonial beer is the drink not of the "wealthy and privileged classes" but of the mass of the population.' It also objected to a tax that required 'inquisitorial inspection', would be expensive to collect and offered great temptation to fraud. Moreover, it was simply bad government to levy a tax on one of the few flourishing colonial industries.[62]

Brewers were no less indignant. At a meeting in Dunedin, Maurice Joel claimed that the tax was 'exorbitant and would press heavily upon the trade' by acting against an important local industry to reduce its competitive advantage against imported beer. There would be no need for a tax on beer if that on sugar had not been reduced. James Marshall added that as brewers must have some profit, some would be inclined to use inferior products in the manufacture of beer in order to keep their margins. This was not a good prospect given that English beer was still competing in the New Zealand market and would grow again if colonial beer deteriorated. William Strachan bemoaned such an excessive tax when duties on Australian light wines had just been reduced while no effort had been made to reduce Australian tariffs to create export opportunities for New Zealand beer. The meeting concluded that the tax would stop competition by closing a number of recently established small breweries and thereby confine the trade to a monopoly of large capitalists.[63]

On 21 August a deputation of brewers from Auckland, Wellington, Nelson, Christchurch and Dunedin met the colonial treasurer. They pointed out that brewing was not as profitable as many imagined; that the tax would interfere with the development of colonial industry; that it was a direct tax on a particular trade and therefore was objectionable; that it would impact far more on the working class if passed on to the consumer; that it would cause more consumption of spirits, which caused more drunkenness than beer; that local brewers objected to being taxed to make up the shortfall in revenue from declining imports of English beer; and that the distilleries had been given twelve months' notice before the removal of their competitive advantage in 1873 and had been given significant compensation thereafter. Finally, they argued, a beer tax had never been introduced in any British colony and therefore should not be introduced in New Zealand.[64]

While Ballance promised to consider the arguments of the deputation, some were less than sympathetic to its claims. To the argument that the poor man's beer was being taxed while the rich man's tea and sugar were given relief, it was pointed out that the real issue only concerned the account of the brewers with the government, as there was no actual need to pass on the increase to the publican or drinker. Further, as the cutting of the sugar duty was of benefit to the whole community, it was reasonable to trade this against the narrower interests of beer drinkers.[65] Others concluded that a publican who bought a hogshead of beer at £6 and sold it at 6 pence a glass for a 350 per cent profit was scarcely suffering.[66]

But the most sustained critique came from 'Anglo-Saxon' [J. W. Jago] in *The Economics of Beer*, which marked his debut as perhaps the most eloquent nineteenth-century New Zealand pamphleteer for the temperance and prohibition cause. Jago reminded his readers that when the brewers had told Ballance that the tax 'must have a mischievous effect and cripple an industry in which large sums of money are invested, and which gives work to a large number of people', Ballance had replied that 'It would not answer the government to cripple an industry.' But what, Jago asked, would be the case if many people were engaged in the distribution and hatching of scorpions and vipers in New Zealand? 'Would it not answer the purpose of government to cripple such an industry as that?'[67] Then followed a detailed explanation of the brewing process to prove that the industry was not essential to the colony and was in fact counterproductive in its consequences. Beer was a waste of barley that could be used for food, and the end of the industry would merely create scope for investment in other areas which would more than offset the loss of jobs. 'I do not think there are many who will plead for the medical virtues of colonial beer, and that phase of the question may at once be disposed of by saying that if no more beer was used in New Zealand than is legitimately taken for medical purposes, any one of the many brewers in the colony would be more than able to supply the entire demand.'[68]

Notwithstanding the claim of one editor that 'Brewers in the House and out of the House have been lobbying in a way never before excelled for pertinacity', it is far from certain how far the eventual demise of Ballance's proposal rested with the specific opposition of brewers and drinkers.[69] As much as anything, and irrespective of the particular commodity, opposition in Parliament stemmed as much from the fact that this was perceived as a class tax and the first attempt in New Zealand to impose a direct tax on local industry as opposed to raising revenue through tariffs on the materials used. Indeed, such was the determination to resist these methods of taxation that beer received support from the most unlikely of allies in the former premier and staunch temperance advocate William Fox. As the *Otago Witness* put it:

> Fox on the beer tax is among the political phenomena of the age. This ardent and uncompromising apostle of temperance has taken a new departure, to the great delight there can be no doubt, of all those who regard him as the champion of great principles. Everyone knows the estimate in which he is supposed to regard alcoholic drinks, their trade and traders. No legislation too extreme to crush the infamous business. . . . And yet the other day he fulminated against a Government that would rob a poor man of his beer. That drink, he said, was not to be regarded as a luxury; on the contrary by the poor man it was considered a necessity of life.[70]

Fox also opposed any measure that would make the government more dependent for revenue on the sale of alcohol.[71]

At around 2 a.m. on 5 October 1878, with at least one-third of the House absent, the Beer Duty Bill was defeated by two votes. Shortly afterwards the government instructed the Customs Department to refund the duty collected since 7 August. As the furore faded from memory, the *Otago Witness* wondered why, despite the victory of the brewers over the government apparently on behalf of the poor man, the poor man in Dunedin had to pay 6 pence for a glass of beer while the same Dunedin beer could be purchased in Wellington or Auckland for 3 pence.[72]

With beer in its sights, the next excursion by the colonial treasury would not be as easily repelled – especially as recession in Britain from 1879 quickly spread to New Zealand. With declining prices for wool exports, a diminishing gold yield and consequent high levels of indebtedness following a decade of sometimes frenetic overseas borrowing to meet the expansionist ideals of the Vogel era, the country entered a depression that would last until the mid-1890s. Amid falling wages and rising unemployment, the Grey Government was replaced by a Hall ministry intent on retrenchment and containment of borrowing. On 8 June 1880 Harry Atkinson, Ballance's successor as colonial treasurer, announced a beer tax of 6 pence per gallon – quadruple that proposed two years earlier. While he claimed to regret the measure, and had helped to throw it out in 1878, Atkinson insisted that 'hard financial times' now made it necessary.[73]

Again, the brewers descended on Wellington to recycle their case and to implore Atkinson 'That in a young colony it is unwise to clog the springs of local industry by taxation' which would lead to unemployment and insolvency among brewers.[74] In Whanganui, licensed victuallers passed a resolution declaring the duty unjust, while a petition from thirty-one Napier publicans pointed out that the whole duty would fall on them and not the brewers, and yet publicans required their bar trade to enable expenditure on suitable comforts for guests staying at their establishments. 'An Ex-Brewer' perhaps justifiably argued that while duty in Britain had increased gradually over a long period, that in New Zealand was coming all at once.[75] A deputation of licensed victuallers suggested alternatives to beer duty, such as a return to previous tea and sugar duties, a special tax on land owners who had benefited by the presence of a railway, an 'education' fee or an increase in stamp duty. To this a widely circulated *Clutha Leader* editorial lambasted the publicans: 'In their opinion the poor man and his wife and family might fairly have their necessities of life taxed to the utmost, educational fees might be imposed, but to tax beer is, in the eyes of the publican, an unpardonable sin.'[76] In Timaru, experienced brewer Francis Franks raged against a *Timaru Herald* editorial declaring that the rich man could afford a price increase and if the poor man could not then 'so much the better for him in ninety-nine cases out of a hundred'. Franks suggested that 'A penny tax upon newspapers would have been quite as reasonable'.[77] Atkinson, not a man renowned for rhetorical flourishes, asked for facts and figures from the trade.[78]

By the time the Bill was again debated in the House on 23 August, worsening economic circumstances had shifted public opinion against the trade. The only point at issue was the extent of the duty. Despite an insistence from Atkinson that the government would need every penny it could get, a motion from Ballance to reduce the amount from 6 pence to 3 pence per gallon eventually carried 41–34. That the former colonial treasurer now accepted a duty at twice the level he had proposed two years earlier was ample testimony to the fiscal difficulties facing the colony.[79] By comparison with the New Zealand duty of 9 shillings a barrel, the Gladstone Government's imposition of an excise in Britain in March 1880 for the first time in half a century amounted to 6 shillings 3 pence on a standard barrel, although with a sliding excise scale depending on the strength of the beer brewed. By comparison, New Zealand got off lightly.[80]

Following the introduction of beer duty in June 1880, stamps confirming that duty had been paid had to be affixed to all barrels of beer for sale. Lion Technical Library

Although brewers were refunded duty that had been collected at 6 pence rather than 3 pence between 9 June and 23 August, this was perhaps little compensation for the changed environment in which they now operated. While the Distillation Act 1868 had caused little real inconvenience, the new measure, to be administered by the Customs Department, came with a good deal more bureaucracy and surveillance. In addition to an expectation that they would 'faithfully comply without fraud or evasion' with the requirements of the Act and any other law relating to the manufacture and sale of beer, brewers had to pay an annual licence fee of £1 and a bond equal to twice the tax required to be paid in any one month. Following a system used in the United States, duty was represented by stamps, purchased by brewers and affixed to the plug or stopper of the tap hole on every cask such that the stamp would be destroyed when the cask was opened for consumption. All purchases of stamps amounting to more than £10 were eligible for a 5 per cent discount, but an absence of stamps was regarded as non-payment of duty, for which the beer was liable for forfeiture and a £20 fine on each cask. There was a penalty of up to £100 and loss of chattels for any bottler who bottled from casks without stamps, and imprisonment for forgery of stamps. Brewers were also required to keep brewing books detailing the kind and quantity of beer produced, the amount sold, and a record of all raw materials purchased for brewing. Entries in the book had to be verified and signed at the end of each month and a duplicate transcript provided to the local Collector of Customs. No beer could be destroyed or run off in a brewery without permission from the Collector, who would appoint an officer to be present. Failure to comply could lead to seizure of a brewery plant and beer and a fine of between £50 and £100. No duty was required on beer for export. Finally, an amendment in 1886 stipulated that one year after its introduction all manufacture of aerated water, cordial, ginger beer and similar drinks was prohibited within a brewery.[81]

If enforcement of the Distillation Act on brewers had been rather lax, that on beer duty afforded no such luxury, as the Customs Department, especially one excise officer, Samuel Jackman, pursued infringers and evaders with vigour. It seems that the first to fall victim was William Strachan of the long-established Victoria Brewery, Dunedin, who pleaded guilty and was fined £60 ($11,300 in 2022) for supplying three casks of beer without the appropriate stamps being affixed to them. However, a charge against a publican for 'knowingly' receiving these was dismissed. Meanwhile, charges at the same court sitting against Joseph Kühtze and another publican were also dismissed when the judge accepted that the stamps may have been washed off in the rain or taken by children.[82] In April 1887 Richard Purchase, an employee of Francis Innes' Standard Brewery in Christchurch, was charged with a breach of the Act for withdrawing beer from two casks for bottling without defacing the duty stamps. In his defence Purchase claimed that a tap could not be put through that part of the cask containing the stamp so he had to use another part. Although Innes claimed that he was not at the brewery on this occasion and insisted he had left clear instructions to his employees regarding tapping, a stocktake by the Customs Department found the brewery to be 5 hogsheads and some gallons short of the required duty payments. It had been 50 hogsheads short in 1882 when Innes had been fined at Timaru and Christchurch. On this occasion Purchase was fined £26 ($6000 in 2022). In all such cases, the prosecuting customs official claimed a reward generally in the vicinity of a quarter of the fine.[83]

In April 1889 Jackman's greatest triumph was the prosecution of Wellington's Junction Brewery on numerous breaches of the Act, resulting in fines and costs of nearly £480 ($110,000) and eventual forfeiture of the brewery plant to the Customs Department. After protracted negotiations a decision by Cabinet was required to return the plant to the owners and reduce the fines to £378 ($87,000).[84] Another to suffer forfeiture was Stuart Monteith of the Phoenix Brewery, Reefton, in March 1897. Although the Customs Department eventually agreed to recommend remission of the forfeiture, a later discovery that the discrepancy in beer duty was due to a clerical error rather than an attempt to defraud did not result in any compensation or apology to the brewery.[85]

*

No sooner had excise duty become a habit than the brewing industry was confronted with the beginning of a potentially greater challenge in the form of the Licensing Act 1881. The winds of temperance that had periodically buffeted New Zealand from the 1830s were by the 1870s gradually evolving to a permanent storm of prohibitionist sentiment committed to nothing more or less than the end of alcohol. As noted already, the dramatic population increase of the 1860s and especially the spectre of large concentrations of freely imbibing males on the goldfields was disconcerting to those with progressive ideals for the future of New Zealand. Post-gold-rush urbanisation, albeit on a fairly small scale, also had a role in stirring the mix of myth and reality that has always shaped perceptions of the relationship between alcohol and society. As Jock Phillips suggests, 'Cities brought new expectations of order, routine and public decorum.'[86] Amid their filth, limited infrastructure and poor housing for the working class, cities also fed fears of a repetition of old-world vices and inequalities in the colony. The increasingly popular British and North American rhetoric of abstinence as the key to moral and material progress found an articulate minority of converts in New Zealand. Contrary to the prevailing non-interventionist economic liberalism of the day, they insisted that the community had every right to protect itself against evils such as liquor. The liberty of those wanting a public house was opposed by the liberty of those not wanting one. Therefore, the moral right lay with the majority. To this end, and because they knew that alcohol controls would only succeed by popular will rather than the blunt instrument of coercive legislation, prohibitionists campaigned for the control of liquor licences to be determined by democratic means rather than the whim of the magistrates.[87] Here they were drawing on strategies developed by the United Kingdom Alliance following its establishment in 1853, but also from the United States where the temperance movement had moved from moral suasion to political action from the 1830s, with thirteen states and territories banning liquor sales during the 1850s.[88]

In 1871 the province of Auckland led the British Empire with a Licensing Act allowing for a two-thirds, male and female, adult majority of a district to veto a licence in a memorandum addressed to the licensing meeting. At the same time an identifiable temperance lobby was emerging in Parliament under the direction of William Fox, who served as premier for most of the period from mid-1869 to late 1873.[89] Against the background of some 100 petitions with 30,000 signatures from various temperance groups to the General Assembly during the years 1871–73, central government moved during the early 1870s to take greater control of liquor legislation from the provinces. After the rejection of licensing bills in 1870 and 1872, success finally came with the Licensing Act 1873, which echoed the earlier Auckland measure by abolishing the issuing of licences by magistrates and creating a new licensing court consisting of a resident magistrate and local commissioners nominated by the governor. Among other things, the Act established two important principles – the right of householders to object to the issue of any licence by means of representation to the licensing court, and the absolute discretion of the court to grant or refuse any licence. An amendment the following year empowered the court itself to make objections to an application or grant renewal or transfer of a licence.[90] Nevertheless, the feeling of the temperance cause was that their real objectives had been 'emasculated' by the legislation. Aside from the politics of nomination to the licensing court, the process of securing the necessary two-thirds opposition to a licence was compromised by the limited means of proving the exact number of adult residents in any district at a particular time. Successive attempts by Robert Stout during the late 1870s to introduce 'local option' bills allowing for licensing decisions by locally elected committees were defeated.[91]

The abolition of provincial government in 1876 highlighted the inconsistency of liquor laws between the provinces. In 1881, no doubt stirred by both prohibitionist agitation and the arguments raised by the beer duty debates of the previous three years, the Hall Government moved to consolidate the numerous provincial ordinances and existing central government measures into a single Licensing Act. This marked the first significant step towards public control of drink and drinking places, as it provided for the creation of small licensing districts such as wards, boroughs or road districts with licensing committees annually elected by ratepayers. Triennial local polls were to determine whether various types of licence in a district could be granted or renewed. Any ten residents could object by petition, and any resident could object personally, to the granting or renewal of any licence. The abolition of Sunday trading, which had been enacted by a number of provinces during the 1860s, was now confirmed across the country, although with an exemption for bona fide guests and travellers. Consequently, all hotels now had to provide accommodation. Hours of sale were restricted to 6 a.m. to 10 p.m. with provision to apply for 11 p.m. or midnight on payment of a further fee, and no alcohol was to be supplied in a public house to any person under the age of sixteen. In a move that signalled New Zealand's sustained determination to divorce alcohol from sociability, public dances and other such festivities were banned from licensed premises.[92] However, these measures were far from perfect. Writing in 1887, Nelson Magistrate Lowther Broad was highly critical of the drafting of the Act as a patchwork of legislation from throughout the British Empire: 'This brilliant idea of arrangement, and the happy thought of striking out here and there words which to the uninstructed mind seemed superfluous, and putting in others which the compiler felt ought to have a particular meaning (although in law they have not) has resulted in this badly-arranged and puzzling statute.'[93]

As economic depression intensified during the 1880s, prohibitionists emphasised a link between alcohol consumption and limited progress for the colony and insisted that the misfortune of many was a moral rather than economic issue. Only the improvement of the individual through abstinence would lead to the development of a settled society and the improvement of the British in New Zealand as a whole. The key to achieving that outcome was to restrict the availability of alcohol through controls both on locations where it could be sold and hours of sale. Prohibitionists also presented 'scientific' evidence of a link between alcohol and insanity, and were inclined to borrow arguments from Charles Darwin to suggest the survival of only the abstinent fittest.[94] The cause gained significant momentum in 1885 with the visit to New Zealand of Mary Leavitt on behalf of the United States-based Woman's Christian Temperance Union (WCTU). Leavitt gave well-attended lectures, and branches of the Union were established throughout the country. The objectives of the WCTU quickly became entwined with the campaign for women's suffrage in the belief that women, who suffered most from the consequences of male drinkers, would vote against alcohol.[95] The following year, diverse prohibitionist elements came together in Wellington to form the New Zealand Alliance for the Suppression of the Liquor Traffic as a national organising body with Sir William Fox as the first president. In the main, prohibition leadership was articulate and middle-class – reflecting those who had the time, means and organisational skill to become involved. But there was also a significant strand of working-class support. While most Anglicans, Catholics and moderate Presbyterians opposed prohibition, more puritanical Presbyterians, Baptists, Methodists and smaller evangelical churches supported it. Moreover, although some suffragists were not preoccupied with prohibition and the Alliance was not fully committed to suffrage, actual and perceived links between the WCTU and the Alliance were strong, and it was certainly the case that the wider suffrage movement benefited from the organisational and campaigning experience gained by many of its leaders in temperance and prohibition causes.[96]

By the early 1890s prohibitionist dissatisfaction with elected licensing committees, and with the 1881 Act more generally, was reaching fever pitch. It was suggested that many committees were 'got at' by the 'liquor power', and, perhaps with more justification, that committees with prohibitionist inclinations were constrained by the prospect of litigation inherent in the complexity of the 1881 Act. Hence the Alliance pushed for a direct vote by all electors in a district, on the basis that a whole community could not be so easily corrupted. This stance was further galvanised in 1891 when the newly elected Sydenham licensing committee, including prominent prohibitionists Leonard Isitt and Tommy Taylor, pledged to close all eight public houses in the borough. A Supreme Court injunction was then taken against the committee on the grounds that it could not close public houses merely because the majority wanted it. Rather, the committee needed to consider the legitimate wants of those who wished to drink. The committee then closed five houses but left three open before closing those as well in 1892. This decision was reversed in the Court of Appeal, where Justice Denniston ruled that the 'incurable bias' on either side of licensing committees prevented them enforcing the law and considering matters in an honest and judicial spirit.[97]

The debate threatened to split the Liberal Government, as its newly elected leader Richard Seddon, a former publican and widely regarded as friendly to the liquor trade, wrangled with the increasingly vocal prohibitionist Robert Stout. Eventually Seddon skilfully negotiated a compromise in the form of the Alcoholic Liquors Sale Control Act 1893. Seddon allowed his MPs a free vote on the Act, thus establishing a parliamentary tradition of 'conscience' votes on liquor matters that has persisted. In return for no increase in the number of hotel licences except in areas of rapid population growth, all voters in an electoral district, and not just ratepayers, were now presented with three options: continuance; reduction of hotels in a district by a quarter; or prohibition. A three-fifths majority of the vote was required to carry prohibition, but only a bare majority for reduction. However, a 50 per cent turnout of the electorate was required for a vote to be valid. No compensation would be paid to any publicans put out of business. While prohibitionists complained bitterly, especially at the three-fifths majority, they undoubtedly took heart from the first poll on 21 March 1894, when Clutha voted to go dry and fourteen electorates voted for reduction, including eight in the Presbyterian strongholds of Otago and Southland. However, thirty-five of fifty-two electorates returned void votes due to low turnout. Thereafter, licensing polls were to be held on the same day as general elections.[98]

In the meantime, another seeming threat to drinkers had emerged when the franchise was extended to women in September 1893, making New Zealand the first country in the world to do so. Accounts of suffrage have followed the broader pattern of New Zealand historiography in tending to take temperance and prohibitionist claims at face value and presenting the specific objectives of the WCTU as an entirely justified response to circumstances.[99] Previous chapters have shown the limitations to that approach. More to the point, it is obvious that women did not flock to the prohibitionist cause in disproportionate droves after the extension of the franchise. While Linda Moore has suggested that women were consistently more likely to support prohibition than men, they did not favour prohibition overall.[100] Had they done so, New Zealand would have been mostly dry very shortly after the 1896 election. In fact, that election and the one that followed were disappointing for the Alliance. Women were no more a cohesive group in society than men, and it is likely that the middle-class domestic ideals and economic status at the core of the WCTU did not sit comfortably with some working-class women.[101] Yet perception created its own reality. An enduring prohibitionist faith that women would eventually unite for the cause was perhaps given more latitude by anti-prohibitionists, to be discussed in Chapter Six, who seem to have done nothing to address their arguments specifically to women until the early 1920s.

Notwithstanding that the women's vote did not play out as most imagined, it was abundantly apparent to anyone connected with the liquor industry in New Zealand during the 1890s that the liberality that had marked the first few decades of European settlement had been significantly eroded. In addition to the multiple logistical and technical challenges facing even the best brewers, there were laws to which careful adherence was required and the potential for reduction of local retail outlets no matter how scrupulously one adhered. Moreover, as the introduction of beer duty had confirmed, the colony was now in the grip of hard economic times from which only the strongest would emerge unscathed.

A bottle from Cowie & Co's brewery in Caversham, Dunedin, which was established in the mid-1870s and closed in 1923. Hocken Collections, University of Otago, 509-262c

AUCKLAN
NEW INDUS
HANC
BISM
LAGE
BISMARK
LAGER BEER
HANCOCKS
"BISMAR
LAGER
HANCOCKS
"BISMARK"
LAGER
BEER
(ICED)
HANCOCKS
"BISMARK"
LAGER
BEER
(ICED)
ON DRAUGHT AND IN BOTTLE

An exhibit for Bismark Lager, the first lager produced in New Zealand, in 1900. It faded from the market around 1905. Auckland Libraries Heritage Collections, AWNS_19011121_5_4

FIVE

'Domineering, Purse-Proud, Scheming Brewers': Concentration and Competition, 1890–1914

During an extensive tour of New Zealand breweries in August 1889, Alfred Lawrence, a leading figure in the Australian liquor industry and correspondent for the *Australian Brewers' Journal*, noted a marked decline in the prosperity of the New Zealand brewing industry during the previous five years. Many brewers were 'sore' at having spent considerable sums on new buildings and plant just as the trade entered a period of decline. The deepening economic depression which had ensured the easy passage of the Beer Duty Act in 1880 certainly had an impact. Equally, as the previous chapter demonstrated, an increasingly trenchant prohibition movement no doubt made many feel nervous for their future. But for Lawrence the worst feature of the New Zealand situation was that the country had twice as many breweries as seemed to be necessary.[1]

In one sense Lawrence caught the New Zealand brewing industry as its wave was cresting. The 1891 census recorded 102 breweries, and there were probably several more, for a population of 626,658. By 1896 there were only eighty-five breweries and by 1911 only fifty-six. A number of those that remained were far larger and more sophisticated establishments than anything seen previously and were well on the path to securing almost total dominance of the industry by the mid-twentieth century. The decline therefore was not in terms of the volume of production but the variety of products available to local consumers. Processes of concentration, merger and technological improvement dominate this period, as do the complex and controversial means by which the larger breweries secured their market by tying the vast majority of hotels in New Zealand. But it is important to remember the human side to this business history. Breweries, and the ever-expanding malting and hop industries upon which they depended, were workplaces providing a variety of experiences to their employees. Further, as the circumstances of beer production changed, the choice of product for the consumer and the way it was presented were also transformed. Buying a beer in 1914 was in many ways rather different from 1840.

*

During the 1880s there was almost no growth in brewery numbers or the average size and capacity of each establishment. Indeed, there was a slight drop in overall output as per-capita beer consumption declined from 43 litres in 1881 to a low of 30 in 1888 before stabilising in the vicinity of 32 litres during the 1890s. Only in 1899 did it again rise above 35 litres, peaking at 44 in 1908 before settling around 42 by 1914.

It seems that the depression did curb drinking from the 1880s, when spending on all 'luxury' goods declined as per-capita incomes reduced by around a quarter. Beer may have been a commonplace with deep roots in the cultural make-up of many people, but technically speaking it was not a necessity.[2] Indeed, contrary to a frequent argument from prohibitionists that those who could least afford to drink drank most, the government statistician for New South Wales, T. A. Coghlan, concluded that throughout Australasia during the late nineteenth century there was a clear correlation between alcohol consumption rates and the performance of the economy.[3] The changing structure of the population also played its part. As a proportion, males aged over fifteen, the most likely drinking group, declined from 47.2 per cent of the population in 1864 to 33 per cent in 1891, while women of the same age increased from 20.6 per cent to 27.3 per cent. In short, New Zealand was becoming a more settled society in which families replaced itinerant, single males as the dominant characteristic. While settled and maturing males chose or were compelled to drink less than before, it was also the case that the most determined drinkers in the population were more likely to be offset by women and children who drank little or nothing. The rise of consumption from the 1890s in inverse proportion to the popularity of prohibition will be addressed in the next chapter.

Tables 5.1 and 5.2 capture a series of dramatic changes and rationalisations in the three decades following the Licensing Act 1881. Above all, there was a 43 per cent reduction in the number of breweries but a 106 per cent increase in the volume of beer produced for a population that increased by 116 per cent. By 1911 the ratio of people to breweries was more than three times what it had been in 1881. The industry also began to concentrate in the leading firms of the four main cities. The nineteenth-century brewing stronghold of Nelson shrank significantly, while few

Figure 5.1: Per-capita Alcohol Consumption, 1881–1914 (including Māori)

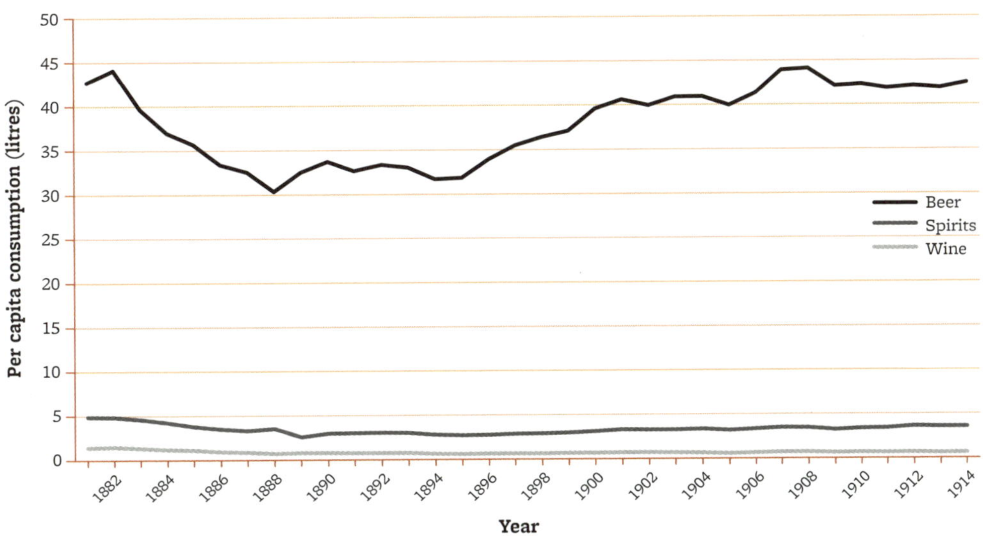

Source: *Statistics of New Zealand*, 1881–1914.

of the goldfields breweries of Otago and Westland made it to the twentieth century. Although average production in all breweries had at least doubled, in Auckland and Otago it increased more than fivefold. Even then, the average capacity of the largest New Zealand breweries, such as Speight's or the Captain Cook, was minuscule by international or even Australasian standards. Tooth & Co's Kent Brewery in Sydney was producing 15 million litres by 1889.[4]

Until the 1880s the vast majority of breweries were individually owned with perhaps two or three employees at most. By the 1890s, a few successful establishments – such as Speight's in Dunedin, Crown, Manning's and Ward's in Christchurch, Staples in Wellington, and Ehrenfried's Albert Brewery and John Logan Campbell's Domain Brewery in Auckland – were able to modernise and expand. All of them had previously distinguished themselves for quality of product and their ability to navigate the numerous pitfalls of brewery ownership and survival outlined in Chapter Three. They were also able to access capital for development. By 1911 seven of the eight breweries in Auckland were registered companies, as were five of ten in Canterbury. There are no reliable figures for Otago. But beyond the four main provinces, sixteen of twenty-five breweries were described as individually owned and seven others were in limited partnerships. Only two had adopted any sort of registered company structure. Consequently, larger breweries were able to take advantage of scale economies in terms of purchasing raw materials, and therefore had greater financial flexibility to undercut smaller breweries on price. In time those breweries able to mechanise and streamline their bottling technology enjoyed an added advantage as their product was able to reach a wider range of retail outlets beyond tied hotels, and they were also able to supply home consumers in the growing number of no-licence districts or those with an inadequate number of hotels following reduction. In 1902 the leading Auckland brewers claimed that they were able to sell beer at two-thirds the price of their provincial counterparts.[5]

Smaller breweries were squeezed by a number of forces. As early as 1870, a *Tuapeka Times* report predicted that if William Theyers kept up the quality of his first brew at the Alexandra Brewery, he would 'defy Dunedin competition and thus keep more money in the district'.[6] Evidently the threat from the metropolis was already apparent, and, as rail networks expanded from the early 1880s, especially down the East Coast of the South Island and throughout the lower North Island, it became very real. Regional breweries were vulnerable to competition from the main centre breweries, which could easily afford transport costs for their beer while still remaining competitive on price.

The second factor, to be considered in more detail shortly, was that smaller breweries were unable to compete in the escalating scramble for tied houses. Few of them were able to extend credit to publicans or to buy expensive freeholds or leaseholds in a market where property values for increasingly scarce hotels were rising fast. The closures of the Milton Brewery and of Daniel Arkell's Gladstone Brewery in Auckland, both in 1906, were specifically attributed to competition from large firms.[7] Other long-standing establishments to fall during the 1890s and the early years of the twentieth century included Pizzey & Arkell's Hokitika Brewery and Charles Innes's Te Awamutu Brewery, both founded in 1865; Theyers & Beck of Alexandra, founded in 1869; George Hally's Cambridge Brewery, founded in 1870; and Cromwell's much heralded and quite sophisticated Swan Brewery, founded in 1874. Francis Franks, who had enjoyed mixed fortunes with a number of breweries, mainly in South Canterbury, finally departed the industry in 1888 to become a town clerk. Others who had entered during the gold boom and were now coming to the end of their working lives were evidently unable to find suitable people to continue their business. It was one thing to run a

Table 5.1: Change in the Brewing Industry, 1881–1911

	Number of breweries	Beer produced per year (L)	Average production per brewery per year (L)	Employees per brewery	Population per brewery
1881	99	20,930,511	211,419	5.31	4,949
1886	98	19,292,132	196,858	4.84	5,903
1891	102	20,765,764	203,585	4.66	6,144
1896	85	24,470,936	287,893	5.47	8,275
1901	74	32,434,800	438,308	9.21	10,442
1906	72	37,817,628	525,244	10.15	12,341
1911	56	43,309,560	773,385	12.35	16,807

Table 5.2: Beer Production by Province, 1881, 1896 and 1911

	1881		1896		1911	
	Number of breweries	Average production per brewery per year (L)	Number of breweries	Average production per brewery per year (L)[1]	Number of breweries	Average production per brewery per year (L)
Auckland	15	288,916	11		8	1,454,910
Taranaki[2]	4	94,429	5		2	292,025
Wellington	11	228,204	12		10	683,545
Hawke's Bay[2]	5	137,802	5		5	292,025
Nelson[2]	14	80,596	12		8	226,486
Marlborough[2]	5	84,000	3		1	226,486
Canterbury	19	281,747	16		10	761,354
Westland	7	66,789	4		5	270,514
Otago	19	371,041	17		7	1,940,205
New Zealand	**99**	**211,419**	**85**	**287,893**	**56**	**773,385**

1 Production figures are not available for 1896.

2 For 1911 the production totals for Hawke's Bay and Taranaki, and for Marlborough and Nelson, were combined in the statistics by dividing the total amount of beer produced in each combined area by the number of breweries.

Source: New Zealand Census, various editions.

long-established brewery or even to continue it out of family loyalty as some widows and sons did, but quite another for an outsider to take on a new business in the uncertain climate of the late nineteenth century with bigger competitors and prohibitionists lurking on the horizon.[8]

The Black Horse Brewery at Wetherstons in Central Otago, founded in 1866, was one of a very few provincial breweries with the resources to supply significantly outside its own area, with its beer available from Bluff to Timaru.[9] It and a few other small breweries were partially protected by specific consumer preference for their products, but there is also ample evidence from generations of fickle beer consumers to suggest that differentiation on taste was closely linked to price. If the price was not competitive, consumer taste would change. Later chapters will also explain that the imposition of measures to restrict beer strength during both world wars, and the consequent reduction of malt, sugar and hops in many beers, effectively blunted the ability of brewers to produce a taste profile that would stand out. A level drinking field on taste was a slippery slope to economic ruin in most other respects.[10]

As certain of the longer-established breweries expanded and others faded out, several new breweries did defy the unfavourable conditions of the 1880s to secure an enduring significance. Stewart's Brewery, established in Greymouth in 1880, survived in one form or another until 1961, making it the longest survivor of this generation of small breweries. In Dunedin the Standard and Union breweries both opened in 1882. The latter, more commonly known as McGavin's, was part of the New Zealand Breweries merger of 1923. The new conglomerate took over and closed the Standard Brewery the following year. Meanwhile, in April 1889 Henry Wagstaff established what would eventually become known as the Tui Brewery at Mangatainoka in the northern Wairarapa. By the end of the year, he was supplying several local hotels and employed a second brewer to cope with demand. The Pahiatua Brewery, set up in competition at around the same time, lasted barely eighteen months, and by February 1891 the Salvation Army had taken the lease of the premises and converted it into a barracks. In January 1896 Wagstaff was fined £75 ($16,500 in 2022) for breaches of the Beer Duty Act and joined the

Founded by the Kühtze family as the Main Trunk Brewery in 1908, this business became the Cascade Brewery in 1935 and continued until acquired and closed by New Zealand Breweries in 1955. Lion Technical Library

list of those who forfeited their business but got it back on appeal. In March the *Pahiatua Herald* confidently reported that the brewery was 'able to bear comparison with any in New Zealand and the liquor turned out by Mr Wagstaff has earned for itself an excellent reputation'.[11] After a brief period in which the brewery was sold to Joseph Isherwood, Wagstaff was back in control by the turn of the century and had also established the Goldfield's Brewery at Paeroa in 1897. In 1903 Wagstaff sold out to the North Island Brewery Company. Sales increased rapidly from 912 hogsheads in 1908 to 2271 by 1914.[12] We will be reacquainted with the Tui phenomenon in a later chapter. Meanwhile, in November 1897, D. J. Barry took over the Gisborne Brewery, which had been founded in 1874 by William Crawford, and began to inject new life into the establishment. It would eventually become part of the New Zealand Breweries merger.[13]

From the mid-1870s a number of brewers sought to generate capital for expansion through the establishment of limited liability companies and share issues. The Albion Brewing Company in Dunedin may have been the first in January 1875, when it took over the premises formerly occupied by the New Zealand Distillery.[14] The following year the Crown Brewery in Christchurch became a limited liability company, and in mid-1881 its local rival, Ward's Brewery, converted, with a capital of £50,000 in 5000 £10 shares, which were soon fully subscribed. Maurice Joel's Red Lion Brewery in Dunedin followed in 1882.[15] By 1889 Staples Brewery in Wellington was severely indebted as it had been forced by competition to purchase hotels to secure a market. A syndicate of mainly Catholic merchants and hotel owners, who had secured

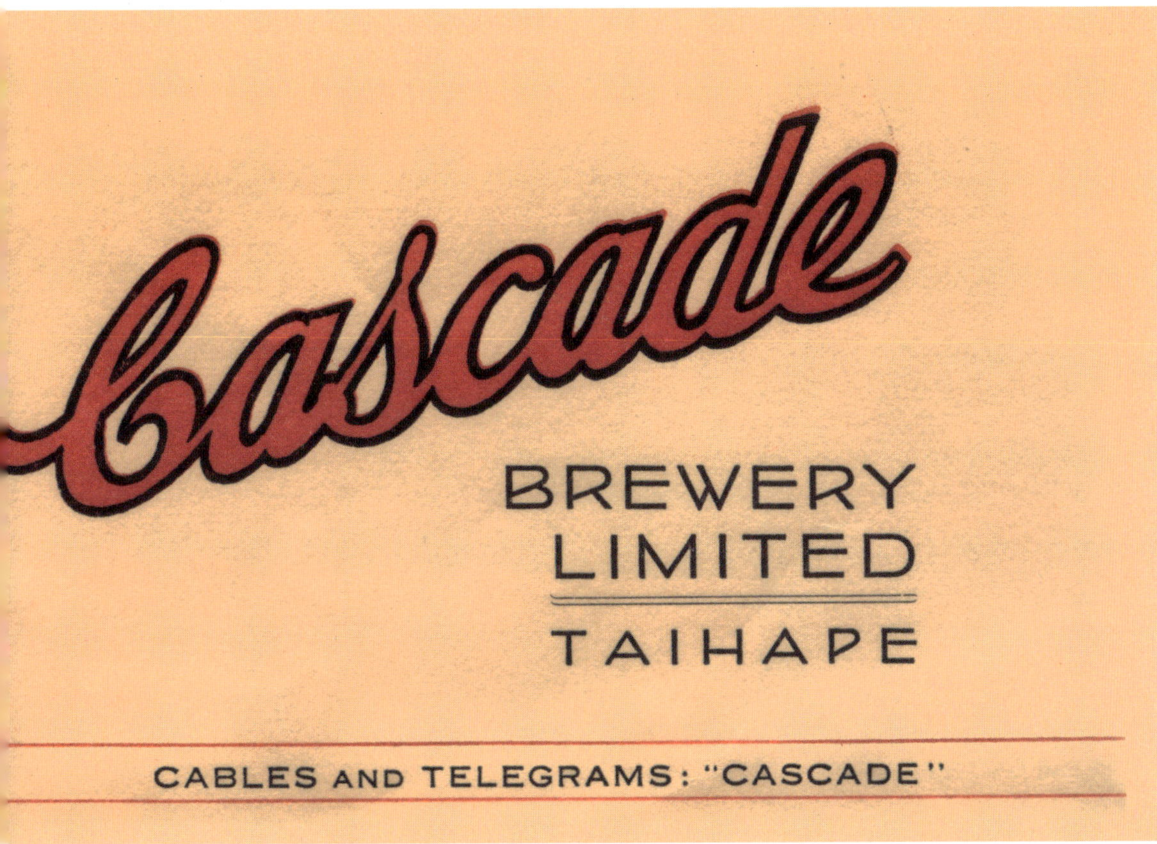

their prosperity during the West Coast gold rushes, offered to purchase the business and pay off its debts, estimated at £10,000. The new company began with 12,000 £1 shares, although it seems that few if any of these went beyond the hands of the syndicate and their immediate associates.[16] Others to take the company path during the 1890s included Speight's in 1897 and Timaru's Alton Brewery as J. Hole & Co. in 1899.[17]

The most significant moves were occurring in Auckland, where the seeds of major brewery mergers were being sewn from the early 1880s. As diminishing gold returns triggered the decline of the Thames population and economy during the late 1870s, Louis Ehrenfried began to turn his attention to the Auckland beer market. With the death of Robert Whitson in November 1883, and no successor in his family to continue his Albert Brewery business, Ehrenfried seized the opportunity for a more comprehensive expansion. After complex negotiations were completed in December 1885, Ehrenfried bought the Albert Hotel on Queen Street and the brewery directly behind it, along with six tied hotels and land to build another. With borrowed money Ehrenfried acquired at least twelve more hotels around Auckland by the early 1890s. Meanwhile the retirement of Thomas Hancock during the 1880s had resulted in his son-in-law Samuel Jagger taking a Nelson merchant, Moss Davis, into partnership at the Captain Cook Brewery. On Jagger's death in 1890 Davis became sole owner. Thenceforth an intense rivalry developed between the Ehrenfried and Davis families.[18]

Running parallel to these developments, John Logan Campbell returned to the Auckland brewing scene in the 1880s with renewed although conflicted vigour. Never a hands-on brewer, Campbell had leased his premises to others from 1851 to the late 1870s. But difficult financial times had prompted a revival of the Domain Brewery firmly under his own control in 1880. By 1888 it was producing 3000 hogsheads annually, and sales were further boosted when Domain Ale collected a first prize at the Melbourne Centennial Exhibition at the end of the year. There were twenty-two tied hotels and eighty others to which beer was supplied, and Campbell was in no doubt that the brewery alone had sustained his fortunes during the depression. Nevertheless, his attitude to the liquor industry was complex and he was determined to sell the business. In part, as he reached his late seventies, he wished to wind up his affairs and retire. But as Russell Stone explains, he was also conscious that his position as one of the most venerable public figures in Auckland did not sit comfortably with involvement in an industry that was now under intense attack from both prohibitionists who despised all alcohol, and others, to be discussed shortly, who regarded large brewers as unscrupulous monopolists.[19] Writing to a business associate in 1884, Campbell described 'The damned grog shop connection [as] simply disgusting, even fettering'.[20] In a letter to William Fox in 1888 he claimed, entirely disingenuously, that the liquor part of his business had developed only during his absence from Auckland and therefore from direct control over the company during the 1860s. In fact, it had been an integral part of the business since 1840.[21]

Despite the prosperity of the Domain Brewery, no buyer could be found in Australasia. In 1890 Thomas Russell, a prominent Auckland lawyer, businessman, and speculator now based in London, suggested the formation of a syndicate, or 'beer trust' as he called it, with the other major Auckland brewers – Ehrenfried, Jagger and Davis, and John Secombe, who now controlled his father's Great Northern Brewery. The single company could then be offered to British investors against the background of a number of large breweries such as Bass and Guinness that were now listed on the London Stock Exchange. As Russell put it, British capitalists were 'on the feed' for brewery incorporations. The resulting capital would allow rationalisation of operations and the acquisition of more tied houses. But the scheme progressed no further when the other brewers responded with what Campbell regarded as 'outrageous' valuations of their businesses.[22]

By the mid-1890s both Campbell and Ehrenfried were more determined than ever to consolidate their brewery interests. After discovering that earlier proposals by 'London capitalists' to buy his business would have seen it merged with Davis's Hancock & Co., Ehrenfried turned to Campbell in 1895 with a merger proposal. Although Campbell was critical of Ehrenfried for his role in the failure of Russell's earlier scheme and admitted to a degree of 'race prejudice' on account of Ehrenfried's German Jewish origins, the two eventually developed a warm understanding. Over the next two years they stepped back and left the complex merger negotiations to the rising stars of their respective companies, Alfred Bankart for Campbell, and Ehrenfried's nephew Arthur Myers. Although Ehrenfried died on 26 February 1897, the new Campbell & Ehrenfried came into existence on 1 April. As well as two breweries, it had forty-six freehold and forty-one leasehold hotels.[23] The following year the Auckland Industrial and Mining Exhibition featured an extensive display for the new company based around a replica two-storey hotel assembled from the bottled products of the brewery and accompanied by a detailed description of the premises and facilities. As the Exhibition catalogue noted, 'The very sight of it is enough to make one sensible of a parched throat, and to long most ardently for the means there presented in great diversity of assuaging a consuming thirst.'[24]

Beyond such formal consolidations there were evidently other less obvious connections between breweries throughout New Zealand. Secombe's Great Northern Brewery was lending money to the Waikato Brewery from the 1870s, and Charles Lewis Innes was apprenticed to the Great Northern soon after his parents took control of the Waikato Brewery in 1889. In turn, when Mary Jane Innes took sole ownership of Waikato following the death of her husband in 1899, Ward's of Christchurch provided loans and arranged a mortgage over the brewery, which then bottled Ward's beer for North Island distribution. The web was further tangled as the Crown Brewery in Christchurch, owned by Alfred, Cecil and Charles Louisson, also had a financial stake in Ward's.[25]

*

The key to brewery survival during the late nineteenth century, let alone to successful mergers, became the control of tied houses. There were essentially two methods of securing ties. As in Britain, brewers frequently arranged exclusive agreements to take their beer by advancing money to publicans, often on a reduced rate of interest. This allowed the publicans to make improvements and refurbishments to their hotels, especially those demanded by increasingly vigilant licensing committees determined to close down any establishment unable to meet an acceptable standard. Sometimes both interest and principal were written off completely. Alternatively, as became common with the larger breweries from the 1880s, a brewery could purchase hotels freehold or leasehold and install its own licensee or lease only to those agreeing to exclusivity. In this instance, standard rents were considerably reduced.[26]

It was noted already in Chapter Two that ties were evident in Nelson and Auckland from the 1840s. The practice gained momentum amid the rapid brewery expansions and lucrative opportunities of the gold rushes. Indeed, Louis Ehrenfried controlled thirty-nine of the fifty-nine hotels in Thames by 1880. When the local licensing committee sought to reduce the total by six on the grounds that the decreasing population of the town did not need so many

A Campbell & Ehrenfried exhibit for a parade in Auckland, 1902.
Auckland Libraries Heritage Collections, NZG_19021018_987_2

outlets, Ehrenfried petitioned Parliament in protest. He lost, but was allowed to choose which six were closed.[27]

From the brewers' perspective tying became more imperative than ever from the 1890s with the rise of prohibition. As hotels began to disappear, especially in rural areas, as a consequence of the Licensing Act 1881 and then the triennial licensing polls held from 1894, the effect was to greatly raise the value of those that remained, to the point where only the larger breweries could afford to control them. They well understood that select closures in an area did not magically reduce the overall consumption of beer. Over ninety licences were lost to the 'reduction' clause in the 1894 poll, along with all of those in Clutha, which became the first no-licence district. Within a year the total number of licensed outlets in New Zealand fell from 1719 to 1579. It slipped further to 1364 in 1902 when Ashburton and Mataura went dry, and to 1218 in 1905 after Grey Lynn, Invercargill and Ōamaru surrendered to the prohibition crusade. As the population increased, the ratio of licences to people went from 1 to 434 in 1895 to 1 to 895 in 1905, and 1 to 1255 in 1916.[28] In theory the number of hotels in a district could be increased if the population increased by at least 25 per cent and a petition signed by at least 25 per cent of electors was presented to the licensing committee, which could then approve a special poll – which would only be valid if more than 50 per cent of electors turned out and more than 60 per cent of those approved an increase. Not surprisingly, there is no record of such conditions ever being met.[29] As later chapters will reveal, an increasing and shifting population would eventually cause very problematic patterns of hotel distribution.

Partially offsetting the decline in hotel numbers were wholesale wine and spirit merchants, which were also the main stockists of imported beer, but could only sell in amounts greater than 2 gallons. Eventually the total number of such outlets was capped at 140 in 1920. Meanwhile, the ability of smaller retailers to sell under 'bottle licences' had been abolished in 1895. The number of chartered clubs that were also licensed to sell alcohol to their members had been capped at forty-six in 1908 and did not increase again until 1947. In short, the remaining hotels also held dominance over the retail takeaway portion of the market.[30]

Before Ehrenfried's entry into the Auckland market, Robert Whitson and especially John Secombe of the Great Northern Brewery had secured a number of Auckland establishments. By 1902 all but two of the fifty-eight hotels in Auckland city were tied to one of the large breweries. Moreover, ties throughout the upper North Island were so comprehensive that the Waikato Brewery, not an insignificant establishment but nevertheless one without sufficient capital to tie, conducted all of its draught beer business through deliveries to private families rather than with hotels. This market was apparently sufficiently large to keep the brewery going.[31] In Wellington, Staples had tied about twenty hotels by 1889 and had supply agreements with as many more. Another twenty were acquired during the next decade, and by the turn of the century the company controlled three-quarters of the hotels in the wider Wellington district. At the same time Manning & Co. had between thirty and forty in Christchurch, and Crown about fifty. The wine and spirit merchants Fletcher Humphrey, with whom the three main Christchurch breweries had close connections, controlled another twenty-five. Speight's had secured a monopoly in Dunedin by 1900 as well as numerous outlets in Southland and Central Otago. Despite hotels being tied to other breweries, Speight's managed to arrange distribution agreements for its bottled beer in Canterbury, Wellington, Taranaki, Hawke's Bay and Auckland. Perhaps three-quarters of all New Zealand hotels were tied by 1902.[32]

There were strenuous objections to these arrangements from several quarters. Aside from the prohibitionists, who naturally questioned anything done by the trade, there was a wider objection amid the New Zealand political climate of the late nineteenth century to anything in the manner of a monopoly which undermined the aspirations of the 'small man', whether he was a farmer, tradesman or publican. As the Liberal Government of John Ballance and Richard Seddon moved to improve working conditions and to open up access to land, restrictive ties became anathema to many. As a correspondent to the radical *Observer* fumed in March 1890:

> If the people will drink (and who has the right to compel them not to drink just what they please?), then let our legislators see that our licensing laws are so administered that the domineering, purse-proud, scheming Brewer shall not be permitted at his own sweet will to run the whole trade, to flood his tied-up houses with such noxious beverages as he now forces on an unsuspecting public; and to this end let some enterprising, honest M.H.R. [Member of the House of Representatives] have again tabled, and if possible passed, a Bill . . . excluding Brewers and wine and spirit merchants from having or holding any interest in or over any licensed public house, or the stock-in-trade or furniture therein, thus separating the Brewer from the Publican. Let the latter buy where and from whom he pleases, using his own judgment as to quality and quantity.[33]

Legislation along the lines suggested had been proposed since at least 1881, but it was only in 1895 that it came to fruition in a section of the amended Alcoholic Liquors Sale Control Act which stated that 'No covenant or agreement, condition, proviso or stipulation' could be made to compel any person or body to purchase alcohol exclusively from any other person or body.[34]

The new measure was totally ineffective. Some clauses in hotel leases were simply replaced with 'informal' discounts on rents in return for continued exclusivity. In the case of Campbell & Ehrenfried, it appears that although the memorandum to lessee always stressed the right of the tenant to a free choice of liquor supplies, an informal discount of 50 per cent on the book rent if a publican took only their product ensured that all were tied. In other instances, especially in the dealings of Staples & Co. in Wellington, hotels owned by the brewery were on-sold to their lessees and the tying clauses were moved from the hotel lease to the sale and purchase agreements. If a hotel could not be sold and a lease was still required, then the lessee could effectively be held to ransom by a clause requiring them to immediately surrender the tenancy on demand of the brewery. By all of these means the law was evaded and the publicans remained compliant.[35]

The increasingly clandestine relationship between brewers and publicans was further complicated by what became known as the speculation in goodwill. Although profit from the day-to-day running of a hotel provided publicans with a secure if not substantial income, there was much more to be made from the transfer of a licence. If a publican sold a lease to another tenant before the lease expired, they could expect a substantial payment in 'goodwill', frequently in the vicinity of £2,000 ($430,000 in 2022) and even as high as £5,000 ($1.1 million) around the turn of the century. These payments reflected not only the improvements that may have been made to a hotel during a tenancy, but also the fact that the ongoing reduction in the number of licences per head of population was making them a particularly valuable asset.[36] The new tenant had every reason to think that hotel values would continue to rise and that they could expect their own substantial goodwill payment in the future. Indeed, Hope Gibbons, a Whanganui brewer, said that many people took over hotels knowing they had little chance of paying off their advances from the brewers unless they sold out for a larger goodwill than they had paid.[37]

Not surprisingly the breweries felt that as they had financed many of the improvements to hotels, it was hardly acceptable for the goodwill to go to the licensee. But as the 1895 Act had also specifically prohibited them from taking any such payment for the transfer of a licence, they were obliged to employ surreptitious measures such as using networks of 'agents' or 'brokers' and convoluted leases to indirectly buy and sell licences in a manner that they could control. Such speculation simply accelerated the value of hotel licences, more than the buildings themselves, to the point where only a well-connected or compliant few could marshal the resources to enter the trade.[38]

There is no clearer illustration of the sometimes volatile relationship between brewer and licensee than the lengthy dispute between Martin Ryan and the Captain Cook Brewery. In December 1898 Ryan took over the United Services Hotel in Auckland and agreed to a rent of £16 7s ($3,500) rather than £30 ($6,200) weekly on condition of taking his draught beer from the Captain Cook Brewery. It was further agreed that all colonial draught supplied should be of 'good quality'. But in April 1900 Ryan sought to leave the hotel and asked a broker to find a purchaser for his lease. His solicitor also wrote to Moss Davis stating that 'the beer which was delivered to him this week was inferior in quality and he desires that better beer should be supplied as some of his customers decline to purchase the beer which he now has in the hotel'. Indeed, Ryan returned some beer in fourteen of the twenty-three months of his lease, amounting to 1624 gallons or one-twelfth of all beer supplied.[39] In a further letter to Davis on 2 October 1900, Ryan wrote, 'I am prepared, however, to give you a further chance and now notify you that I desire you to supply me with beer of good quality and of proper measure. If you are not prepared to supply me with Speight's Dunedin export beer, I shall order it myself.'[40] True to his word he offered Speight's and Ward's to his customers within a fortnight, and claimed to be selling three times more beer than in the previous month.

Inevitably, Davis rejected claims about the quality of the Captain Cook Brewery's beer and duly presented Ryan with an account for full rent from 1 October 1900, and in March 1901 initiated legal proceedings for recovery of the unpaid rent. Ironically the case was heard by the ardent prohibitionist Chief Justice Sir Robert Stout, who reluctantly upheld the claim for full rent on the basis that although the 1895 Act was supposed to abolish tied houses and exclusive purchase agreements, it did not cover this 'device' of rent reduction and the court was powerless to act. 'I am therefore compelled, though I think the spirit of the statute has been ingeniously evaded, to hold that the letter has not been violated.'[41] It was little consolation for Ryan that a jury agreed that bad beer had been supplied. For when he repeated his determination to order good beer from wherever he saw fit, Moss Davis replied in the presence of Ryan's solicitor 'daring him to do so, and saying that by the time he had done with Mr Ryan the latter would be walking the Auckland streets barefooted'.[42] It was obvious, according to a trenchant pamphlet summary of these events published in 1904, that the supply of bad beer did not constitute a breach of contract. 'According to this the brewer may send barrels of anything that will pass as beer, however inferior or bad it may be. The unfortunate tenant has the alternative of accepting it, knowing he cannot sell it as beer, or rejecting it and finding himself situated as Mr Ryan is at present' with a declining custom and increased rent. As the law seemingly allowed brewers to shut out all beer superior to their own, there was no incentive to improve quality or have competitive prices. 'This is nothing less than commercial serfdom in the twentieth century and in New Zealand.' Ryan remained bound to his lease until December 1904.[43] His career thereafter is unknown.

These and related matters came to public notice in 1902 when the Hon. John Rigg introduced a Bill aimed at the enforcement of the 1895 Act and ultimately 'extinction' of tied houses. In this he was presumably inspired by an investigation in New South Wales during the previous year that revealed 90 per cent of hotels to be tied in some way.[44] A select committee was established to examine the New Zealand case, but from the outset it encountered a parade of reluctant and evasive brewery and licensee witnesses who were either unwilling to reveal the nature and extent of their business dealings or determined not to antagonise those with control over their livelihoods. Moreover, the hearings were dominated by brewery interests to the extent that one of the legislative councillors on the committee was Charles Louisson, the managing director of Christchurch's Crown Brewery. As the Wellington lawyer and prohibitionist Arthur Atkinson informed the committee, 'The majority [of those in tied houses] are merely tenants at will and they are absolutely at the mercy of the brewers; and therefore it is not to be expected that they who know most about the evils of the system can come and testify before this Committee concerning those evils.'[45]

For the brewers, Arthur Myers argued that if they and the wine and spirit merchants did not provide capital for the hotels, no one else would. There was certainly no prospect of banks lending against a background of triennial licensing polls and clear evidence of widespread hotel closures. Here Myers echoed a point made by Auckland's *Licensed Victuallers Gazette* shortly after the Campbell & Ehrenfried merger in 1897. While this trade paper declared that it 'hated monopoly' and wished every publican could operate with freedom, it was also pragmatic:

> [W]hen we realise that it takes £5000, or more, to furnish and stock some of our best hotels, we ask how many men in the retail trade would have gained the position they now occupy as keepers of first-class houses if they had not received assistance at the start of their hotel career from the brewer and the spirit merchant, and how could they have obtained such assistance if those who advanced the money saw no way to secure interest on the advance made and eventually a repayment of the principal itself?[46]

Myers and others also insisted that the firm hand taken by the brewers ensured both quality of product and good conduct of the premises.[47]

One of the few to challenge the powerful brewers before the committee, albeit in veiled terms, was a small Kaiapoi brewer, Frank Egan, who said he had once had ten hotels taking his beer but had lost these to the Crown Brewery because he could not match their price. Although he declined to comment on specific instances where brewers had pressured tenants for a stake in goodwill payments, he did observe that 'I come from a country [Ireland] where, if such action were taken in raising rentals, you would get flattened out with a bullet from behind a hedge. You must not interfere with a sitting tenant.' Egan's testimony suggests that the 'interference' was not just with 'good-will' payments, but with the actual ability to hold a licence. In terms of the provision of the licensing law that any licence 'endorsed' for a breach of the law could be cancelled, and probably would be in the hands of many an aggressive licensing committee, Egan implied that brewers were sometimes involved in arranging such endorsements as a means of getting rid of a difficult tenant in order to both replace them with one more malleable and pocket some of the goodwill from the transfer of the licence. As Egan put it, 'the best-conducted house in the world may get an indorsement at any time, perhaps through the carelessness of a servant or some employee', and publicans were at the mercy of any who might want to force them out.[48] In response, and in an effort to deflect attention away from the brewers, Charles Louisson

An early-twentieth-century Speight's advertisement.
Lion Technical Library

seemed particularly determined to secure an admission from the police that endorsement of licences was initiated by them. This they denied.[49] As the last witness before the committee, Louisson also claimed that exclusive beer supply and reduced rental agreements were merely an 'option' offered to tenants rather than a 'condition', and that the only tied houses in existence were some few secured before the passing of the legislation in 1895.[50]

After hearing witnesses over a fortnight, the select committee quickly declared that 'From the evidence submitted to them they are satisfied that the provisions of the Bill would not further the interests of hotel-keepers or of the general public. They recommend therefore that the Bill be not further proceeded with.'[51] The Legislative Council readily agreed, and further attempts by Rigg to introduce the Bill in 1903 and 1904 also failed.[52]

*

Whereas the relationship between brewery managers and licensees attracted much public scrutiny, rather less is known about the working experiences of brewery employees. Until the late nineteenth century almost all worked in very small establishments with consequently close interaction between employer and employee. Wages, hours and conditions were a matter of individual agreement. But as the economic strain of the long depression continued and the brewing industry began to consolidate from the 1890s, some signs of a division between capital and labour emerged. On 27 June 1890 around 120 brewery and cordial manufacturing workers, including coopers and carters, met to form the Dunedin Brewers' and Cordial Manufacturers' Union. The meeting noted that while brewery workers were generally satisfied with their pay, conditions and working hours, cordial workers were far less content. Despite these differences, workers in the two drinks industries frequently joined forces. Compared to most of New Zealand, Dunedin was highly industrialised by 1890 and the city was also a stronghold for the Maritime Council, which linked seamen, watersiders, miners and other groups in an industrial alliance. Indeed, as the maritime strike escalated from late August 1890, employees at the Dunedin Brewery and Speight's donated £9 15s and £7 3s, respectively, to the Maritime Council's Strike Defence Fund. But the collapse of the strike in early November, and difficulties in gaining recognition from cordial manufacturers, undermined the influence of the union and it appears to have faded soon after.[53]

In a curious case in Christchurch in August 1890 the Maritime Council threatened action against Whitcombe & Tombs, printers and publishers, for running its business on non-union lines and refusing arbitration. When it emerged that one of the shareholders in the company also held shares in a prominent, though unnamed, local brewery, workers in Lyttelton determined to boycott establishments stocking the firm's beer. One publican's Saturday takings were reported to have fallen from £16 to £3. The manager of the brewery quickly explained that the shareholder from Whitcombe & Tombs held very few shares in the brewery, the brewery was not in sympathy with Whitcombe & Tombs, and they sent their own printing to a union establishment. Thus reassured, union representatives called off the boycott.[54] But the ease with which the matter was resolved caused some bemusement to the *Auckland Star*.

> It is a curious fact that all through the labour movement, we have never been treated to a boycott of any brewery or distillery. There has not even been a dispute about wages. This speaks well for the way in which persons engaged in the liquor business treat their workmen. Or, perhaps it shows that the British workman has not yet got to the stage when he can complacently do without his beer. He will suffer a good deal before he will boycott the man who supplies his pint. The grocer and the butcher may be boycotted, but never the brewer or the publican.[55]

It was also suggested that in the ongoing agitation for shorter working hours, scarcely a word had been said on behalf of overworked publicans, hotel servants or barmaids.[56]

Around September 1892 another effort was made to form a brewery workers' union, apparently aimed at covering New Zealand as a whole. Although rules and regulations were laid out, brewers reportedly declined to recognise the proposed body in any form as brewery pay in New Zealand was above average.[57] A sporadic Otago Breweries Employees' Industrial Union operated around the turn of the century, but it was only in 1912 that a more permanent Dunedin Brewers, Bottlers, Bottle-washers and Aerated Water Workers' Union was registered with the Labour Department. In early 1913 an Otago and Southland Breweries and Bottling Houses' Employees Award was negotiated with coverage for thirty-three manufacturers and/or bottlers. Meanwhile, an Aerated Water Workers and Other Bottlers' Union registered with the Labour Department in Wellington in April 1905 and lasted to at least mid-1908. But not until mid-1914 was a Wellington Brewers, Bottlers, Bottlewashers and Aerated Water Employees Union registered. In Canterbury a Maltsters and Brewery Employees' Union emerged in 1906 and spread to South Canterbury by 1909. Auckland brewery, bottling plant and aerated-water factory workers were unionised in late August 1907. Under the influence of a Captain Cook Brewery cellarman and future Labour prime minister, Michael Joseph Savage, who became president in 1909, the union gradually, although not unanimously, became aligned with socialist objectives. In 1911 it joined the militant 'Red' Federation of Labour, although unlike most other Federation affiliates it did not deregister from the arbitration system. In 1914, following the defeat of the Red Feds, the union joined the United Federation of Labour.[58] Union developments beyond the main centres seem to have been non-existent. There was no union in Nelson until 1921, or Westland until 1938.[59]

In the years immediately preceding the Great War, the unions secured local awards along fairly similar lines and preference clauses in favour of the employment of union members. It seems that negotiations between employers and employees were entirely amicable, with those in Auckland in June 1913 being described as 'conducted in a tone of utmost good feeling'.[60] Although the Auckland union joined the general strike later in 1913 and stayed out for three weeks, this action served more than anything to expose divisions over the militant direction initiated by Savage, and dissident members registered a breakaway union which lasted throughout 1914.[61] There is no other evidence of industrial action of any kind until much later in the twentieth century.

As with variations in other industries, a forty-eight-hour working week was the norm in Dunedin, but forty-five or forty-six hours elsewhere, and forty-four at Campbell & Ehrenfried. Paid public holidays ranged between five and seven-and-a-half days per year, with some employers providing a further week's paid holiday by the 1920s. All other aspects of pay and conditions were governed by one's role within the brewery, with separate clauses covering brewers, bottlers, bottle or cask washers, cellarmen, coopers, corkers and wirers, maltsters, packers, and various bottling-house, malthouse and brewery labourers. Within these responsibilities there were also marked variations in pay based on age, with only those over twenty enjoying the full rates. In 1903 rates for adult employees at city breweries ranged from 25 shillings a week for labourers to at least £6 for brewers, and these rates were reportedly similar in rural areas.[62] A decade later, when the Auckland Conciliation Council heard a lengthy case between the Auckland union and the brewers, it was resolved that leading cellar hands should be paid at least £2 15s for a forty-six-hour week, although the three large breweries jointly resolved to pay not less than £3. Pay for brewery and malthouse labourers increased from £2 8s to £2 12s; chilling-room men from £2 12s to £2 16s; night-cellar men from £2 10s to £2 14s; and bottling-house workers from £2 5s to £2 10s.[63] The shilling or so an hour for these jobs and perhaps 10½ pence an hour for bottling workers compares with an average of 1 shilling and 3 or 4 pence an hour for unionised tradespeople in other industries, and at least a shilling an hour for other labourers and carters.[64]

Running parallel to changes in the brewing industry, hop growing and marketing also developed its own sophisticated work practices by the late nineteenth century. After hesitant beginnings in various parts of New Zealand during the 1840s, hops became firmly established in the Nelson region as it provided a suitably warm climate with sufficient rainfall but a lack of wind.[65] Aside from repeated efforts, especially by local horticultural societies, to encourage cultivation in areas as diverse as Otago, Manawatū and Waikato, Nelson hops soon dominated the local brewing industry. As one report noted in April 1871:

> The cultivation of hops, both for local use and export, is now an industry of considerable importance in the town and neighbourhood of Nelson. The locally grown hops, indeed, although selling at a price (2s per lb) above that of the imported article, have by their superior excellence quite driven the latter out of the local market.[66]

By the end of the decade the price had shifted in favour of Nelson hops as weak harvests in England produced import scarcity, and the Grey Government, in response to the emerging economic depression, doubled the duty on hop imports at the end of 1879. By the late 1880s the export volume of New Zealand hops, mainly to Victoria, was as much as eight times greater than the amount imported, as brewers now relied almost entirely on locally grown hops.[67] Imports were only used for particular beers. At the beginning of the twentieth century Campbell & Ehrenfried imported Californian, English and Bavarian hops for blending with local hops, while Manning's in Christchurch used Nelson hops for their draught ale but imported Kentish for the bottled equivalent.[68] By 1888 there were 519 acres in hops throughout New Zealand, an area that more than doubled to 1050 acres by 1901 before a gradual decrease over the next decade as the yield per acre improved. At any time at least 90 per cent of the crop was in Waimea and to a lesser extent Collingwood in Golden Bay. Initially the crop was based on Kentish varieties such as Bumford, Green Bine, Cluster and Golding, but by the late 1890s a Californian import began to dominate, which soon assumed particular local characteristics under the name Nelson Californian.[69]

Prosperity was not straightforward. In 1884 the cost of trenching, digging, providing poles for an acre of hops and tending them until harvest was estimated at £200 ($43,000 in 2022). Brewers also periodically complained that hops were spoiled by poor drying. Their moisture content needed to be removed slowly in a hop kiln or oast house, but both required capital to establish and were only used for a short period once a year. Hence in the early years many growers resorted to more primitive methods of drying, including by the sun, with kilns relatively common by the mid-1880s.[70]

The hop harvest, unlike most things in the brewing industry, was very much the preserve of women and children. Hops were planted in September and trained on poles

Baling hops for Nelson merchants the Bisley Brothers, *c.* 1940s. Nelson Provincial Museum, NPM 15 T_181244_8x10_229_Bisley_Bros_MM

TOP: A hop kiln or oast house near Motueka – for drying hops. Nelson Provincial Museum, NPM 10 LHB_49

BOTTOM: Hop pickers in Motueka during the early twentieth century. Nelson Provincial Museum, NPM 17 T_181873_3_8x10_111 or strings in October and November. The harvest, which began in late February and could last until early April depending on the crop, was very labour-intensive, as all of the picking was done by hand until at least the mid-1950s, when machine picking began. While men cut down the hop vines, constantly tended the fires in the drying kilns and baled the hops, all of the picking was done by women and children. Farm women also had the responsibility of growing and preparing sufficient food to sustain the annual influx of pickers. In the mid-1880s, Thomas Tanner, who established a briefly flourishing hop garden near Hastings, required as many as 250 pickers who were brought by coach from the town. Although most of the Nelson picking was done by locals, other pickers regularly came from around the country, especially Wellington. The wreck of the steamer SS *Penguin* off Cape Terawhiti in February 1909 claimed the lives of a number of women and children returning to Wellington after a season hop picking.[71] In April 1907 a report from the Nelson School Committee noted that more than a hundred children were absent hop picking, and it was difficult to carry out the terms of the School Attendance Act as many families left town for another district. As it was recognised that they were dependent on the money from picking, it was suggested that city schools should follow those in hop-picking districts in having a shorter holiday at Christmas and close for a month or more at the height of hop picking during March and April.[72] Nevertheless there were periodic complaints from growers that they made losses due to a shortage of pickers able to harvest the crop in time.[73]

From the late 1870s, E. Buxton & Co. of Nelson began to act as hop brokers, buying from growers and selling throughout Australasia. Their correspondence from 1901 provides a good illustration of a sophisticated business in which quality and timing was everything and urgent business was conducted by telegraph. In April they reassured the Crown Brewery in Christchurch that all hops provided were sufficiently dry. 'We don't think they are the least damp, but should any of them not keep thoroughly sound we will make good any loss. We think that if they were any harder dried they would be too brittle and easily broken for the purpose you require them.' In May they wrote to the secretary of the Commerce and Industry Department regarding samples of Nelson hops that were being forwarded for a representative of the Department to take on his visits to 'dangerous places' in the hope he could open up new markets, especially in South Africa. In June 140 bales of hops were dispatched to Speight's with a note that the usual cash discount of 2.5 per cent would be deducted for 'the best lot of hops shipped from here this season'. By July they were informing New Zealand brewers that although the price was being lowered to clear stock from the previous season, it was best to order quickly before an enthusiastic Australian market took the surplus.[74]

Growers and brokers alike also appreciated that the breweries had an ever-advancing understanding of what to do with the hops as brewing science moved forward rapidly in the wake of Pasteur, Hansen and others. As mentioned in Chapter Three, it is uncertain how and when knowledge of various scientific and technical developments reached New Zealand. At the very least, and as an unforeseen benefit of the requirement of the Beer Duty Act to keep brewing books, brewers now kept regular analytical data, such as precise quantities of raw materials and resulting specific gravities of the brew. But many of the important advances were European, especially German, and yet New Zealand brewers and the consumer preferences they satisfied were more closely aligned to the British brewing industry, which was slower to absorb change. On the other hand, the *Australian Brewers' Journal*, a comprehensive Australasian-focused monthly trade and technical journal launched in 1882, provided up-to-date information and advice to brewers in the manner of several such publications that had begun in Europe and Britain during the 1860s. Although the New Zealand climate was closer to Britain's and did not present the same challenges that prompted Australian brewers to embrace some European innovations before they were common in Britain, especially in the case of pure yeast strains, the influence of the journal ought not to be underestimated.[75] Indeed, in August 1903 it offered considerable praise for the quality of the library at Staples Brewery. 'Not a few old dusty volumes of brewing lore thrown on a corner shelf, written in the days when they tested the quality of beer by its resisting power against leather breaches, but valuable books on the theory, science, chemistry and practice of brewing, and by the most able of modern teachers.'[76] At a more speculative level, the New Zealand brewing industry may also have been influenced by the level of technical and scientific expertise emerging in the local meat and dairy industries following the development of refrigeration and the pursuit of international markets during the late nineteenth century.[77]

Almost certainly the most formally qualified brewer in New Zealand at this time was Henry Duncan of J. R. Dodson & Son, the successor to Hooper & Co. in Nelson. Duncan visited Britain in 1890 to examine brewing techniques and especially to study scientific analysis. He received a glowing reference after three months with Ansell & Sons Brewery of Birmingham, stating that

> you have applied yourself with great zeal and real effect to the task of studying the scientific and practical details of English brewing. You have acquired considerable skill in chemical analysis upon the most recent scientific authors. . . . I can say without hesitation that you are the most diligent worker that I have as yet met with and I very much regret your leaving me.[78]

Thereafter Duncan visited Europe and spent three months in Copenhagen studying pure yeast culture. He made a further visit to Britain and Germany in 1893 and became a member of the London Institute of Brewing, obtaining an honours certificate in brewing from the London City and Guilds Institute, a certificate probably not held by any other Australasian brewer.[79] William Dawson of Speight's was another who kept up to date with brewing developments. From 1895 he took six to nine months on half-pay every second year to travel overseas for both pleasure and to explore the latest brewing technology in Europe and the United States. Information from his first trip led directly to a significant overhaul of the brewery in 1896. In 1905 Dawson took his son Reg overseas and found a two-year course for him at a brewing college in Birmingham.[80]

Others sought to import expertise rather than obtain it themselves. In 1902 Staples Brewery engaged English brewer Percy Phipson, described by Pat Lawlor as 'a brewing technologist able to separate real facts about brewing from nostrums and superstitions that had come down through centuries of tradition'. Phipson revolutionised the running of the brewery, but was highly critical of New Zealand barley and read a paper on the subject to the Wellington Philosophical Society in November 1905.[81] Two years earlier the Crown Brewery in Christchurch had brought out another experienced English brewer, C. J. Sloman, who immediately introduced a bottle-conditioned pale ale that was still in production during the 1940s.[82] But on more than one occasion the *Australian Brewers' Journal* claimed that English brewers engaged in Australasia were not a success because they did not know how to handle colonial raw materials. Many terminated their contracts and established other businesses.[83]

ONE OF THE MALT KILNS.

TOP: The bottling and labelling room at the redeveloped Captain Cook Brewery, Auckland, in 1900 when the brewery briefly produced New Zealand's first lager. Lion Technical Library

MIDDLE: Conrad Breutsch was brought to New Zealand from Switzerland to oversee the establishment of the Captain Cook Lager Brewery in 1900. He later worked at the Gisborne Brewery. Lion Technical Library

BOTTOM: A brewery employee tending the malt in one of the drying kilns at the Captain Cook Brewery, Auckland, 1900. Auckland Libraries Heritage Collections, AWNS_19000223_12_2

Perhaps the most significant embrace of European brewing science and expertise, if not apparent until sometime later, was the development of New Zealand's first lager brewery under the guidance of a German, A. F. Metzler, and a Swiss, Conrad Breutsch, which opened in 1900. Lager, from the German *lagern* – 'to store' – emerged in Bavaria, where brewers had for centuries stored their beers in cold caves and cellars to mature. By accident rather than design, a yeast strain had developed which thrived in cool conditions and settled at the bottom when fermentation was complete – as opposed to the more common ale yeast, which required warmer conditions and rose to the top. The result was a very crisp and clear beer, and one with greater stability and durability at lower alcohol levels. Following an extensive brewing tour through Europe and Britain during the early 1830s, the Munich brewer Gabriel Sedlmayr the Younger began to perfect lager brewing and gradually spread the technique throughout Europe. While lagers in the Bavarian style were initially dark beers, in 1842 in the Bohemian town of Plzeň a Bavarian brewer, Josef Groll, began to transform the region's soft water and local barley, with its low protein, into the first golden-coloured pilsner. This new style spread rapidly across Europe and with German migrants to North America, being greatly assisted by new industrial technology, especially refrigeration machines which enabled lager to be produced throughout the year and in a variety of climates. The large-scale manufacture of glasses to replace stoneware and pottery also assisted lager, in that the preference for lighter rather than darker-coloured beers was accentuated among drinkers who could now see what they were drinking.[84]

But the new beer trend was slow to take hold in Britain, where consumer taste and perhaps a patriotic resistance to continental innovations ensured that ale remained supreme. Consequently, lager was not in the repertoire of the brewers who shaped the New Zealand industry during the nineteenth century, and seems not to have been tried by the several German and other European brewers, and Ehrenfried in particular, who established themselves in the colony. Aside from taste preferences, few brewers had the capital to transform their ale breweries to lager production, and especially to afford the ice machines necessary to keep cellars cool while the beer was conditioning.[85]

One of the few who did have sufficient capital to contemplate such a venture was Moss Davis of the Captain Cook Brewery. Following travels in Europe during the 1890s he apparently concluded that there was potential for lager in New Zealand, although there was seemingly nothing in the existing market to justify such a belief. Imports of beer from Germany, the vast majority of which was likely to have been lager, reached a peak of perhaps 30,000 litres per year during the mid-1890s, or 1 to 2 per cent of all imports and 0.1 per cent of all beer available for consumption. But by 1898–99 the annual German import had fallen by as much as two-thirds. Imports from the United States, a significant portion of which is also likely to have been lager, peaked at more than 54,000 litres in 1898 before falling away sharply. Nevertheless, Davis sent his head brewer, Charles Spooner, to visit breweries in Britain and Germany during 1899, and Breutsch and Metzler were immediately engaged to establish a lager brewery in Auckland. Amid considerable festivity attended by a large number of Auckland notables, the Captain Cook Lager Brewery was opened on 13 February 1900.[86] In opening the new plant, the Hon. Edwin Mitchelson, a former minister of public works, declared that he was 'pleased to learn that the new brand of lager beer was not so intoxicating in its effects as the ordinary beer – this was a great feature in its favour', and that it was 'to all intents and purposes a temperance beverage'.[87] The *Auckland Weekly News* pursued this theme with a sweeping and xenophobic dismissal of prohibitionists and much of the rest of the world.

> The temperance party would not be such opponents to the drinking of beer if they realised the whole of its results. Our race has made beer its popular beverage, since it emerged from the obscurity of prehistoric times, and who dare say that our forefathers, drinking even to excess their national beverage, were not stronger, more intelligent, than more temperate nations? It is a curious fact but true, that the beer drinking nations rule the world in wealth, in military power, in mechanical and scientific skill. The prohibitionists can not point to one great nation which is strictly abstemious in the matter of drink. The more abstemious the nation as a rule the lower it is in the social scale. The Hindus, the Chinese, the Latin nations of Europe drink no beer and scarcely any intoxicant. The British people, the Americans, the Germans, Scandinavians, and Dutch are great beer drinkers and the pick of the world's peoples.[88]

Davis added that he hoped the public would recognise that by buying the new beer they would be assisting to retain their money in the colony as it was made from colonial-grown hops and barley. He claimed that there were significant imports of lager to New Zealand which he intended to 'obliterate'.[89]

Such is the rarity of the Bismark Lager label from around 1900 that an undamaged example seems not to have survived. *Brewnews*, Lion Technical Library

Yet the New Zealand palate remained more attuned to British tastes than those of continental Europe. The new beer, variously advertised as Bismarck or Bismark [*sic*] Lager Beer, supposedly named for the former German Chancellor Otto von Bismarck, was only a minor contributor to the New Zealand market. It certainly drove German imports out as they immediately plunged to 6000 litres in 1900 and only 2500 in 1904. But, if the end of its advertising campaign is any indication, production seems to have ceased by May 1905. Thereafter the Germans reclaimed their tiny niche market, with imports increasing to over 23,000 litres by 1908 and around 69,000 in 1913 on the eve of the Great War. But by then, with the sharp increase in overall New Zealand production during the first decade of the twentieth century, this was still only 0.1 per cent of all beer available. Anti-German sentiment from 1914 further undermined the prospects for lager.

As the twentieth century opened, New Zealand consumption was mainly of ale, broadly defined. Certainly, some brewers had their specialties such as Daniel Arkell's AK Bitter Ale or XXX Mild Ale, Campbell & Ehrenfried's special Strong Ale and Amber Ale or McGavin's Gold Medal Strong Ale, which was frequently advertised in the *Pelorus Guardian and Miners' Advocate*, but not at its source in Dunedin.[90] The specialties created by William Dawson at Speight's have been noted already. While the 1911 census, the only one of the period to report on the type of beer produced, revealed that one-third of the production of the four small Southland breweries was stout, this was only 2 per cent of production in neighbouring Otago and no more than 7 per cent elsewhere in the country. Harvest ale, a staple of the rural workforce during the nineteenth century, entirely disappeared from the advertised offerings by the mid-1890s.

Surprisingly, amid these scientific, technical and taste advances, the breweries were less successful in addressing one of the fundamentals of beer supply – suitable bottles. With no bottles manufactured in New Zealand until the 1920s, brewers were dependent on imports and more especially on recycling of bottles from imported beer. In turn, many of these were repurposed champagne bottles.[91] From the 1840s discounts were offered to those providing their own bottles or casks and to those returning them after use. In 1879 Louis Ehrenfried advertised for the return of 10,000 ale and porter bottles in Thames. It was not uncommon for breweries to employ a dedicated bottle collector, while others paid for the return of bottles. In 1911, D. J. Barry of the Gisborne Brewery noted that several men in the town made their living by this means, and numerous children significantly boosted their pocket money during the summer.[92] One, possibly apocryphal, explanation for the bottle shortage was that many of them were being purchased and then sent into dry districts where draught beer could not be sold.[93] Any brewery going out of business was quickly targeted, although there was also stiff competition from cider and cordial makers for any spare bottles. Barry imported bottles from Sydney and as far away as Germany, but a significant number arrived broken despite waterside workers being inclined to handle them with greater care than many other items. The shortage became so severe that bottles were put on the free import list in 1903, meaning that they could be imported from beyond the British Empire.[94]

Corks for bottles were also a challenge. Most were imported, although when a new cork-cutting factory was established in Auckland in 1877 it was noted that they were able to make the corks to sizes that local brewers and bottlers wanted, whereas many bales of imported corks frequently contained sizes unsuited for local needs.[95] By the 1890s there were also cork factories in Dunedin and Petone, although limited duty on imported corks meant that these were not especially profitable ventures.[96]

Bottling itself was for a long time a labour-intensive and somewhat fraught process, as it created opportunities for contamination of the beer if bottles or corks were not scrupulously clean. While many smaller breweries did their own bottling, others contracted to specialist bottlers. For example, Powley & Keast became the most prominent bottlers in Dunedin from the mid-1890s after Richard Powley lost the licence of the Apollo Hotel following a reduction vote in 1894.[97] Towards the end of the nineteenth century more established breweries were able to invest in efficient bottling machines. In 1892 McGavin's invited interested parties to view their 'Vacuum Bottling Machine' and the 'Triumph Corking Machine' imported from Denmark. It was noted that 'the machine saves considerable time and labour, as one man only is required to do the bottling, and a machine with five spouts will easily fill 1200 bottles in an hour without spilling any'.[98] Three years later, Powley & Keast's float in the Dunedin Labour Day parade apparently fascinated the crowd by including 'trollies in which the process of bottling their favourite brands of liquors was being gone through'.[99] Into the twentieth century bottles were commonly green with concave bottoms, corked, wired and labelled by hand, but only labelled if destined for hotel retail rather than home delivery.[100] Those breweries with a wider distribution, such as Speight's, sent bulk beer around the country to have it bottled on contract by specialist firms and sometimes smaller breweries. Speight's had eighteen such contracts by around 1900. Consequently, there could be quite different labels for the same beer in different parts of the country.[101]

The short-lived production of Bismark Lager did herald one significant change. New Zealand began to follow British and more immediate Australian moves to counter the lager threat and at the same time meet a growing consumer demand for a lighter, less intoxicating, more carbonated and clearer or 'bright' bottled beer. Following a visit to Australia in 1907, Moss Davis's son Ernest, who was gradually establishing his own reputation as a leading advocate for brewers and a scourge of prohibitionists, oversaw the Captain Cook Brewery's shift to production of its first artificially carbonated, filtered and pasteurised beer. Unlike bottle-conditioned ales that contained an amount of sediment in the bottle from secondary fermentation and were prone to become cloudy unless decanted carefully, these 'no deposit' beers, as they were often referred to in New Zealand, were effectively ale on lager principles. The common method was to place the bottles in a tank of warm water and raise the temperature to around 130°F for twenty minutes, and then slowly cool. The Captain Cook Brewery's advertising for its new Imperial Ale made explicit reference to the process by emphasising the 'steam and hot water bottling machine . . . ensuring, as far as humanly possible, clean bottles to put a good article into'.[102] Following his work in Auckland, Conrad Breutsch moved to the Gisborne Brewery in 1908 and began to produce Gold Top ale by the new method. Percy Phipson followed suit with Staples' Red Band Ale in 1913, which became perhaps the most renowned and enduring of this style. Even then, the process of carbonation remained somewhat approximate and rudimentary and Staples' bottling staff were prone to regular soakings with high-pressure beer.[103]

By around 1900 the New Zealand brewing industry was very much in transition – a blend of commercial and technical sophistication with elements of the primitive and makeshift. At one end were a few larger and increasingly monopolistic breweries in the main cities, at the other a group of determined smaller players thus far protected by local geography or consumer taste preferences. But this was also an industry under threat, especially from the prohibition lobby, and the first quarter of the twentieth century, including the multiple challenges of the Great War, was to bring the most fundamental changes in its history.

STRACHAN'S
KAKA
ALL BRIGHT
DUNEDIN ALE
TRADE MARK REGISTERED
KAKA
BREWED & BOTTLED IN NEW ZEALAND BY
W. STRACHAN & Co.
Drink success to the
Best L
KAK
AL
"CLEAR TO THE LA
Brewed & Bottled in New Z

Dunedin's Strachan & Co. advertisement for Kaka Ale during the First World War. Alexander Turnbull Library, Eph-A-ALCOHOL-1914-1918-01

SIX

'A Red Hot Nest of Liquor Dealing Intriguers': Beer and Prohibition, 1900–1919

In many respects, the first two decades of the twentieth century were the most critical in the history of beer and brewing in New Zealand. The industry was fighting for survival as support for prohibition swelled. The anti-liquor vote peaked at 55.8 per cent in 1911, nearing the 60 per cent threshold needed for success. Prohibition nearly won in December 1919, when only a simple majority was required to turn New Zealand dry. While the conventional portrayal of this period is of a tide of prohibition that nearly succeeded, it is equally valid to ask why some people voted against it. The brewers in particular deployed a range of strategies to boost their own standing with the public and to put obstacles in the way of the prohibition campaign. Yet as the liquor industry fended off the prohibitionist threat, especially during the dramatic upheaval of the Great War, it was forced to make significant concessions in the licensing laws. These fundamentally altered drinking culture and the brewing industry throughout the twentieth century. The closure of hotels due to no-licence or reduction votes and other legislative change altered where, and later when, people could drink. By 1914 drinking was rather more covert and divorced from public sociability than it had been a decade earlier.

The rise in support for prohibition from the 1890s to the end of the Great War was startling and remained very strong throughout the 1920s, as Table 6.1 demonstrates. The reasons for its decline by the 1930s will be explored in the next chapter. Licensing polls, except for a special referendum in April 1919, were now held on election day and the 'liquor question' was fiercely contested. The first polls were local, with voters deciding on the licensing status of their own electorate from three options. In 1911, the poll became national, with voters deciding on the status of alcohol for the whole country and facing a choice between continuance and prohibition. In 1919, a third option – state purchase and control – was added to the poll, and these choices remained until their removal in 1989.

In the 1896 to 1908 polls, voters could choose any two options, and 30–40 per cent generally did. The figures are therefore beyond 100 per cent. While we can assume that some of those who voted for no licence cast a second vote for reduction in the hope of achieving at least some change, we can be less certain as to how many of those who voted for continuance may also have voted for reduction. Voting patterns were surely also influenced as prohibitionists

Table 6.1: Votes at Licensing Polls, 1896–1938

Local

	No licence (%)	Continuance (%)	Reduction (%)	Percentage required for change
1896	37.8	53.1	36.4	60
1899	42.1	50.7	38.3	60
1902	47.5	46.6	41.5	60
1905	50.1	46.1	38.1	60
1908	52.5	44.6	38.5	60

National

	National prohibition (%)	Continuance (%)	State purchase and control (%)	Percentage required for change
1911	55.8	44.2		60
1914	49.0	51.0		60
1919 (1)	49.0	51.0		50
1919 (2)	49.7	44.5	5.9	50
1922	48.6	48.6	5.8	50
1925	47.3	44.4	8.3	50
1928	40.1	51.1	8.8	50
1935	29.6	63.4	7.0	50
1938	29.0	60.4	10.6	50

Source: *New Zealand Official Yearbook*, various issues.

could see the fruit of their labours in the form of twelve districts that voted for no licence by 1908 (Table 6.2). While the initial success was in the south of the South Island, especially where a more conservative Protestantism held sway, the Auckland victories in Grey Lynn and Eden and those in Wellington undoubtedly suggested that the main centres would go dry before long.

These dry districts led to the closure of 141 hotels. Successful reduction votes in other places claimed another 343. For example, while the decisions of Grey Lynn and Eden to go dry prompted the closure of the five hotels in these districts, the reduction vote in Auckland Central in 1908 resulted in another five of fifty-eight closing in June 1909. Meanwhile, in Ōhinemuri, fourteen hotels were closed ostensibly because the miners of Waihī followed through on a threat to vote for no licence unless the local publicans reduced the price of beer from 6 pence to 3 pence a glass.[1]

The increasing momentum of prohibition was due to better organisation, better resourcing and more effective lobbying rather than to any arguments substantially different to those first articulated during the 1870s.[2] Indeed, there were a wide range of views within the prohibition movement about what was desired from legislative reform and the best strategies to achieve it. Prohibitionists gained much confidence from the 1894 poll result, especially the success in Clutha. They were convinced that, once tried, prohibition would remain. During 1894–95, efforts were made to federate various temperance organisations together under the umbrella of the New Zealand Alliance and to emphasise the value of a national organisation to secure national prohibition

Table 6.2: Districts Voting Local No-Licence, 1894–1908

1894	Clutha
1902	Ashburton
	Mataura
1905	Grey Lynn
	Ōamaru
	Invercargill
1908	Bruce
	Eden
	Masterton
	Ōhinemuri
	Wellington Suburbs
	Wellington South

legislation rather than merely change at a local level.[3] Opponents, such as the Liberty League, which emerged in Christchurch during 1896 as a non-political anti-prohibitionist organisation focused on the personal liberty of the individual, were abruptly dismissed as 'a red hot nest of liquor dealing intriguers whose only political motto is "Our trade, our politics."'[4] But in reality Clutha was an artificial poll result. Many supporters of continuance had not voted at all in an effort to void the poll due to turnout of less than 50 per cent – a risky strategy that worked in thirty-five districts. The failure to gain any significant ground in the 1896 or 1899 polls also confirmed that the support of women for prohibition could not be taken for granted and a more extensive education campaign was needed before further progress could be made.[5]

A pattern emerged whereby prohibitionists tended to spend two years after the licensing poll engaged in agitation for legislative reform, and then a year devoted to public education as the election approached.[6] In this they were assisted by voluminous pamphleteering and public lectures drawing impetus from the rise of prohibition in the United States, in particular where the Anti-Saloon League had quickly achieved national prominence following its formation in Ohio in 1893. By 1913 nine US states had complete prohibition and thirty-one others had local option laws, placing more than half of the United States population under some form of prohibition. Numerous British and American temperance and prohibition campaigners toured New Zealand from the mid-1880s, and in addition to the WCTU other American organisations such as the Sons and Daughters of Temperance also gained a foothold.[7]

Prohibitionists tended to recycle well-rehearsed arguments and seldom engaged in toe-to-toe debate with their opponents. But there were periodic exceptions and trenchant attacks on the conduct of the trade. In a contribution entitled 'The Story of the Clutha', Frank Isitt, brother of the most articulate and enduring prohibition leader, Leonard Isitt, claimed that the dry district was not a complete success as, the moment the hotels were closed, one magistrate rapidly issued more wholesale licences to suppliers in neighbouring districts.[8]

In 1898 the Christchurch politician and ardent prohibitionist Tommy Taylor published *The Shadow of Tammany* as a counter-narrative to a Royal Commission that he maintained had entirely failed to address widespread evidence of corruption between the police, publicans and brewers.[9] Taylor, after interviewing various witnesses throughout the country, insisted that police were not enforcing the licensing laws and that complaints made to senior officers, including Commissioner of Police Arthur Hume, were ignored. Police, especially in country districts, were apparently afraid to do their job, and in particular to prevent Sunday trading, due to the need to be aware of who held influence in the area. Inspector John Cullen of Greymouth, who would become commissioner in 1912, informed the inquiry that he had been transferred from Timaru at the behest of publicans who resented his enforcement of the licensing laws.[10] In a chapter entitled 'The Brewer and Two Wanganui Friends Succeed', Taylor reported the case of a Hāwera sergeant who initially escaped sanction for being asleep in a public place during the day as he was deemed to be ill rather than drunk. When he was later dismissed for disturbing the peace when he mistakenly tried to enter a house that was not his own at 11.30 p.m., a dismissal endorsed by Premier Seddon, the sergeant immediately went to Wellington and approached Martin Kennedy, managing director of Staples Brewery. Kennedy wrote to Seddon stating that he had known the officer on the West Coast, could attest to his good conduct, and asked for his reinstatement. He was soon reinstated to the force, although reduced in rank and transferred to Lyttelton. However, he was soon promoted again following further pressure.[11] No doubt the spectre of Kennedy's intervention was accentuated by his machinations for Staples & Co. around tied houses outlined in the previous chapter.

Staples was again in direct confrontation with prohibitionists in 1903 when the brewery sued the Wellington printers Wright & Carman for libel following allegedly defamatory doggerel before the 1902 election. The first three verses will suffice.

A LAY OF THE TRADE.

Staples and Co. were brewers,
And they plied a roaring trade;
A hundred tied house tenants
Their sovereign will obeyed;
'Sell our beer,' said they, despite the law
'And likewise pay our rent;
'And if you don't, well out you go!'
And out they surely went.

Staples and Co. the brewers
Had a covenant with death
To be his zealous agents
As long as they had breath;
They served their master faithfully,
And did his work so well,
That the track of their beer-carts round the town
Was marked like the road to hell.

Staples and Co. were brewers,
And they raked the money in;
What matter to them that it bore the stain
Of misery and sin,
Of the tears of widow and orphan,
Of the harlot's bitter shame?
Staples and Co. were brewers,
So they took it all the same.[12]

TOP: A 1910 depiction of the challenge facing the prohibition campaign. Alexander Turnbull Library, B-K693-3

BOTTOM: In 'Brewer Rabbit & the local option poll', *c.* 1910, Trevor Lloyd depicts the threat to the trade, as twelve districts had gone dry. Alexander Turnbull Library, C-109-008

OPPOSITE: 'The Octupus of No License' represented New Zealand's increasingly complex and restrictive licensing legislation by 1905. Hocken Collections, University of Otago, 509-262g

THE OCTUPUS OF NO LICENSE.

The "No License" monster, crawling along its pathway of fanatical superstition, is endeavouring to claim New Zealand as its victim; its bigoted ignorance of mankind preventing it recognising the greater evils it would bring to bear by the introduction of its hypocritical fallacy—refusing to see in its blindness, of crass stupidity and narrow-mindedness, that the spending of a few superfluous shillings across the bar in search of knowledge of mankind blended with the conviviality of good-fellowship is a much lesser evil than the vices its ignorance would otherwise introduce.

In subsequent verses, Staples were accused of trying to bribe Christian voters with large donations to both the Anglican and Catholic churches, of which more below, and of unscrupulous intervention to overturn an initial no-licence vote in Newtown in 1902. In defence, Wright & Carman argued that the anonymous composer did not target Staples specifically but all brewers, and merely 'wanted a little poetic discussion on the evils of the drink traffic generally'. Further, as there was public opposition to tied houses, and as Staples had these, it could be taken as typical of the trade. The jury deliberated for fifteen minutes and awarded one farthing damages to the brewery – rather less than the £1000 claimed. The judge also awarded 20 guineas costs.[13]

There was no such levity in the Newtown licensing case. The Wellington suburb of Newtown went dry at the 1902 election by a margin of seventeen votes, reduced to four after a recount. A petition was immediately lodged alleging 'gross irregularities' in at least one voting booth and claiming that ineligible people had voted. Rumour at least suggested that any ballot tampering had been done to favour continuance. In February 1903 a magistrate declared the licensing poll void, a decision that caused much bitterness among prohibitionists. Their revenge came later in the year when the local licensing committee, which they dominated, refused to renew five hotel licences on the grounds that, as no decision on the licensing status of the district had been confirmed by the electors in 1902, the committee had no jurisdiction to renew or grant licences. This decision was appealed to the full bench of the Supreme Court in July 1903, which, by a 3–2 margin, refused to intervene on the same grounds as the licensing committee.[14] To resolve the deadlock, it was agreed that the case of Frank Smith, licensee of the Park Hotel, would be taken to the Privy Council in London as a representative case. While the case awaited resolution, the five hotels continued to trade until the police, on advice from the Crown Law Office, secured convictions for illicit trading and each was fined £5. The hotels were closed from September 1903.

However, in May 1904 the Privy Council determined that the clear intention of the legislation in any case where none of the licensing proposals was carried was to continue with the status quo. As there had been no public objection to Smith's application for a licence, the licensing committee had to hear his case and had to pay the costs of the appeal to the Privy Council.[15] These were estimated at £800–1000 ($160,000–200,000 in 2022). On behalf of the New Zealand Alliance, Frank Isitt described the decision as a 'triumph of trickery' and complained bitterly at the improper electoral process that had triggered it in the first place. Within two weeks of the decision from London, the five licences had been reinstated, no doubt to the satisfaction of Staples, which held ties over all of them and had financed the progression of the case to London.[16]

These events, combined with Seddon's friendship with Kennedy, and the fact that one of his daughters was married to the son of another prominent brewer, Hamilton Gilmer, fed claims that the premier was an apologist for the brewers and for the trade more generally. Certainly, he orchestrated a more moderate Licensing Bill in 1893 and continued to thwart demands for legislative reform throughout the 1890s, not least in never having to concede to a national prohibition poll or a simple majority threshold.[17] But Seddon's constant outmanoeuvring of prohibitionists is best understood as both a reflection of his own personal moderation and an effort to maintain unity and hence the wider reformist agenda of the Liberal Party by ensuring that his Government would not rise or fall on the single issue of prohibition. Having said that, it is unlikely that a single-issue group would have been able to mount a threat as a broader political party.[18] More to the point, it is evident that Seddon was not merely a puppet for the needs of the trade. He told Kennedy and others that they should expect some reduction in hotel licences to satisfy prohibitionist agitation. He also warned them against too frequent public statements and their own demands for licensing reform and refused any suggestion to increase the three-fifths threshold to two-thirds or three-quarters.[19]

Following Seddon's death in June 1906, and with the Liberals now under the less decisive leadership of Joseph Ward, prohibitionists saw more scope to capitalise on the rising tide of support that had seen the total vote for no licence exceed that for continuance in 1902. The key to securing national prohibition was to limit voter options and capture the moderate vote. A target was those who saw local no-licence as a means to help the poor and vulnerable and instil an aspiration to middle-class respectability by removing their access to drink, but saw no need to curb their own drinking. These were people who voted no-licence but freely imported alcohol to dry districts for their own use. The Alliance also campaigned relentlessly against the three-fifths requirement and pointed out that the twelve dry districts by 1908 would have been fifty with a simple majority.[20] On the other hand, the trade was also looking to protect its interests by bringing about decisive licensing reform that would end the perennial debates. Therefore, in late 1909, under the direction of Minister of Justice John Findlay, secret negotiations were initiated between Alliance representatives and the trade. The Alliance proposed a single ballot in which all votes for local no-licence would be counted as votes for national prohibition. A 55 per cent majority was also proposed as a compromise between the simple majority and the existing three-fifths.[21] To the surprise of most in Parliament, Sir Joseph Ward announced the existence of the agreement in December 1909. If carried, prohibition would remove production and importation of alcohol for all except scientific and medicinal purposes. It would take effect five years after the poll and no new poll would be held until prohibition had been in force for three years.[22]

Although some Members of the House were happy that agreement had been reached, others, and many Alliance members, were very dissatisfied with a process in which they felt that the Alliance leadership's single ballot proposal had alienated the moderate voter who wanted local no-licence but not national prohibition.[23] Deep divisions emerged within the Alliance at national conventions in late 1909 and again in mid-1910. Naturally, the trade refused to release the Alliance from the agreement over the single ballot clause. Eventually the Alliance retreated from the agreement and rather surrendered the initiative.[24] The fragile Ward Government, still wanting compromise to avoid a split in the Liberal Party, then passed a Licensing Amendment Bill that allowed for a national prohibition poll but retained the three-fifths majority and kept local option as a separate ballot.[25] Following a very intense Alliance campaign in 1911, the national prohibition vote peaked at 55.8 per cent – higher than the original 'compromise' threshold negotiated in 1909 and only 16,000 votes short of the three-fifths required. This was a moral victory that offered real hope for the 1914 campaign, but perhaps no real consolation.[26]

The 1910 licensing legislation contained other victories for the prohibitionists that had far-reaching consequences for brewers and the trade generally. From its New Zealand foundation in 1885, the WCTU mounted a strenuous campaign against the employment of women anywhere in the liquor trade. Within the broader campaign to preserve home and family and elevate the domestic role of women, attractive barmaids were seen as enticing men to drink. Prohibitionists also tended to associate licensed premises with brothels and argued that alcohol created desire and reduced self-control. It was therefore imperative to remove the temptation of women. From 1893, no new publicans' licences were to be issued to women except for widows and wives covered under provisions of the Married Women's Property Protection Act.[27] Following the banning of barmaids in Glasgow in 1902 and significant restrictions emerging in several Australian states during the same year, in July 1903 a private member's Barmaids Abolition Bill put forward by Harry Ell failed as it was thought too drastic to throw all barmaids out of work. It was becoming evident, however, that some sections of the trade also supported the phasing out of barmaids. Whether this was driven by their own moral conviction or a sop to the prohibitionists is not entirely clear.[28] In late 1908, the three main brewers in Nelson declared that future leases on hotels that they owned would prohibit the employment of barmaids. In retort, 'very indignant' local barmaids

THE BARMAID WINS THE DAY.
King Dick: Well, ladies, if I must choose between you, give me the barmaid.

Despite the message of this cartoon from 1903, Premier Richard Seddon was unfairly portrayed as being under the thumb of the alcohol industry. While barmaids survived his tenure, they were banned in 1910 and not reinstated until 1961. *New Zealand Observer*, September 1903 tried to dissuade people from drinking local beer. 'In one free house, the landlady cut off the pipes from the casks of local beer, cancelled orders for aerated waters and cordials from a local brewer, and drew only a Southern ale.' While the boycott soon faded, prohibitionists were reportedly 'hilarious over the brewers' manifesto'.[29] Faced with such rare agreement between the trade and the prohibitionists, the fate of barmaids was inevitable. Under the Licensing Amendment Act 1910, it became illegal, from 1 June 1911, for women to be employed on licensed premises during opening hours, with the exception of those already employed before the Act, any woman still holding a publican's licence, or the immediate family of any publican. A national register of barmaids was established to aid enforcement of this measure.[30] This combination of legislation and moral suasion ensured that the pub became very much a male preserve.

The pub was also a very white environment, as further legislative restrictions sought to curtail access to alcohol by Māori. The Outlying Districts Sale of Spirits Act 1870, which superseded the 1847 Ordinance, established twelve Native Licensing Districts in areas where at least two-thirds of the population were Māori. Liquor licences could only be issued to Pākehā with written permission from local Māori assessors. Further amendments to the legislation during the 1870s encouraged communities to elect temperance committees and empowered the governor to proclaim dry districts in areas of at least two-thirds Māori population if one-third of the adult Māori population or ten local chiefs petitioned for it. Against a sharp decline in the Māori population, which reached its lowest point at just over 42,000 in 1896, and with many predicting eventual extinction, temperance and later prohibition campaigns were as much a product of Pākehā paternalism as Māori initiative.[31] During the 1890s, the Te Aute College Students' Association, with the emerging influence of Āpirana Ngata to the fore, actively discouraged the use of alcohol at meetings and tangi. Similarly, when the Tūhoe prophet Rua Kēnana established a community at Maungapōhatu in 1908 he also banned alcohol.[32] Some publicans adopted their own measures to restrict Māori drinking. In 1903 those in Ōpōtiki formed a 'No Quart Trust' in which Māori would not be served with a quart of beer for consumption on the premises. They could purchase beer by the gallon but only if bringing their own buckets and glasses and drinking elsewhere. 'The object is to discourage drinking habits among the Maoris, who don't know when they have had enough.'[33]

While South Island Māori retained equal drinking rights with Pākehā, the 1910 legislation further strengthened dry districts in the North Island, especially the King Country. It became an offence to supply alcohol 'to an intoxicated male Native' or 'any female Native, not being the wife of a person other than a Native'. North Island Māori males could only buy alcohol for consumption on licensed premises, while Māori women were prohibited from drinking in pubs unless they were married to a Pākehā. Although provisions existed between 1912 and 1931 for Māori to renounce their native status and obtain equal access to the purchase, sale and consumption of alcohol, fewer than a hundred ever exercised this option.[34]

*

Faced with such unrelenting pressure from prohibitionists, the trade, with Martin Kennedy and other prominent brewers in the lead, also grasped the necessity to organise. In 1890 Kennedy led the formation of a Brewers', Wine and Spirit Merchants' and Licensed Victuallers' Association – commonly referred to as the Brewers' Association. He served as permanent chairman of various subsequent iterations of this body until his death in 1916, and meetings were frequently postponed when he could not be present.[35] Among the objectives of the Association were:

> (a) the general protection of the trade throughout the colony, and to endeavour to obtain compensation in the event of adverse legislation, (b) to diffuse, by means of papers, statistics, and letters among members, particulars of all matters affecting the general interests of the trade, (c) to ensure more unanimity and the cooperation of the trade in the work of the Association, especially with reference to all proposed Parliamentary legislation, and in general licensing matters.[36]

While wine and spirit merchants paid a flat fee of £2 5s per annum, brewers paid a levy of 3 pence per hogshead of beer produced, an arrangement later attacked by prohibitionists as a secret 'trade defence fund' which was supposedly used to interfere with the electoral process. It is clear, however, that the fund was used for normal administration and publication costs, and its existence was certainly no secret.[37] During the first year there were fourteen meetings of the executive and £75 was given to the Christchurch branch to assist legal action in the Sydenham licensing case. The Association also had a permanent secretary, but the position was dispensed with in August 1891 after non-payment of the levy by Auckland and Dunedin brewers, and the latter, for reasons unknown, briefly withdrew from the Association.[38]

The broad strategy of the Association was to protect the trade through supporting the election of candidates favourable to its interests. Some suggested selecting its own candidates to directly oppose prohibition candidates, but it was deemed more useful to support particular candidates and in so doing reinforce the perception that prohibitionist candidates were single-issue fanatics. Yet the problem with such an approach was that some electorates contained several candidates sympathetic to the trade, while other candidates were unwilling to display too much public commitment in case it undermined their chances of election.[39]

A second strategy was to make representations to the premier and other parliamentarians and to draft alternative legislation to that proposed by prohibitionists. In mid-1893 Gilbert Stringer was contracted to travel to the United States to collect relevant information about the harmful effects of prohibition in various states. Shortly afterwards, a deputation of brewers to Seddon prior to debate on the crucial Alcoholic Liquor Sale Control Bill pointed out that local no-licence would result in a large number of people connected with the trade becoming unemployed. The loss of licence fee revenue would also mean an increase in local rates. With regard to tied houses, the brewers insisted that although they advanced money to publicans in return for taking their beer, publicans were otherwise free to act as they saw fit in their business.[40] In July 1896 a proposal for national prohibition was described as one that would 'allow a minority of ultra-extremists to bring ruin and desolation upon hundreds and thousands of families who had done no wrong and committed no crime, and would swell the already overcrowded ranks of the unemployed'.[41] However, in July 1900, Seddon informed Kennedy that further deputations were not advisable as new legislation was being drafted and the views of the Association were well known by ministers.[42] In Seddon's customarily shrewd judgement, this was not a time to further antagonise the prohibition lobby.

Ultimately, the response from the trade was relatively indecisive and not helped by the fact that brewers, publicans and wine and spirit merchants had varying priorities. The leading brewers had different political leanings; the smaller provincial brewers were sometimes hostile to their larger counterparts; brewers generally had an uneasy relationship with wholesale merchants; and both had a volatile relationship with publicans, especially as breweries competed with each other to secure control over the limited number of licensed premises. Nor did the trade possess skilled orators and public debaters of a calibre to match the likes of Leonard Isitt and Tommy Taylor. Many smaller brewers objected to the levy to finance the Brewers' Association. Others did not think a centralised organisation was necessary, while others, certainly during the 1890s, simply did not believe the prohibition threat would ever materialise, despite the rising vote.[43] In early 1899, the *Licensed Victuallers Gazette* observed that prohibitionists were making mistakes, but the trade was too apathetic to take advantage of them. Three weeks later, it added:

> Just so long as the different members of the trade allow personal feeling and petty jealousies to interfere with united action, just so long will they pose in the peculiarly anomalous position of an industry that pays the largest proportion into the consolidated revenue of the State, and yet has no means of making its latent power felt, and, furthermore, just so long will it run the imminent risk of injury from the guerrilla warfare carried on by its opponents who are leaving no stone unturned to give effect to their ideas and desires.[44]

The nub of the problem according to Arthur Myers was that while publicans were interested in issues around the local option poll, 'on all other questions a supreme apathy was very much in evidence'.[45] Six months later, the secretary of the Brewers' Association, H. J. Williams, complained of 'insidious attempts being made by some hotelkeepers to sow discord between the brewers and their customers'.[46] By the turn of the century, membership of the Association comprised mostly large breweries from the main centres with only sporadic representation from others. In June 1902, the bank balance was only £24, with eleven subscriptions, all of them brewers and most of whom did not pay on time.[47]

As centralised opposition to prohibition faltered, various local initiatives emerged. In Christchurch by early 1894, Manning & Co. typified a determination to ensure that brewers were scrupulous in adhering to the licensing laws. A resolution was passed that in order to combat Sunday and after-hours trading, the brewery would reduce rents on all hotels over which they had freehold if satisfied that the tenant was strictly complying with the law. Responding to this report, a *Press* editorial noted that any move to ensure respectability of the trade was a good one and that brewers who owned or controlled most hotels were best placed to do it. It was doubtful, however, whether these actions would in any way satisfy the 'extreme section of prohibitionists'.[48] The following year, Patrick Burke resigned the presidency of the Canterbury Licensed Victuallers' Association, allegedly in response to the hostility of some who objected to his insistence that the trade should be conducted in a respectable fashion. While *The Press* did not wish to comment on the enemies Burke may have created, the principle of his position was clear.

> If we are to have the shameless disregard of law and morality which for years past have distinguished the conduct of not a few of our public houses, then we must make up our minds to it that the sober and temperate portion of the population will get sick of the whole business and give way to the prohibitionists, for a season at all events. We fear the latter state would be worse than the first, but still that is what an ill-regulated trade will bring us to.[49]

At the end of the decade, during a luncheon to celebrate extensions to the Crown Brewery, Charles Louisson was critical of the requirement for publicans to obtain a certificate of respectability as no other trade required this. Yet he also emphasised that there had been necessary improvements during recent years and 'He would advise the licensed victuallers as a class to so conduct themselves as to earn the respect and esteem of the majority of the public.'[50] Meanwhile, in Auckland Moss Davis was another who felt the need to advertise his integrity. As his entry in the *Cyclopedia of New Zealand* noted in 1902: 'Mr Davis has assisted to improve the hotels in Auckland, and has also done much to raise the status of the trade by getting a good class of licensed victuallers into the houses, and securing a close observance of the licensing laws.'[51]

Brewery advertising also continued the long tradition of emphasising the medicinal and medical qualities of beer in contrast to pernicious spirits. Typical of thousands was a 1902 Great Northern Brewery advertisement which claimed: 'The company's bottled ales and stouts are especially suitable for invalids and for all who require an invigorating tonic, as well as a delicious, wholesome beverage. They are strongly recommended by the medical faculty, being the purest brewed in the world, next to the celebrated Burton-on-Trent ales and Guinness's stout.'[52]

Numerous breweries also advertised non-alcoholic beverages to attract a wider customer base and perhaps deflect from their main activity. James Paul of the Egmont Brewery produced 'Paula' as 'a bracing tonic', 'Kola Champagne', and 'Ginger Stout' made with 14 litres of Egmont Stout to 120 litres of lemonade.[53] Many others produced low-strength 'hop beer' or 'hop ale'. Questions frequently arose as to what could be defined as non-alcoholic beer. In 1889 James Hamilton, a Nelson cordial maker, produced a 'non-alcoholic beer', but as it was 4 per cent alcohol by volume he was required to desist or take out a brewers' licence.[54] Finally, an amendment to the Beer Duty Act effective from 1 January 1908 defined non-alcoholic beer as containing no more than 3 per cent alcohol by volume. While this still required a licence to produce, it was not subject to duty. However, some brewers were inclined to flirt with the law. Innes & Co. of Hamilton accumulated a long list of convictions after 1904 for failing to record all production of hop beer in their brewing books and for selling 'Innes & Co Non-Intoxicating Hop-Beer' to local fruiterers when it was above 3 per cent alcohol by volume. Finally, in February 1916 they were warned that any further problems could lead to cancellation of their licence to brew.[55]

Others took a riskier path to counter prohibition by testing the complexities and contradictions of the licensing laws in the courts. Here the cause célèbre was certainly the Gore Brewery, which found itself within a no-licence district after Mataura voted to go dry in 1902. At one level there was no difficulty for the brewery to keep operating as it was only the sale of alcohol within the district that was illegal. On the other hand, the brewer, John Whittingham, maintained that his brewers' licence, granted by the commissioner of customs under the Beer Duty Act 1880, allowed him to sell beer from the brewery in quantities greater than 2 gallons. In mid-1903 Whittingham's lawyer wrote to the inspector of police at Invercargill stating that for the purpose of a test case his client was prepared to admit:

> (1) having sold beer to persons resident within the Mataura electorate . . .
> (2) that he has a bonded store at Mandeville (just outside the no-license area) which is supplied from his brewery at Gore, and where orders are received, and such orders are supplied from Mandeville; and (3) that he has an office and store at Mandeville where he receives orders from within the Mataura electorate and from there supplies such orders.

When the police replied that they needed time to consider the matter and decide whether they could take a prosecution on 'abstract grounds', Whittingham promptly invited a constable to the brewery and then sold 2 gallons of beer to a customer in order to force a prosecution. The case quickly found its way to court, where the magistrate responded that the effect of no-licence was not to revoke brewers' licences, but to impose conditions on sales. Whittingham was fined for selling in a no-licence district, but only £5 as he had invited the prosecution. In October 1904, he was convicted for selling beer at Mandeville without having a retail licence to do so.[56] However, after Invercargill also went dry in 1905, Whittingham and the two Invercargill breweries quickly worked out how to comply with the law by obtaining retail licences for depots just outside the boundaries of the no-licence districts. Indeed, there is evidence that they profited as a consequence, because traditional city competitors such as Speight's were not inclined to challenge them for the no-licence district market. The Gore and Southland breweries thrived into the 1950s.[57] Meanwhile, hotels close to the boundary of a no-licence district also thrived. The Lorneville Junction Hotel, just to the north of Invercargill, apparently had the longest bar in New Zealand during the early twentieth century.[58]

Suffice it to say, these developments, and the decision of some breweries to relocate, greatly antagonised prohibitionists. When the prime minister, W. F. Massey, toured the south in 1914, members of the Invercargill Temperance Reform League complained about breweries that had been established since no-licence was passed. Massey stated that the government would do what it thought right in relation to the matter. Subsequent communication from the comptroller of customs to the minister of justice reaffirmed provisions of the Licensing Amendment Act 1910 which prohibited new licences to breweries within no-licence districts or within 5 miles of the border of one, but allowed renewal of the licence for an existing brewery or successor brewery. Both breweries were in existence before the passing of the 1910 Act; therefore no action could be taken against them.[59]

Other brewers sought to counter prohibition by ingratiating themselves with their communities through acts of public service or philanthropy. Among prominent brewers, Charles Louisson was elected mayor of Christchurch from 1888–89, followed by Samuel Manning in 1890. Louisson was elevated to the Legislative Council in 1900. John Logan Campbell was mayor of Auckland in 1901, followed in 1905 by Arthur Myers (1905–09), who was MP for Auckland East from 1911–21 and a Cabinet minister during the Great War. In Whanganui, Hope Gibbons made numerous contributions to public life, including serving on the committees for Queen Victoria's jubilee celebrations in 1897 and the South African War memorial in 1906. He served as mayor during the early 1920s.[60] In Dunedin, William Dawson of Speight's was a city councillor from 1885, mayor in 1887, and MP for Dunedin Suburbs between 1890 and 1893. His campaigns to regain the mayoralty in 1894 and 1901 both fell foul of the Customs Department. When Wellington head office queried an excessive amount of Speight's beer written off as 'waste and consumed on the premises', it was informed that Dawson had supplied this at the brewery on the day of the election. In due course he was warned for attempting to woo electors with free beer and for not paying duty on it.[61]

Dawson's business partners took a more conventional approach to public engagement. In addition to his many other business interests, Charles Greenslade was instrumental in floating a company operating Kaikorai cable cars, and was a member of the Dunedin Chamber of Commerce and involved in the administration of bowls, rowing, sailing, the model yacht club and the Caledonian Society.[62] James and Mary Jane Speight were foundation members of the Moray Place Congregationalist Church in 1864, and an obituary for Mary Jane in 1896 described her as 'widely known for her unostentatious charities and kindly help to the poor and afflicted'.[63] Her son Charles was extensively involved in other businesses, sporting and charitable organisations and was a militia volunteer and member of the organising committee that raised two contingents for

A beer barrel race at an Auckland location around 1906.
Auckland Libraries Heritage Collections, NZG_19060317_30_5

the South African War from 1899.[64] The following year, Speight's contribution to the war effort, along with other Dunedin breweries, was to arouse the ire of prohibitionists by providing the Otago–Southland contingent with 200 dozen bottles for the trip to South Africa. A month later Speight's celebrated the relief of Ladysmith by giving free beer to any who called at the brewery to drink it. In February 1901, a public debate as to whether the banquet to farewell the Sixth Contingent from Auckland should include beer or tea was eventually resolved in favour of beer. Campbell & Ehrenfried provided 600 dozen bottles for the ship at very short notice. Troops lined the rails and gave three cheers for beer when it arrived on the wharf.[65]

Of the hundreds of instances of breweries making contributions to community, charitable and patriotic causes, Staples appears to have been the most philanthropic. In May 1897, when it was noted that the brewery 'subscribes liberally to charities', they contributed £25 to the Wellington hospital operating theatre fund.[66] In mid-1899, they gave £500 ($106,000 in 2022) towards construction of Wellington's Catholic Cathedral, and shortly after gave the Anglican Church the same amount for a similar purpose. 'Actions like these,' according to the *Australian Brewers' Journal*, 'should surely be a little proof to the most rabid prohibitionists that, after all, there are still a few good qualities left in those interested in the brewing trade.'[67]

To the contrary, they prompted howls of indignation from prohibitionists throughout the country. As one Ōamaru correspondent fumed: 'This outburst of generosity on the part of the wholesale liquor vendors shows unmistakeably that the trade realise, to some extent at any rate, that if their business were to go into the dock, and the jury of public opinion were unfettered, Staples and Co, with their army of retail slaves, would soon be a thing of the past.'[68] In retort, the *New Zealand Tablet* had no doubt about the real issue. 'Even in the case of those who would seek to raise a political cry over the affair, we have a strong suspicion that the root cause of their hostile comment – as of that of the other strong sectarians – was the fact that the Catholic Church has benefited by the generosity of the Staples Company.'[69] The brewery was not deterred by the controversy. For many years they made an annual donation of 40 tons of coal for distribution by the Wellington Benevolent Institution. In 1912, they gave £100 ($19,000) towards a new Wellington Art Gallery.[70]

Among many smaller contributions, the Greymouth and Kumara breweries each provided 10 gallons of ale towards the refreshments for the hospital ball at Māori Creek in September 1900, while the Awaroa Bowling Club expressed gratitude in 1913 for a donation of one guinea by the Great Northern Brewery and a pair of bowls from Campbell & Ehrenfried. Shortly afterwards, the Great Northern contributed £140 ($26,000) to the staging of the Auckland Industrial, Agricultural and Mining Exhibition. Around the same time, Ernest Davis of the Captain Cook Brewery, who will loom larger in the next chapter, paid bail allowing striking Waihī miners to be released from Mt Eden prison.[71]

*

In broader terms, beer and brewers sought to counter prohibition through voluminous pamphlet and press campaigns from the 1880s to the 1930s. The main arguments involved distinctions between drinking and drunkenness; issues of personal liberty and morality; the trade as a legitimate business with economic benefit; New Zealand alcohol consumption in international context; and much more besides. Contributions came from brewers, publicans and consumers with obviously vested interests, but also from theologians, academics and avowed non-drinkers who distinguished between true temperance freely pursued and prohibition achieved by legislative coercion. Without exception anti-prohibitionists, whether representatives of the trade or other perspectives, firmly condemned excessive drinking. As anti-prohibitionist journalist John McKeague stressed, 'the evil consequences of drunkenness are serious and beyond dispute. It is the duty of every true man and every true woman to see that excessive drinking is effectively dealt with.'[72] Further, it seems from the total absence in anti-prohibition tracts of any reference to drinking by either women or Māori, that they were also willing to concede these groups to the prohibitionists.

While some contributions were narrow in focus, such as the close scrutiny of the no-licence district of Ashburton, others attempted to assemble every conceivable argument in one document. It seems that all of the material was written by men, but there is no discernible pattern, whether of socio-economic origin or political affiliation, among those who can be identified. Religious and theological objections to prohibition ranged across the spectrum from Presbyterians to Catholics who frequently agreed on little if anything else. While all anti-prohibitionists advocated for 'liberty', some conceived of it as 'free will' and others as 'free trade'. It is impossible to quantify the impact of anti-prohibition literature as a whole or of any particular argument within it. Circulation figures have not survived. Undoubtedly many individuals shaped their feelings and their vote on prohibition from personal experiences and the information and prejudices of friends and family. Given that the New Zealand Alliance clearly believed that its considerable expenditure on publications and touring speakers did yield results in its favour, it is only logical to assume the same of its opponents.

One of the first to enter the anti-prohibition field was Speight's, which published the sixty-four-page *The History of a Glass of Beer* in 1893. In detailing the long history of brewing from antiquity to the present, explaining the brewing process and the raw materials used, it argued that the trade was as legitimate as other businesses and made a significant contribution to the New Zealand economy that would be lost under prohibition. 'A drink that has been for centuries the staple beverage of the nation and has not yielded even to tea, coffee or cocoa in its rank in the dietary scale is not going to give way to the clamour of the rabid prohibitionists, who see only the drunkards and ignore the masses to whom the daily consumption of beer is a benefit and a pleasure.'[73] An important corollary of breweries was the communal value of hotels.

> At the present day hotels have undoubtedly advanced, commercial and social. They are convenient for entertaining travellers, for transacting business, and for spending a social hour, and their suppression would leave a blank in colonial life that would inevitably be filled with something worse. The humbler classes would be deprived of the opportunity of getting wholesome liquor at all, because they could not keep a cellar, and probably, nay almost certainly, drinking would be resorted to on the sly with all the attendant evils of secrecy.[74]

Indeed, as prohibition made gains with the establishment of no-licence districts, these provided much fuel to anti-prohibitionists through case studies of the consequences. Drinking was now done out of public view and without the social regulation of a group setting. An increase in drinking in the home also exposed women and children to a masculine world that had previously been contained elsewhere.[75] Gisborne lawyer and businessman Douglas Lysnar's particular target was the consumption of spirits in no-licence districts. He suggested that spirit drinkers rather than beer drinkers prevailed because beer was too bulky a commodity for illicit, or 'sly grog', sellers to keep a sufficient stock on site without fear of being detected. Further, the prohibitionists' blanket condemnation of all alcohol in New Zealand had alienated public sympathy from the problem that did exist.

> If the Prohibition Party would for the present wage war against the sale of spirits, and concentrate their forces against that enemy, they would gain headway much quicker and more effectually, and educate the rising generation, who are not abstainers, to prefer light wines and ale which seldom or ever intoxicate and are not ruinous to the constitution. If they can succeed in routing the spirits demon, the educating of the next generation from temperance to abstinence will be a much easier task.[76]

Others claimed that no-licence districts created a class distinction in drinking habits in that the wealthy could afford to transport and stockpile alcohol in a district while the poor did not have the resources to do so.[77]

Anti-prohibitionists fervently believed that legislative coercion did not produce a temperate society. A correspondent to *The Press* in 1899 argued that prohibitionists had no right to force others into their preferred form of self-denial. It was precisely because they were failing by argument to persuade the people of the virtues of teetotalism that they now resorted

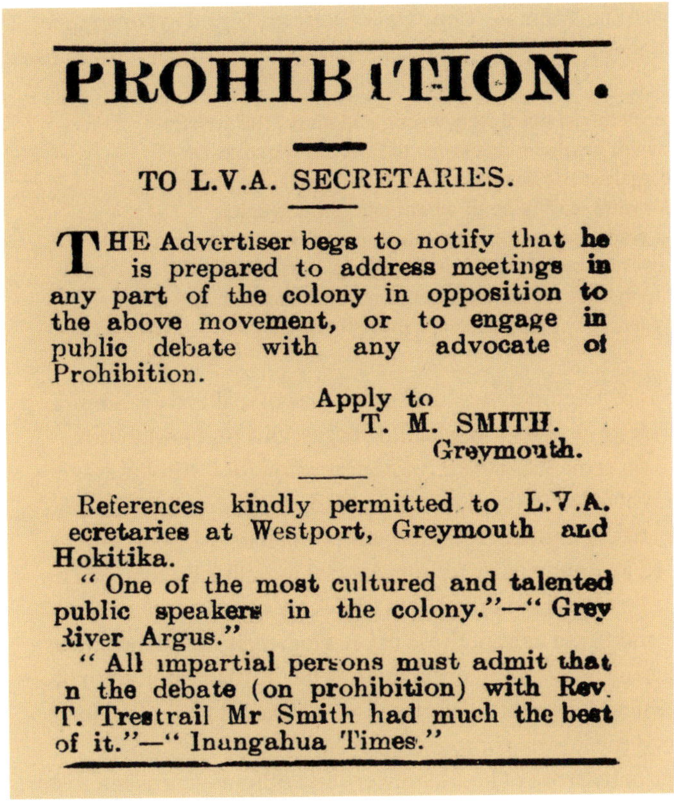

PROHIBITION.

TO L.V.A. SECRETARIES.

THE Advertiser begs to notify that he is prepared to address meetings in any part of the colony in opposition to the above movement, or to engage in public debate with any advocate of Prohibition.

Apply to
T. M. SMITH.
Greymouth.

References kindly permitted to L.V.A. Secretaries at Westport, Greymouth and Hokitika.

"One of the most cultured and talented public speakers in the colony."—"Grey River Argus."

"All impartial persons must admit that in the debate (on prohibition) with Rev. T. Trestrail Mr Smith had much the best of it."—"Inangahua Times."

A 1903 advertisement for an anti-prohibition campaigner. *Licensed Victuallers Gazette*, 12 November 1903

to a 'system of coercion which is as much opposed to the instinctive British law of liberty as it is to sound ethical principles'.[78] 'H. F. B.', in a lengthy letter to the *Otago Daily Times*, suggested that the United States, Canada and the no-licence districts of New Zealand offered abundant proof that prohibitive legislation did not work. It 'outrages the sense of justice of large numbers of people' in that it was the law most frequently circumvented and treated with sustained contempt.[79] The active agent in producing self-control was not compulsion but moral suasion. 'The temperate man,' according to Rev. William Thomson, 'is he who has realised the value as well as the dignity of right living.'[80] The case for self-control and moral suasion was most frequently articulated with reference to scripture. Paul's exhortation to Timothy to 'Drink no longer water, but use a little wine for thy stomach's sake and thine often infirmities' was at least confirmation of the use of wine for medicinal purposes.[81] There were also clear statements throughout the Bible that wine was associated with conviviality and sometimes intoxication. Here criticism of intemperance was directed at the individual drunkard and not at drink per se or its purveyors. Not only did Christ indulge in wining and dining, but his first miracle as outlined in the Gospel of John was the transformation of six stone jars of water into wine at the marriage feast of Cana.[82]

The implications of prohibition posed a particular challenge for Catholics. Until 1911, the church generally stayed aloof from what it regarded as a never-ending prohibition debate. It held firmly to traditions of moderation and regarded prohibition as illegitimate. Father Pierre Regnault of Waimate was one who condemned prohibition and the intemperate tone of prohibitionist clergy in a long letter to the Waimate Temperance Reform Union in November 1899. He stressed that drink was not an evil of itself, but rather in the way it was sometimes used.

> I cannot, on moral grounds, condemn as absolutely wrong and sinful the strictly moderate and well regulated use of alcoholic drinks by the proper persons. Did I do so, I should set myself up against the Saviour of the world and His apostles, all of whom took wine in moderation, and none of whom were in this sense Prohibitionists. Neither, by the way, were any of the Reformers of the sixteenth century, and least of all the most notable of their number, Dr Martin Luther. I would then – and do – insist in every case on moderation as a bounden duty, for excess in drink, as well as in speech, is against all laws, human and divine.[83]

Accordingly, Regnault declined to assist a 'temperance' union that was in reality a prohibitionist cause. But the hand of the Catholic hierarchy was forced amid suggestions that prohibition would ultimately extend to a ban on sacramental wine.[84] Another tension for Catholics was that they were disproportionately involved with the liquor trade. Hotel-keeping, especially during the gold-rush era when entry to the industry required only a small investment but returns could be significant, provided an opportunity for rapid social mobility. By 1921, when Catholics made up 14 per cent of the New Zealand population, they constituted 28 per cent of those working in licensed premises and 18 per cent of those in brewing, malting and bottling. Although similar census data correlating religion and occupation is not available before 1921, it is reasonable to suggest that the nineteenth-century pattern was similar, as it was for North America. Publicans and police formed the largest components of New Zealand's Catholic 'bourgeoisie'. The powerful Wellington network of hotel, brewery and other business interests revolving around Martin Kennedy, Hamilton Gilmer, Thomas Macarthy and other Catholics is the clearest example.[85]

Beyond the specific focus on alcohol consumption, critics also challenged prohibitionists on the broader logic and consistency of their crusade. If they were to ban alcohol on the basis that some consumers of it caused problems, then they ought also to ban sexual relations or anything that caused lust because these sometimes produced abuse.[86] As for the liquor trade as a business, 'It Is' railed against condemnation of the many for the failings of the few. 'Will anyone assert that among the clergy, lawyers, auctioneers, or hawkers, there are not scoundrels and blackguards of the deepest dye, who have caused the ruin and misery of many innocent, unoffending individuals?'[87] Prohibition was also condemned as the thin end of a wedge that ultimately desired the eradication of such 'luxuries' as tobacco, horse racing, theatres, public transport on Sundays and other things seemingly repellent to the evangelical churches.

Where prohibitionists deployed various rather speculative methods to calculate the national drink bill in terms of expenditure on alcohol and the costs of mitigating its harm, representatives of the trade were able to respond with greater certainty regarding the numbers employed and the contribution of excise duty to national revenue. In 1899, the flamboyant *Prohibition Fallacies* predicted that no-licence districts, let alone total prohibition, would cause a great deal of economic suffering to employees of breweries, hotels and related industries.[88] By late 1914 the revenue from the New Zealand trade was estimated at £1.17 million per annum ($205 million in 2022) or enough to pay all of the 16,000 old-age pensions, the total costs of running hospitals, all of the charitable aid, and the total outlay on police and prisons. By contrast, prohibitionists were depicted as contributing nothing to the revenue of the country through their actions, and in fact creating significant costs through the triennial licensing polls and enforcement requirements in no-licence districts. They also failed to provide any meaningful explanation of the fate of some 12,000 employees connected with the trade who would lose their jobs under prohibition. Moreover, all workers would face a higher tax burden to offset the loss of excise duty at a time when New Zealand's national debt was increasing and war expenditure looming.[89]

The longest and most eloquent contribution to the anti-prohibition cause came from a seemingly unlikely source. In February 1911 William Salmond, ordained Presbyterian minister and professor of mental and moral philosophy at the University of Otago, published *Prohibition: A Blunder*, a trenchant, multi-dimensional critique of prohibition with a particular emphasis on arguments from scripture and on free will and moral suasion rather than coercion.[90] Assisted by £1127 ($216,000 in 2022) contributed by the New Zealand Brewers' Association, including £350 from Speight's, Salmond's pamphlet ran to five editions in seven weeks. This left the Association with a liability of £133 and an extra levy was asked of breweries to clear the debt.[91] This unlikely alliance of interests suggests that the commonly assumed binary between the non-conformist churches and prohibition ought not to be taken for granted. Moreover, to emphasise the appeal of anti-prohibition across sectarian boundaries, Salmond's work was later substantially paraphrased by 'Liberty and Regulation', almost certainly the volatile Catholic priest, editor and Irish nationalist J. J. Kelly.[92]

*

By mid-1914 prohibitionists were surely more confident than ever of victory in the December licensing poll. The no-licence vote had increased at every election since 1896, and the first national prohibition poll in 1911 had fallen only 16,000 votes short of the 60 per cent threshold. They had also conducted a customarily vigorous election-year campaign, including visiting international speakers such as British Labour politician and future Chancellor of the Exchequer Philip Snowden.[93] But the declaration of war on 4 August decisively altered the course of the campaign. On 10 December, the prohibition vote fell sharply to 49 per cent. For the New Zealand Alliance, Rev. James Cocker claimed that uncertainty as to whether the election would be held at the usual time or postponed had prompted many prohibitionists to suspend their activities. Moreover, some voters were perhaps influenced by the idea that revenue from excise duty would be necessary to help pay the cost of the war, 'forgetting that four million pounds have to be worse than wasted before one million can be secured for the revenue'.[94] To the contrary, others condemned prohibitionists for attempting to use the war to assist their cause. Ultimately, it seemed that voters were not willing to disturb existing patterns of business and finance by adopting such a radical measure as prohibition during the war.[95]

Meanwhile, brewers had acted quickly to assert their contribution to the war effort. By mid-August Christchurch's Crown Brewery had donated £250 ($44,000 in 2022) to the National Patriotic Fund and Staples contributed their obligatory £500 ($88,000). Among many other contributions, from early 1915 Speight's employees made weekly collections for the Otago Patriotic and General Welfare Association while the Tui Brewery at Mangatainoka contributed £100 to Belgian relief.[96] Such gestures prompted an Invercargill correspondent to ask what contribution had come from the considerable funds of the prohibition movement and to conclude that 'If the measure of their contributions is the measure of their loyalty they might as well belong to Germany.'[97]

From mid-1915 moves were afoot to ensure that brewers made a more systematic contribution to war finances. Remarkably, beer duty had not been increased since its imposition in 1880. A suggestion in 1913 to double it from 3 to 6 pence per gallon had fizzled out.[98] Now, via the Finance Act, the single rate of duty was replaced, on the model used in the United Kingdom, by a sliding scale based on the strength of beer calculated from the specific gravity of the wort. The minimum duty of 3¾ pence per gallon was payable on beer under a gravity of 1047 – roughly 4.5 per cent alcohol by volume, although impossible to determine exactly without knowing the final gravity at the end of fermentation. The next category charged an extra one-sixteenth pence for each unit up to 1055, roughly 5.5 per cent alcohol by volume, while an extra two-sixteenths pence was charged for each unit above 1055, roughly any beer over 6 per cent alcohol by volume. Brewers were responsible for keeping detailed production records and providing correct utensils, primarily a hydrometer, to enable customs officers to assess the duty.[99] Given that these measures were progressing through Parliament during July and August 1915 as lengthy New Zealand casualty lists were beginning to emerge from Gallipoli, it is scarcely surprising that there was virtually no sign of debate or opposition. Nor would any emerge when duty was further increased in 1917. Indeed, Charles Speight was instrumental in persuading brewers not to reduce the strength of beer as a consequence of the Finance Act. He warned that if brewers reduced strength to reduce duty, the government would likely increase duty again during the next session of Parliament.[100] However, as the next chapter suggests, it seems that the average strength of beer did fall during the post-war years.

The war presented other changes for beer drinkers and brewers. Almost immediately, imports plummeted, never to return to former levels. While 1.3 million litres were imported in 1914, only 145,000 arrived in 1918 and 86,000 by 1922. A slow recovery to 550,000 in 1938 was soon curtailed by the outbreak of another war. The shortage of imported goods also extended to casks and oak for local cooperage. An alternative Russian supply only lasted until the 1917 revolution.[101] The shortage of men was another potential issue, although it is not clear how breweries navigated possible labour shortages as numerous brewery employees volunteered or were balloted for military service.

The greatest challenge, however, came from prohibitionists in conjunction with a government increasingly sympathetic to calls for 'national efficiency'. Immediately following the outbreak of war, the Alliance were in correspondence with the minister of defence, James Allen, regarding alcohol restrictions in military camps and for soldiers generally.[102] By February 1915, additions to the initial War Regulations Act of November 1914 prohibited the sale of alcohol to soldiers in uniform unless it was for consumption on licensed premises. In November, it became an offence for any person to take alcohol on board a troop-train or into a military camp. In August 1916, an 'anti-shouting' regulation was passed making it an offence for anyone to buy alcohol for a soldier. Women were also prohibited from entering or remaining near a bar after 6 p.m., and police were given wider powers to search establishments for breaches of the law.[103] By late 1916, Allen was able to assure the absent Prime Minister Massey that the anti-shouting regulations were vigorously enforced and the police reported a number of prosecutions. 'It is generally admitted that the effect of the regulations has been good in clearing out the bar sponger and certain other objectionable features of the trade especially in the evenings. The regulations have undoubtedly had a good effect upon the soldiers and less men are to be observed about the towns under the influence of alcohol.'[104] Others, though, regarded the measure as a sham that was easily evaded.[105]

An *Auckland Weekly News* depiction of the battle over alcohol during the First World War following the release of the National Efficiency Board report recommending six o'clock closing and a vote on national prohibition. Auckland Libraries Heritage Collections, AWNS_19170809_37_7

Meanwhile, the Alliance repeatedly highlighted abstemious military commanders such as Lord Kitchener and Admiral Jellicoe, and seized on the observation of British Prime Minister David Lloyd George in March 1915 that 'We are fighting Germany, Austria and drink; and as far as I can see, the greatest of these deadly foes is drink.'[106] Of greatest value to the prohibition cause was the report in early April that King George V would not drink during the war. Efforts were then made to arrange 'Follow the King' meetings at Trentham and other camps before troops departed overseas, and troopships were provided with copies of the Alliance's *Vanguard* magazine.[107] There was also a shift in the focus of the Alliance campaign during 1915, from prohibition as a matter of morality to a vital component of national efficiency for the war effort. Alcohol was blamed for inefficiency and lost productivity; it was also an inefficient use of limited sugar supplies and shipping space would be enhanced due to less alcohol cargo. Soldiers drinking less or not drinking at all would be less likely to fall victim to vice and especially to venereal disease.[108]

With Britain, France, various Canadian provinces and Australian states restricting hours and terms of sale, the Alliance began to campaign vigorously for early closing of hotels. This was not a new measure as evening hours had already retreated to 11 p.m. in 1893 and 10 p.m. in 1910. Several substantial petitions presented to Parliament during 1915 and 1916 revealed growing public support for 6 p.m. closing, following the example of four Australian states that adopted the measure during 1916.[109] Leading a deputation of Auckland businessmen to the prime minister in May 1916, avowed prohibitionist and former Cabinet minister George Fowlds insisted that they only desired to bring publicans into line with the hours of other businesses. The potential loss of revenue would be no different than when shops and offices were required to maintain shorter hours from the 1890s. Moreover, many of those called up for military service would also sacrifice careers and professions without compensation. 'Surely in this twentieth century it is absurd to demand sixteen hours a day for trading in that which is unnecessary for human well-being, when only ten hours are allowed for trading in the necessities of life.'[110] In an echo of Lloyd George, Fowlds concluded: 'It would be impossible to estimate how many thousands of lives have been unnecessarily sacrificed, and how many thousands more will be sacrificed before the war is finished, because of the failure of our Empire to adequately grapple with this greater foe than Germany and Austria combined.'[111] Massey replied that the government would consider the matter, but that New Zealand was a democracy and not a single hotel had been closed as a result of the 1914 licensing poll. The deputation replied that the three-fifths majority requirement was not democracy and that the war was creating a different environment.[112]

Momentum towards early closing did not go unchallenged. In October 1915 *NZ Truth*, in a rebuke to motions of efficiency, mused that Germany as 'the nation that drinks more alcoholic liquor than any other nation in the world' was fighting all of the other powerful nations and 'that beer-drinking nation is fighting with such skill, daring and audacity that there are times when even optimistic Britons feel perplexed; and wonder whether we are not in for another war of seven years, or even thirty years' duration'.[113] In short, if beer was acceptable for Germans, it was surely acceptable for the allies. At Parliament's Public Petitions Committee in mid-1916, M. Myers observed on behalf of various licensed victuallers' associations:

> Here are these men going away. A great many of them who are drinkers, who like their glass in moderation, go away to fight for their country – they may never come back. And who are we, who is Mr Isitt, who is a member of the committee or anybody else, that we should say that that man is not going to have a little pleasure before he leaves – if he thinks a little pleasure can be obtained from a glass of liquor?[114]

Shortly afterwards, Colonel H. R. Potter, commandant of Trentham Camp, addressed soldiers in unequivocal terms.

> I am strenuously opposing the proposal. The reason I do so is because I trust every man and until a man is found wanting, I do not want him penalised. It all depends on you men by your conduct as to whether six o'clock closing is to be carried into law. We have not asked for it, and we do not ask for it. It is a big temptation, but I say that if you men are fit to go into countries where the control of alcoholic liquors is not so drastic as in New Zealand, you should first be capable of looking after yourselves here, by showing you are able to exercise self-control under New Zealand conditions.[115]

Lieutenant-Colonel Noel Adams, commandant of Featherston Camp, expressed similar views. Yet proposals to establish 'wet canteens' in the camps to keep soldiers out of city hotels were rejected by military authorities.[116] In mid-July 1916, Alexander Herdman, a leading Cabinet minister and an autocrat renowned for his unrelenting support of the war effort, accused prohibitionists of using the war to further their own ends, not least by allowing 'their great champion himself', Leonard Isitt, to present the early closing case before the Petitions Committee. Herdman pointed out that voters had endorsed continuance at the 1914 licensing poll and since then the trade had made business decisions on that basis. 'Under these circumstances it would be a gross injustice if Parliament were to curtail those rights.' They should not consider anything else but the safety of the state, and other matters such as prohibition could wait until after the war.[117] Attempts to amend the War Regulations to include 8 p.m. or 9 p.m. closing or a national referendum on the matter were narrowly defeated in July 1916.[118]

While the campaign for early closing lost some momentum during the second half of 1916, it was revived by the end of the year amid a pronounced intensification of the war effort. In July, with opposition from only four MPs, the government had passed the Military Service Act introducing conscription. Those deemed to be 'shirking' and who had not been exempted military service on economic, religious or other grounds were liable to imprisonment.[119] In January 1917, Allen, acting as New Zealand's de facto war leader on the home front during the prolonged absence of the prime minister and deputy prime minister in London, established the National Efficiency Board. Its brief was to consider industrial efficiency and manpower allocation and to reconcile recruiting with other labour needs. The Board, consisting of four prominent and generally conservative businessmen representing each of the four military districts, one farmer, and Allen as government representative, was also given a mandate to consider broader issues of social and economic organisation relating to post-war reconstruction. It was to continue for one year beyond the end of the war.[120] Perhaps anticipating the inclination of the Board and the public regarding early closing in particular, Allen wrote to Massey on 28 April 1917 following his meeting with a deputation of Dunedin businessmen, 'There is no doubt a very strong feeling in the South that the Government has not given a strong lead to the country on this question. I feel sure we shall have to face the problem in Parliament.'[121]

It is evident that such directive measures from the government possessed a deeper objective than simply winning the war in the shortest possible time. There was also a

An anti-prohibition parade in Newmarket, Auckland, in 1918 at the height of prohibitionist agitation before two close referenda in 1919. Lion Technical Library

determination to use the war effort to exceed the level of commitment of other dominions of the British Empire to strengthen New Zealand's 'special relationship' with Britain and to establish a 'moral debt' in British minds that would endure beyond the conflict. Fears that the introduction of military conscription would undermine the image of a voluntary and willing war effort were skilfully offset by an emphasis on equality of sacrifice and the pursuit of national efficiency. Most revealing was Allen's reply to a question about the level of reinforcements during a parliamentary debate on 7 August 1917: 'I do not want New Zealand to be in the position of Australia and Canada. We have a higher aim, a higher purpose and I hope at the end of the war we shall reap a higher reward.'[122] A very competent administrator, but a somewhat dour and uncompromising personality, Allen was likely influenced by the death of one of his sons at Gallipoli. Moreover, he was less able than Massey to subtly manage the various cleavages in New Zealand's wartime politics – and not least that around alcohol.[123]

At its first meeting in February 1917, the National Efficiency Board indicated that it would consider industrial relations, business efficiency, amusements and alcohol consumption among its first tasks. By May, the Board was recommending compulsory direction of labour and industrial conscription to counter perceived under-utilisation of available resources.[124] The Board subsequently heard evidence from more than sixty witnesses – brewers, publicans, wine and spirit merchants, chartered clubs, employees in hotels and clubs, hop and barley growers and merchants, vine growers and winemakers, the New Zealand Alliance and other temperance advocates, churches, and protection societies for women and children. Judicial, magisterial, medical and military evidence was also considered.[125]

The Board's final report, released in July 1917, stated that it had not been influenced by moral issues but only by questions of efficiency. It recommended national prohibition but acknowledged that this was a decision for the people to make and therefore a licensing poll should be held as soon as possible. Reasonable financial compensation, later estimated at £4 million ($485 million in 2022), should be provided to the trade if prohibition was introduced. Regarding the loss of revenue from excise, it declared: 'That the increased national thrift which is most desirable and which would follow the diversion to productive trade channels of the money spent upon liquor would, in the opinion of the Board, as evidenced by world-wide experience where the abolition or reduction of the use of alcoholic liquor has operated, more than counteract the loss of public revenue now received from this source.' In the meantime, the Board recommended that hours of sale should be 8 a.m. to 6 p.m. on four days of the week, until 9 p.m. on one day of the week and until 1 p.m. on the statutory half-holiday. All alcohol sales should be for on-premises consumption only and the strength of spirits reduced.[126] In a confidential memo to Allen, the chairman, William Ferguson, noted that the Board originally considered issues purely from the perspective of war conditions, 'yet as the enquiry proceeded it became apparent that in the interests of National Efficiency the Board had to consider the liquor question also from the point of view of the state and of the individual both during the war and afterwards, and it has therefore submitted its recommendations so as to promote permanent National Efficiency.'[127] The Alliance naturally praised recommendations made by 'business experts' without a prior view on the liquor question. The release of the report also coincided with the presentation of a petition for early closing with 177,000 signatures.[128]

The government now moved quickly to introduce a Sale of Liquor Restriction Bill with a recommendation for 8 p.m. closing and to remain in force for the duration of the war and six months beyond. In reply to a question from Robert Fletcher, MP for Wellington Central, Allen confirmed that 'no complaints have been made by the military authorities since the first year of the war that overindulgence of liquor by New Zealand forces has affected training' and 'It is a fact that on repeated occasions compliments have been paid to the physique of our reinforcements, and not only to their physique but also to their discipline and sobriety.'[129] Indeed, convictions for drunkenness had declined sharply, from 13,000 in 1914 to 8500 in 1917, and would fall to 7000 in 1918. George Witty, MP for Riccarton, in addition to repeatedly claiming that many signatures on early closing petitions were fraudulent, questioned the efficiency of Australian states and Canadian provinces that had early closing yet contributed proportionately fewer men to the war than New Zealand, which had not adopted the measure.[130] Alexander Herdman raged that 'the time has arrived in New Zealand when the liberty-loving section of the community should rise in rebellion against a despotic sect in the community who think of nothing else but no-license, who believe in nothing else but no-license, and will have nothing else but no-license. Who is behind this movement for the early closing of hotels? The no-license party.'[131] Yet the die was cast. Supporters of early closing believed that it would reduce overall alcohol consumption. This belief seems to have been shared by their opponents, who frequently pointed to a loss of revenue for the trade if it was introduced. Consequently, a belated proposal by Massey to replace 8 p.m. with 6 p.m. closing passed by forty-four votes to twenty-six. Six o'clock closing became law on 1 December 1917 for the duration of the war and six months beyond.[132]

Neither the Great War nor the alcohol war was over. Throughout 1918, pressure mounted for the special licensing poll recommended by the National Efficiency Board. Allen received numerous deputations from the likes of the Businessmen's Efficiency League, which included former Board member W. D. Hunt, who had no reservation in now revealing his prohibitionist inclinations. Allen became convinced that the issue for many was really efficiency and not just prohibition. As he wrote to Massey in September 1918, 'They were very much in earnest. They spoke as men who were convinced and who are determined to act up to their convictions.'[133] The Alliance also pressured Allen with claims that returning soldiers were vulnerable to the trade and in need of protection from it. As the *Vanguard* insisted in January 1918: 'The greatest kindness that can be shown to the men who have risked their lives for us will be to close up every liquor bar in the country, and cut out the sale of liquor altogether. The liquor, which is bad enough for everybody, is especially dangerous to those whose nerves have been racked and strained in going through the hell of war.'[134] Three months later it asked: 'When will the defence authorities wake up to the fact that the fathers and mothers who have given up their sons for the war look to them for some measure of protection for their boys from needless injuries and especially at a time when those boys are hardly in a position to protect themselves?'[135] The Alliance also protested against drinking on returning troopships, called for the removal of alcohol at all ports on the return journey to New Zealand, and demanded that all bars stay closed for a day when troops arrived home. Local 'vigilance committees' were also formed to ensure that the early-closing and anti-shouting regulations were enforced.[136]

In due course, the War Regulations were amended to allow military authorities to order the closure for a maximum of twenty-four hours of hotels within 25 miles of the ports of Auckland, Wellington, Lyttelton and Port Chalmers when troopships arrived. A 242,000-signature petition was then presented to Parliament calling for the special licensing poll by the end of 1918. This represented a scarcely credible 95 per cent of all of those who eventually voted for national prohibition in April 1919. The trade's 306,000-signature petition calling for deferral of the poll until the 1919 general election represented an even less credible 116 per cent of those who eventually voted against prohibition.[137] An even more decisive victory for the Alliance followed with confirmation from William Ferguson that the recommendation of the National Efficiency Board was that the special licensing poll should be decided by a simple majority rather than the customary three-fifths. Finally, on an assumption that prohibition would be well supported by soldiers, the Alliance insisted that all of those in camp, returning or still overseas be able to vote in the licensing poll.[138]

Ironically, the total closure of hotels, albeit temporarily, had nothing to do with prohibition agitation. As the global influenza pandemic gripped the country during November 1918, tension developed between the government and those directing relief efforts. The deputy chief health officer, Dr Joseph Frengley, ordered the closure of bars from 13 November on the grounds that men appeared more susceptible to the virus than women and that it was fatal to those prone to excessive consumption of alcohol. He was overruled by the minister of health, George Russell, a known opponent of prohibition, who evidently supported the view of brewers and publicans who insisted that 4 p.m. closing was sufficient. Following an outcry from relief workers and the chairman of the Auckland Hospital Board, who publicly accused Russell of being influenced by political considerations, the minister relented and ordered the closure of all bars within 3 miles of any city or borough from 18 November. The only alcohol now available was by a doctor's prescription. While most accepted the decision, aside from a few convictions for illicit trading, some complained about the inefficient supply of alcoholic stimulants to flu sufferers.[139] Consequently, the official inquiry following the pandemic featured intense debate as to the role of alcohol. One doctor insisted that Scotch and soda in moderation had assisted his own recovery. Others said patients recovered despite, not because of, alcohol. The Otago Medical School declared its support for alcohol 'judiciously administered' as a stimulant for severe cases and a tonic during convalescence.[140]

There is a further irony in the lifting of the hotel closure on 4 December. That evening as Parliament debated the Licensing Amendment Bill, Charles Wilkinson, MP for Egmont, moved an amendment removing the clause stating that six o'clock closing would remain for six months beyond the war. By that deletion, passed by one vote, six o'clock became the permanent closing hour for hotels. In the Legislative Council, Sir Francis Bell subsequently proposed that the matter be put to a referendum at the licensing poll, but this was quickly rejected.[141] For a measure that would have far-reaching consequences, there was little comment. The *Otago Daily Times* was one of the few to editorialise, and then only favourably.

> We venture to think that the general public will, as a whole, view the prospect with perfect composure. Six o'clock closing of bars has now been the rule in the dominion for a period sufficiently long to demonstrate that as a practical measure of reform it possesses many recommendations. The community has survived remarkably well the change that the introduction of this system of early closing involved, and has suffered no apparent hardship in the curtailment of the hours during which the law permits hotel bars to be open. On the other hand, it has enjoyed the benefits, not entirely negligible so far as the maintenance of order and decency is affected, that have accrued from the reduction of the facilities for the sale of drink. We have little doubt that the popular verdict would go strongly against a reversion to the old order of things.[142]

At the time the more significant outcomes of the Bill were that a special licensing poll would occur in April 1919; that prohibition, if introduced, would include compensation to the trade; and that local option votes were abolished. However, the customary general

WEATHER FORECAST FOR ALL NEW ZEALAND.

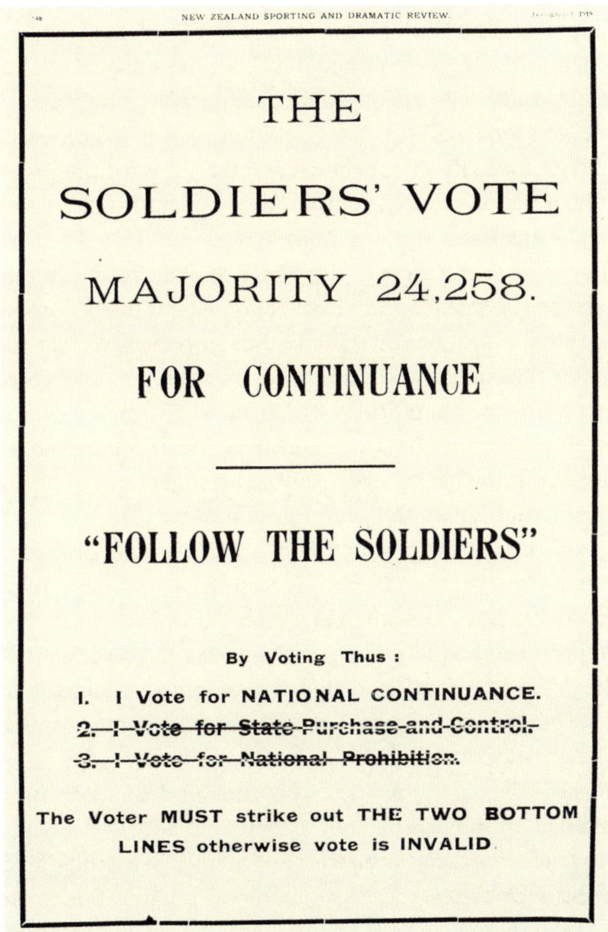

NEW ZEALAND SPORTING AND DRAMATIC REVIEW.

THE

SOLDIERS' VOTE

MAJORITY 24,258.

FOR CONTINUANCE

"FOLLOW THE SOLDIERS"

By Voting Thus :

1. I Vote for NATIONAL CONTINUANCE.

~~2. I Vote for State Purchase and Control.~~

~~3. I Vote for National Prohibition.~~

The Voter MUST strike out THE TWO BOTTOM LINES otherwise vote is INVALID.

TOP: Nobody took the outcome of the two referenda on prohibition in 1919 for granted. In the end, New Zealand avoided going dry in November 1919 by 3262 votes. Auckland Libraries Heritage Collections, AWNS_19190313_33_3

RIGHT: As the licensing poll approached at the end of 1919, citizens were urged to follow the example of soldiers, who had voted overwhelmingly against prohibition in the April 1919 referendum. *Sporting and Dramatic Review*, December 1919 licensing poll would also be held in conjunction with the election at the end of 1919 – an election that had been deferred from 1917 due to the war. This would include three options: continuance; prohibition without compensation; and state purchase and control.[143] Although it was rumoured that the third option was added to the ballot paper in return for conceding to the prohibition demand for a simple majority, state control had been promoted by bodies as diverse as the Anglican Church and the New Zealand Socialist Party since at least 1905 as a reflection of those wanting temperance but not prohibition. The Moderate League ran a strong campaign for state control from 1914 and the newly formed Labour Party passed a motion in support of it in 1916. Undoubtedly the trade also favoured it because they imagined it would split some of the moderate vote from prohibition. If neither national prohibition nor state control secured 50 per cent of the vote, then continuance would remain as the status quo.[144]

Throughout the very intense campaign leading to the special licensing poll on 10 April 1919, the trade was repeatedly criticised for a lack of strategy and visibility against reports that the New Zealand Alliance spent as much as £100,000 ($11.6 million in 2022). Indeed, when the initial result was declared in New Zealand, the criticism perhaps appeared vindicated as prohibition held a majority of 13,000. But the trade, which had been 'jeered' for suggesting during the campaign that 80 per cent of the soldier vote would go against prohibition, was duly rewarded. With Moss Davis of the Captain Cook Brewery coordinating from London, lengthy articles were placed in soldier newspapers such as the *New Zealander* and cars were made available to enable soldiers to get to voting places. Although a committee representing the New Zealand YMCA and prohibition organisations supported by like-minded bodies in the United Kingdom endeavoured to present the case for prohibition, 32,000 of the 40,000 soldier votes favoured continuance, which ultimately prevailed by 264,189 votes to 253,827.[145] As the *Australian Brewers' Journal* reported the soldier verdict: 'Their own statement is that the threat to rob them of their beer on their return home, after fighting the Empires' battles, was the last straw upon a load of grievances heaped upon them by the Government.'[146] From the other side, London's *Labour Leader* presented two explanations for the soldier vote. 'The first is that the New Zealand soldiers who came to Europe mainly as teetotallers have fallen victims to the temptations which have surrounded them. The second is the unwillingness to sacrifice the revenue which New Zealand at present gets from the liquor traffic at a time when the Dominion has to face heavy additional taxation due to war expenditure.'[147]

Paul Christoffel emphasises, however, that the notion that the soldier vote 'saved' New Zealand from prohibition in April 1919 ignores the matter of the general licensing poll in conjunction with the December 1919 election. Now, after another frenetic campaign, the margin for continuance was only 3262. Prohibition secured 49.7 per cent of the vote, continuance 44.5 per cent and state purchase and control 5.9 per cent. While the Alliance feared the power of the third option and campaigned exhaustively against it, the reality was that virtually all of the state purchase and control vote had shifted from continuance and not national prohibition. The question with both polls in 1919 is not why did prohibition do so well, but why did it do so poorly? Given that the majority of the war dead were young men likely to have voted against it, and that prohibition had now become a reality in the United States, Norway and some Canadian provinces, there was every reason to expect a considerable improvement on the vote of 1914. Christoffel suggests that the failure was possibly due to more intense Catholic opposition in response to increased sectarianism, especially the activities of the Protestant Political

Association, during the war. But, aside from the prohibition efforts of Bishop Henry Cleary in Auckland during 1918–19, it is doubtful whether prohibition had ever attracted a significant Catholic vote that could have deserted it in 1919.[148]

Regardless of the explanation for the failure of prohibition in 1919, the brewing industry and the trade as a whole had survived the most tumultuous quarter-century in its history and a genuine threat to its survival. Prohibition would never again come so close. Yet it would continue to cast a long shadow over New Zealand politics, and beer drinking, for at least the next fifty years. Moreover, the survival of the trade came with significant concessions, especially six o'clock closing. The brewers now set out to adapt to these changes, and eventually did so in ways that were not always palatable to the consumer.

HARLEY
XXX ALE X
15
HARLEY & SONS

A Harley's Brewery float in a Nelson procession during the 1930s. Nelson Provincial Museum, NPM 7 FNJ_half_29

SEVEN

The Big Ten and a Rival: Mergers, Competition and Depression, 1920–1940

The two narrow victories for continuance at the polls in 1919, on the back of the immense dislocation of the Great War and the introduction of 6 p.m. closing, had severely tested the fortitude of brewers and drinkers alike. The inter-war decades were scarcely less challenging. While the threat of prohibition eventually retreated and the New Zealand Alliance became a shadow of its former self, this outcome was not apparent for much of the 1920s as brewers and the trade more generally engaged in defensive campaigns by turns legitimate and shadowy. No sooner had some relief from prohibition arrived than severe economic depression gripped the country, especially the pockets of the ordinary drinker who increasingly found solace in home-brewed alternatives of frequently dubious legality and taste. However, amid these struggles the two most significant brewing forces of the twentieth century, New Zealand Breweries and Dominion Breweries, also took shape and set about transforming the local production of beer into a sophisticated industrial duopoly.

The next chapter will reveal that the greatest legacy of 6 p.m. closing, in the form of the notorious 'swill', did not emerge until the mid-1940s. For now, the most evident adaptation to early closing among determined drinkers was to ignore it. A thriving after-hours trade emerged throughout the country during the 1920s, especially in rural areas where enforcement was more lax, as local police understood that country hotels had few if any customers during the day and it was difficult for rural workers to travel to some of them in time after work.[1] Irrespective of the hours, per-capita beer consumption, which had remained under 45 litres since the early 1880s, spiked at nearly 55 litres in 1920, and 50 the following year amid the return of servicemen from overseas, the quest to resume normal life, and a brief post-war economic boom. By 1921 total annual production had also reached 71 million litres compared with 45 million in 1911. Thereafter production and consumption entered a gradual decline through the remainder of the decade – no doubt aided by a doubling of beer duty in November 1921, which led to a price rise.[2] As with the duty increases in 1915 and 1917, there was little reaction, aside from the complaint of Gisborne MP Douglas Lysnar during the Budget debate that beer was the poor man's drink and, unless the government had assured itself that it would continue to be retailed at 6 pence per pint, that is to say the price would not be passed on to consumers, then the duty should not be increased. In turn, publicans later complained that economic circumstances forced them to pass the increased duty on to consumers as a price increase.[3] But most accepted the necessity of the doubled duty as a sharp economic recession took hold from late 1921 with the end of Britain's wartime commandeer of New Zealand farm exports, a flooded global primary produce market and a sudden decline in the value of wool exports. As unemployment and industrial unrest grew, the Massey Government slashed state spending and public service wages. It saw beer duty as a convenient boost to the revenue.[4]

Amid these fluctuating post-war fortunes and the ever-present spectre of prohibition, there was no real warning of the mergers and industry concentration to come. Although Campbell & Ehrenfried and the Great Northern Brewery in Auckland had merged in April 1915, this did not seem to indicate a trend. Aside from long-standing amicable relations between Arthur Myers and the Great Northern management, common ground emerged when both breweries wished to capitalise on the developing taste for lager, or at least for a filtered and chilled 'no deposit' beer. With a cost of £17,000 ($2.9 million in 2022) to install the necessary brewing equipment, the decision was made to combine resources under the name of the Lion Brewery Ltd. Beer production was shifted away from the Albert Brewery in Queen Street to the Great Northern site in Khyber Pass Road, Newmarket.[5] A conference of brewers early in 1922 emphasised the need for unity in the trade as another licensing poll approached. There was also a rise in support for various 'corporate control' proposals, whereby the breweries and hotel interests would be merged into a single corporation with the government holding 20 per cent of the shares and the remainder in the hands of former owners.[6] But the *Australian Brewing and Wine Journal* described the merger plans announced in May 1923 as a 'bolt from the blue' that had only been known to an inner circle of leading brewers who had met in Wellington at the instigation of Alfred Bankart of Lion.[7]

By mid-June it was confirmed that ten breweries would combine as New Zealand Breweries (NZB) from 1 July: Auckland's Captain Cook and Lion; Barry's of Gisborne; Staples of Wellington; Crown, Manning's and Ward's from Christchurch; and Speight's, McGavin's and Strachan's from Dunedin. Alfred Bankart and Charles Speight presented the amalgamation as advantageous on several fronts. Firstly, it would enable a more concerted and united campaign against prohibition. Secondly, it would hopefully broaden the base of the industry through a public stock offer, deflecting criticism that it was the preserve of a few greedy beer barons.[8] Indeed, a very sophisticated advertising campaign was launched to persuade the public to invest in an offer of £1 million of debenture stock, paying a tax-free interest rate of 10 per cent per annum, twice the contemporary going rate. Amid claims that the share offer would likely be significantly oversubscribed, it was restricted to smaller share allocations to enable as many people as possible to become involved. 'That,' according to the *Australian Brewing and Wine Journal*, 'would be good policy, for it would also strengthen the position of the trade against future attacks of prohibitionists, who will be staggered when they find the popularity of the proposal with the public.'[9] Thirdly, and perhaps of most importance, the merger would remove at least the perception of underhand practices and tied-house corruption that had dogged the trade since at least the 1880s. The various companies would transfer their brewing interests to the new entity, but would each retain their existing hotel interests. The need for tied houses was gone. In a later summary for its own staff, NZB claimed that competition for hotels had resulted in both inflated values for some of them and other very marginal concerns being financed to secure the trade for a particular brewery. 'In order to make a living, a number

Gold Top, produced by the Gisborne Brewery, first appeared in 1909 and was still popular in the late 1930s, long after the brewery had become part of the New Zealand Breweries merger in 1923. Lion Technical Library of hotelkeepers found it necessary to adopt practices which were not desirable in the public interest. This position resulted in the industry being in disfavour and was a main contributor to the large prohibition vote.'[10] An official company history later noted, 'There was no question of any company taking over another. It was the banding together of a group of brewery companies who were protecting themselves against distorted economic conditions resulting from what has been described as "prohibition-appeasing legislation", particularly the artificial limitation of hotel licences.'[11]

While the merger was described as being open to others of the forty-six currently operating breweries to join at a later date, most showed no inclination to do so, although apparently some offers were not accepted. In the South Island, those that remained outside the merger were small breweries with little distribution beyond their locality. More competition remained in the North Island, with the likes of Macarthy's Phoenix and City breweries in Wellington, and the Taranaki and Tui breweries. Moreover, there was little real attempt to rationalise and challenge them by the new company. Multiple NZB branches remained in each city and there was initially no sign of sharing of resources or technical information between them or rationalising of such things as distribution and advertising. Instead, profitability was mainly to be found in removing the uneconomic competition for tied houses, and in the fact that the monopoly of hotel outlets in some areas tended to allow higher beer prices.[12] More than anything, the new arrangement was closer to a cartel than a single firm. Further additions to NZB at this stage were limited to the purchase and subsequent closure of the long-established Tuapeka (formerly Black Horse) Brewery at Wetherstons in October 1923, Christchurch's Union Brewery in March 1924 and Dunedin's Standard Brewery in November the same year. The next move to expand would not come until 1936 when the Timaru Brewery was acquired and kept in production.[13] While Speight's was the largest brewery in the new entity and Charles Speight remained integral to developments, and for many years NZB had its head office in Wellington, it became apparent that real power resided with the Auckland directors, who possessed a large combined shareholding.[14]

New Zealand Breweries certainly prospered. Although the initial debenture issue was significantly undersubscribed, indicating perhaps a lack of public faith that prohibition would be kept at bay, the first annual general meeting in August 1924 reported an £85,000 ($9.5 million in 2022) net profit for the first nine months and paid a 10 per cent dividend to shareholders. The following two years produced net profits of £103,000 ($11.5 million) and £109,000 ($12 million) and 5 per cent dividends. The healthy amounts at which brewery shares changed hands in local markets were also taken as an indication that investors, if not the wider public, did not seriously believe prohibition would come. Indeed, trading was especially brisk after its defeat at the 1925 licensing poll. Nevertheless, Alfred Bankart seldom missed an opportunity to attack the polls and the prohibition movement at annual general meetings.[15]

In July 1927 a much smaller merger occurred, albeit one of some significance to the later corporate reimagining of New Zealand, and specifically Monteith's, brewing history. Westland Breweries Ltd was formed from Mandl's Brewery and the Crown Brewery, Hokitika; Pearn's Brewery, Kumara; and Monteith's Phoenix Brewery, Reefton. All breweries continued in operation, although a new brewery and bottling plant was also opened in Greymouth in 1929. An advertisement at the beginning of that year also made it clear that the focus of the operation was firmly local.

> Summed up, the reason for this wide popularity is that Westland Beer and Stout is the best that money can buy. Everyone who wishes to look well, feel well, and keep well for 1929 must begin by asking expressly for Westland Ale, or Stout which is noted for its purity, quality, and flavour. It is a good wholesome brew that will put life into you. Stocked by all leading Hotels and Bottle Stores. Make your 1929 motto 'Westland Please' and keep Westland money in Westland.[16]

By the end of the 1931/32 financial year Westland Breweries had a paid-up capital of £31,000 ($4.1 million) and reported a net profit of £701 ($93,000).[17] However, it later emerged that its financial performance was not all that it seemed. Investigation by the Customs Department during 1939–40 revealed that brewery books were not correct at one brewery for nine years and twelve years at another. Based on reconciliation of raw materials used and the quantity of beer sold, and with allowance for wastage, the brewery was eventually presented with an assessment of £9000 ($980,000) for deficient duty and a £1000 penalty.[18]

The opening of the Coutts' family brewery at Ōtāhuhu in 1929, the forerunner to Dominion Breweries, attracted strenuous protest from prohibitionists. DB Breweries

*

The next move in a busy decade of realignment for the industry developed from low-key and unpromising beginnings, and was in many respects quite remarkable given the brewery concentration occurring elsewhere (Table 7.1). In November 1929 William Joseph Coutts, of a long-established brewing family that had anglicised its name from Kühtze during the Great War, moved with his sons from the Main Trunk Brewery in Taihape and sought a licence for the new Waitemata Brewery in Ōtāhuhu, Auckland. Although the licence was granted by the minister of customs, this decision was questioned by the prime minister, Sir Joseph Ward, and predictably opposed by NZB.[19] On the day the new brewery opened, about seventy members of the Women's Christian Temperance Union gathered outside in protest. It was rumoured that the buses to get them there had been secretly paid for by Ernest Davis of NZB.[20] After hymns and prayers, Rev. T. H. Eccersall addressed the meeting:

> Let it be understood that we have no quarrel with the owner of the brewery. I have never met that gentleman, and desire here and now to express my best wishes for his well-being . . . I do not know whether he is here, but if so he is just as welcome as anyone. He may be a generous, genial, good and moral person, but – I say this deliberately – he has a hell of a job. If it is not so, our presence here is useless and our protest is vain. . . . The Prime Minister has submitted that the licence for the brewery was granted without the serious consideration it should have received. We, therefore, submit that an injustice has been committed and urge that the licence be cancelled even if compensation be given, and hope that the present brewers will put their money into a better form of investment that shall help forward the prosperity and sobriety of the people of the district.

Bessie Lee Cowie, president of the WCTU, added, 'We hope that God will turn this place into a flourmill, a woollen mill, a dairy factory, or a church – something good for the people. We have had no help from Parliament or man, so we have come to ask Him.'[21]

Against these objections, Coutts began brewing and secured a toehold in the Auckland market by early 1930. However, the brewery was undercapitalised and lacked distribution outlets for draught beer due to the dominance of hotels owned by companies linked to NZB. At this point Coutts came into contact with Henry Kelliher, a Gallipoli veteran and entrepreneur

Table 7.1: Change in the Brewing Industry, 1911–1936

	Number of breweries	Beer produced per year (L)	Average production per brewery per year (L)	Employees per brewery	Population per brewery
1911	56	43,309,560	773,385	12.35	16,807
1916	55	51,139,536	929,809	13.4	20,895
1921	56	71,100,731	1,269,655	20.2	22,708
1926	45	64,737,590	1,438,613	20.7	31,292
1931	46	44,398,559	965,186	19.8	33,102
1936	44	52,282,286	1,188,233	19.7	33,768

Source: New Zealand Census, various editions.

J. R. Dodson & Son's Nelson Brewery in 1926.
Nelson Provincial Museum, NPM 16 T_181499_3_8x10_258

who had accumulated a fortune from varied business interests during the 1920s. He owned Lever & Co., wine and spirit merchants and bottlers, and also held a financial interest in the Auckland Bottle Company, which had been established during the early 1920s to finally address the bottle shortage that had plagued the New Zealand brewing industry since its beginning.[22] Kelliher, who was well travelled in Europe, agreed with Coutts on the potential market for clear, sediment-free lager and ales in the continental style. But these could only be produced with extensive investment in new plant. Consequently, a new public company, Dominion Breweries (DB), with a capital of £75,000 ($8.5 million in 2022), was floated on the New Zealand Stock Exchange in April 1930 with Kelliher's financial backing. Its initial share offer, despite the deepening economic depression, was oversubscribed.[23] Kelliher's considerable business acumen and genius for marketing and the technical skill of the Coutts family enabled the business to thrive in a difficult market. By July 1931 the capital of the company had been increased to £250,000 ($31.5 million) and there was a 40 per cent increase in turnover during the 1931/32 financial year. The initial struggle for hotel outlets was offset by about half of the product being bottled during the mid-1930s, and therefore available for home consumption, compared with perhaps a quarter for the industry as a whole. By the mid-1940s, DB produced around 40 per cent of Auckland beer output and 20 per cent of national output.[24]

An initial consequence of the emergence of DB was the government's suspension of new brewery licences throughout the 1930s, seemingly in response to the protest over DB's original licence. One exception was Ballins in Christchurch, which, for reasons unknown, converted from licensed wholesalers with hotel interests to brewing in 1936.[25] But various attempts to launch lager breweries to compete for the market identified by DB amounted to nothing. An application to establish a New Zealand Pilsener [*sic*] Syndicate based in Auckland was declined in November 1930. A similar venture apparently being established in Pakuranga in 1931 with advice from a 'lager expert' from Copenhagen progressed no further. Viking Lager Ltd, which had £20,000 subscribed by investors, was wound up in 1939 without ever brewing, and a judge was highly critical of a company being established without any prospect of success.[26] Meanwhile, NZB did produce some lager from a new Auckland plant from the late 1920s, and suggested that it was important to have a light beer to counter the German and Dutch products that were being imported in large quantities. Given that imports, from any source, constituted less than 1 per cent of all beer available for consumption at the time, one suspects that they were actually attempting to stifle DB at its inception.[27]

A more serious consequence for NZB came in the revival of the scramble for outlets. While DB had some success in establishing 'agencies' to take orders and arrange delivery of their bottled beer to customers, they were frustrated by the NZB dominance of hotels. In April 1936, by means of a newspaper advertisement, DB launched a scathing attack on the situation in Wellington.

> CERTAIN Brewers who own or hold the Leases of certain Hotels in the Wellington District, have warned their Tenants that Leases will not be renewed, or will only be renewed at an increased Rental, if they sell any bottled beer other than that of their own manufacture.
>
> Are these Brewers to be permitted to dictate to the public as to what it shall drink? Hotels obtain their Licenses on condition that they meet the Requirements of the People, and such an Act of Intimidation is repugnant to the British Sense of Freedom and Fairplay.
>
> A List of Hotels which refuse to be dictated to, and are free to supply the Public with any Brands of Bottled Beer they demand, more especially DOMINION BITTER and VITA-STOUT (products of the WAITEMATA MODEL BREWERY) will be published at an early date should this Objectional Practice continue.[28]

While there is no evidence that the promised list of 'free' hotels was ever published, three months later L. J. Stevens, managing director of DB, stated at the annual general meeting that although the tied-house system was undesirable and possibly illegal, the brewery had no choice but to become involved in order to secure outlets. 'So long as the present system remains in existence, this company must, of necessity, adopt the same policy as its competitors, but the directors would certainly prefer a system under which our products would be sold on their merits and in open competition with the products of our competitors.' At this point, DB owned or controlled twenty-four hotels.[29] In a later analysis, NZB acknowledged that while it had stood aside for a period in the hope that DB would not make any impact, from 1933 it found it necessary to re-enter the business of hotel acquisition and financing in order to protect its trade.[30]

There was now a renewed and extravagant battle for hotel assets. In 1936 DB purchased hotels in Huntly and Taihape for £45,000 ($4.9 million in 2022). The company spent another £28,000 on furniture, fittings and advances to hotel owners, an amount increased to £153,000 ($18.7 million) the following year. In 1938 DB purchased the Market Hotel in Auckland for £31,000 ($3.6 million), far beyond its government valuation of around £9,000. Meanwhile, NZB spent £496,000 ($60 million) in 1937 to purchase the Wellington hotel interests of Staples & Co. and another £280,000 ($34 million) building and fitting out the Hotel Waterloo in Wellington. As NZB directly acquired hotels, old established companies such as Staples, Speight's and Ward's, which had continued after the 1923 merger only to manage hotel interests, were now wound up as they no longer served a purpose – although their names endured in various of NZB's regional beer brands.[31]

Addressing the Royal Commission on Licensing in 1945, P. B. Cooke KC, for the National Council of the Licensed Trade, insisted that New Zealand tied-house arrangements were different from those in Britain. While it was usual for hotels to carry only one brand of draught beer, they also generally carried bottles from other breweries not subject to the tie. Nor, he claimed, was there evidence of prohibitive prices for items not within the terms of the tie.[32] There is also an argument, taking the 'Bible' kept by Harley's of Nelson as an example, that relations between publicans and breweries contained a degree of care and benevolence and were not merely an economic transaction. In February 1920 the Black Horse Hotel was purchased by W. E. Trevett with a loan from a returned soldiers' settlement scheme. In 1934 Mrs Trevett approached Harley's with concerns regarding debts, but seeking to purchase another, presumably more lucrative, hotel. The brewery counselled against this and advised her to decrease her own spending. In June 1934 it was noted, 'She has now learnt to live within her means', but there was evidently some 'domestic trouble'. This was apparently resolved by 1935 when Harley's noted that Mrs Trevett says 'she is now quite reformed'. But on 12 August 1940 it was reported that Mr Trevett was 'fed up with pub life' and wanted to open a market garden. After a meeting with Mrs Trevett, it was noted, 'Husband now on wagon, hates handling drunks after hours.'[33] The couple finally left the hotel business in 1948.

Meanwhile, the publican of the Brightwater Hotel frequently complained about business and his inability to make a profit. His rent was reduced from £8 to £6 in 1932, but the following year, after requests for a further rent reduction, Harley's recorded 'Brownlie not cooperative re return of empties, taking larger casks etc. Advised better for his own interests to get out or try another pub.' In March 1945 the publican of the Commercial Hotel, Murchison, was called to Nelson by J. A. Harley to be informed '1. He is the cause of 8pm closing put on hotels by policeman; 2. That his pub is reputed to be a brothel; 3. That his business is slipping.' He denied all charges. Around the same time, it was reported to Harley that the new publican of the Globe Hotel had 'Domestic trouble and that his wife had skipped with a Yank'.[34]

*

Beyond the seeds of the rivalry between NZB and DB that would eventually squeeze out or swallow up every other brewery in New Zealand, the usual repertoire of challenges remained for large and small alike. Although fire was not the scourge it had been in the nineteenth century, it still periodically exacted a toll. Blenheim was especially unlucky. In March 1920, fire completely destroyed the Commercial Hotel, owned by the Marlborough Brewery. Four years later the brewery itself was badly damaged by fire. Exactly six years after that, another hotel belonging to the brewery was also destroyed by fire.[35] In July 1921, fire significantly damaged Cowie's Brewery in Dunedin; in April 1927 a man described as 'mad through drink' was refused a bottle of whisky by a publican and took revenge by setting fire to stables and other buildings of the Napier Brewery. Early the following year fire gutted the packing and bottling department of the NZB premises in Gisborne and a large stock of beer was ruined. Six months later the small St Albans Brewery in Christchurch was completely destroyed by fire.[36]

Other challenges were less accidental. In April 1921, a well-planned burglary at the Lion Brewery in Auckland netted £500 ($49,000 in 2022) in wages from a safe in the strong room.

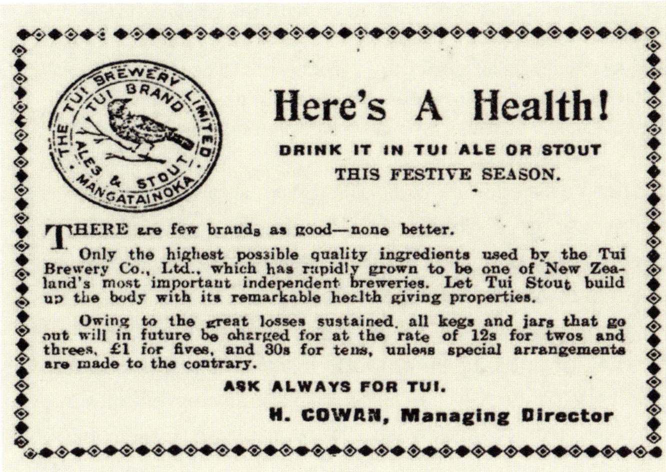

The Tui Brewery at Mangatainoka in the northern Wairarapa could trace its origins to 1889. In the late 1920s it was still a small local brewery. It was taken over by Dominion Breweries in 1969. *Pahiatua Herald*, 3 October 1929, 2

Less successful were those who entered the Southland Brewery early in a morning in June 1928 and used gelignite to blow the safe open with such force that its door was propelled 12 feet across the room, through a table and then a brick wall. The safe contained only 44 shillings of petty cash and the 'marauders' left some of that behind.[37] At the petty end of the offending scale, two McGavin's employees were convicted in February 1923 for theft of four dozen bottles of ale and stout from the brewery in what was apparently not an isolated case. They were given two years' probation on condition that they abstained from alcohol. A decade later two Greymouth men pleaded guilty in the Magistrates' Court to theft of beer from the Westland Brewery.[38]

Some brewers were caught out by the regulatory framework that controlled most aspects of brewery operations. In 1922, Jakob Bertelsen of the Paeroa Brewery was fined and had his licence cancelled for failure to keep a register of casks as required under the Finance Act. He was also the subject of an unfavourable police report for the manner in which he conducted his brewery. Similarly, in 1930, William Lawson of Wakefield had his licence cancelled for selling his beer from a location not approved for the purpose by the Collector of Customs.[39] For others, problems stemmed from what they put in their beer. Although Charles Speight informed a Parliamentary Committee in October 1922 that no 'bad beer' was sold in New Zealand, MP Allen Bell claimed the following year that New Zealand was almost unique in not testing the quality of beer in bars. Although farmers were fined for adulteration of milk, there was no sign of testing of beer for adulteration – a situation apparently due to powerful brewery interests.[40] But with the establishment of the Department of Scientific and Industrial Research in 1926, analysis of the quality of food and beverages was improved. The report for 1927 stated that although the permissible amount of salt in beer had been fixed at fifty grains per gallon, a quarter of samples analysed throughout the country were in excess of this level. In December 1927, C. L. Innes & Co. of Hamilton, a brewery that was no stranger to breaches of regulations, became the first in New Zealand to be fined for salt in beer. Despite the brewer saying he had difficulty in adjusting the quantity of salt when brewing, they had been previously warned and were now fined £5. A year later, and after a previous warning, the Tui Brewery was fined £5 in the Palmerston North Magistrates' Court for salt at the level of sixty-five grains per gallon. The brewery admitted to using salt and hardening crystals in beer, but its lawyer stated:

> I am astonished that the department found it necessary to go on with the prosecution. Fifteen extra grains got into the brew somehow, and if a pinch like that was added to a person's porridge in the morning he would never notice it. It was certainly not the wish of the company to breach the regulations. The Tui Breweries have a reputation for producing one of the best light beers in New Zealand, as those who have sampled it know.[41]

The Taihape Brewery, however, could not be saved from what was reported as a 'tragedy'. Although a Customs official was present as required to watch the transfer of stout between storage vats, neither he nor the brewery staff noticed that a tap had been left on and 3000 litres of beer had run across the brewery floor and into the Taihape drains.[42]

The vexed matter of low-strength hop beer also surfaced periodically. In 1932, the Customs Department made a claim to DB for unpaid beer duty. DB insisted that the product was hop beer. Customs pointed out that the product was brewed with similar ingredients to normal beer, but less sugar, and DB labelled this as 'lager' with a clear intention to sell it in hotels where there would be no demand if it was labelled 'hop beer'. Customs then forwarded two DB brews to the dominion analyst, who determined that the alcohol content was 2.38 per cent and 1.88 per cent, respectively, within the definition of hop beer. Nevertheless, Customs felt obliged to be present the following year to supervise the brewing of DB's hop beer.[43]

The inter-war decades also provided major challenges, and some small victories, for the supply of raw materials to the industry. Barley growers had long complained that they were never able to quickly settle on growing contract prices with brewers. In 1926, they also accused brewers of playing growers off against each other to keep prices down by telling those in Marlborough that Canterbury could grow all that was required. Hence it was important for growers to stand together.[44] But it was evident that Canterbury could not meet requirements, as barley was being imported during the mid-1930s. In late 1936, in the context of wide-ranging reforms under the First Labour Government, Minister of Agriculture Lee Martin initiated a review of the barley industry to counter imports. He pointed out that 'some of the buying interests were not adopting systematic steps to thoroughly exploit the possibility of locally grown supplies. To remedy this, and at my instigation, buyers of malting barley will in future adopt a system of contract growing for the whole of their anticipated requirements.'[45] Fixed prices and standard grading for quality were established and permits to import would be given only to those who could show that they had taken reasonable steps to source local barley and that any shortage was due to unforeseen circumstances.[46] While there were still periodic complaints about quality, and this became a justification for some continued imports, most tensions seem to have been resolved by late 1938.[47] By the 1950s the Canterbury Malting Company provided an advisory service to growers, including field work on farms, lectures, demonstrations and laboratory testing. A small experimental malting plant was also installed for pilot testing.[48]

Issues confronting hop growers were not so easily resolved. As hops were a supplementary crop for many farmers, grown in addition to such things as potatoes, the industry was always volatile and not especially systematic. Growers entered and exited as they saw fit. In times of prosperity some would return or increase their acreage and consequently depress the market with an oversupply. In hard times they departed and scarcity ensued. Unsurprisingly, the acreage of Nelson hops fluctuated markedly from year to year. Further, the yield was always subject to fluctuation depending on weather and soil conditions.[49] The 1937 season was especially wet compared with other years and posed significant difficulties for the harvest. However, the greater threat came from the appearance in the early 1930s of *Phytophthora*

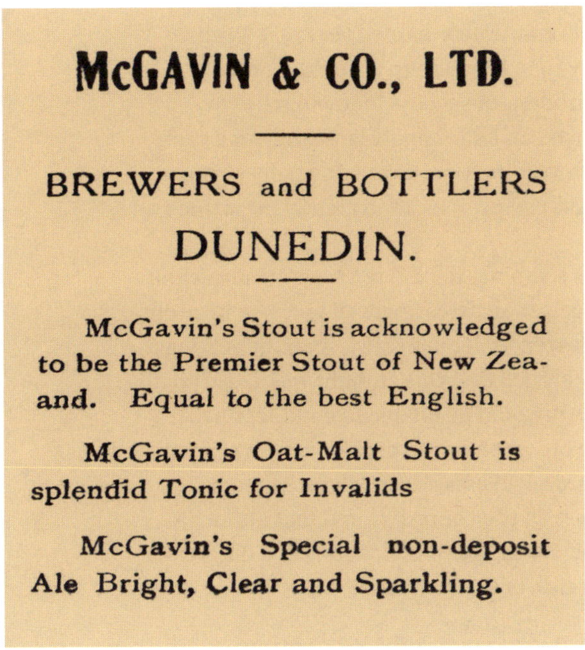

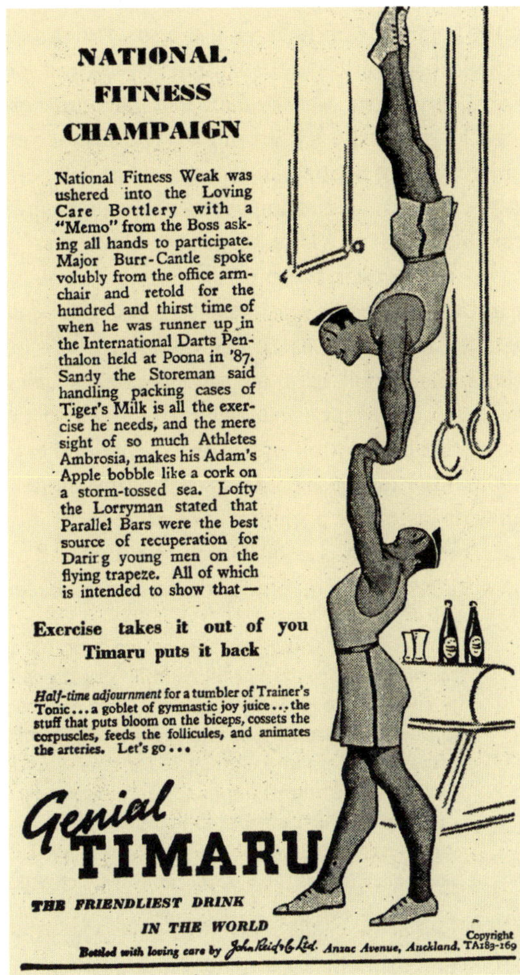

TOP: During the inter-war years, no doubt in response to the emergence of New Zealand Breweries, Harley's of Nelson encouraged continued support for local business. Nelson Provincial Museum, NPM 1 AG_83_E3_1

ABOVE: A May 1920 advertisement for McGavin's Stout from Dunedin. Lion Technical Library

RIGHT: The advertising campaign for Timaru Ale, usually with various sporting themes, was perhaps the most comprehensive throughout New Zealand during the 1930s. *New Zealand Herald*, 2 March 1939, 18 *citricola* or black root rot, a soil fungus which killed hop plants during the winter. While some traditional and lower-yielding English hop varieties such as Fuggles and Bumford were resistant, the higher-yielding Nelson Californian variety, which had become common from the 1890s, was not. In August 1936, Kathleen Curtis, head of the Mycology Department at the Cawthron Institute, which was already engaged in sustained research on the problem, described black root rot as a 'menace to the industry'. Growers who did not abandon hops in despair were forced to continually replant their gardens or to revert to lower-yielding varieties.[50] Only a more systematic approach to brewing science, to be discussed in the next chapter, would begin to arrest the problem from the late 1940s. Meanwhile, the volatility of the industry was addressed through the Primary Products Marketing Act 1938, with Hop Marketing Regulations and the establishment in 1939 of the Hop Marketing Committee, comprising one government official and five growers' representatives. The Committee now regulated the production, pooling, storage, export and marketing of hops. No hops could be sold except through the Committee, which in turn used the long-established E. Buxton & Co. as their official broker. With consolidation and price stability, hop growers were better able to navigate their natural challenges.[51]

*

At the beginning of the inter-war years, the threat of prohibition remained very real. With national prohibition in the United States from January 1920 providing a stimulus for the global movement, New Zealand campaigners were confident of success following the two very close polls in 1919. From 1923, the Alliance employed an increasingly large full-time staff, with fifty-eight by 1925, and a permanent head office in Wellington. It was financially strong, sponsored various touring overseas speakers, published *Vanguard* as a fortnightly magazine and was able to call on a large cohort of volunteers.[52]

By contrast, the efforts of brewers to resist this seeming tide did not have an auspicious start to the 1920s. In mid-1921 Ward & Co. appealed against an assessment of taxes on its profit, but the Court of Appeal determined that it was not allowed to deduct £2123, spent in fighting the 1919 licensing poll, from the assessment. The brewery argued that this amount was a legitimate and reasonable expense in order to produce income as it would be unable to conduct further business if prohibition won. The expense had been forced upon it by a decision of government, and in other business contexts the commissioner of taxes had established a precedent in allowing insurance premiums as just deductions. However, the court declared that although the expenditure was undoubtedly 'prudent and proper' for the business, there was a difference between producing income and preventing the 'extinction of the business'.

> It might appear illogical that, while legitimate expenses incurred in the production of income were deductible, similar expenses incurred for the much more important purpose of keeping the business alive were not deductible, and it might also appear inequitable that the Legislature should on the one hand force a certain class of traders into a struggle for their very existence, and, on the other hand, treat the reasonable expenses incurred in connection with that struggle as part of the profits assessable to income tax. Such questions, however, were matters which might reasonably be considered by the Legislature, and were not for the consideration of the Court, which was concerned solely with the law as it stood.[53]

Ward's was given leave to take its case to the Privy Council. The Law Lords concluded that 'The expenditure in question was not necessary for the production of profit, nor was it in fact incurred for that purpose. It was a voluntary expense incurred with a view to influencing public opinion against taking a step which would have depreciated and partly destroyed the profit-bearing thing.' While the expense may have been 'wisely undertaken', the law did not allow it to be exempted from profit upon which tax must be paid.[54]

Renewed momentum against prohibition soon emerged at a national level. The original Brewers' Association had been wound up in May 1920 and was soon replaced by the National Council of the Licensed Trade of New Zealand. From June 1922 the council periodically published *Cheerio*, with the first issue announcing that it was

> frankly published for the frustration of the prohibitionists and in the interests of continuance. Its purpose is to present the case for sane and rational acceptance of all the good things of life; for reasonable regulation and improvement of conditions surrounding the supply of liquor to the public; for temperance and self-restraint with liberty.[55]

In August 1922 a 10-guinea prize was offered for the best continuance slogan, although the winner appears not to have been reported.[56] Ironically, the introduction of prohibition overseas assisted the cause. While prohibitionists had long made claims about the evils of drink, the trade was now able to draw on evidence of the problems of prohibition enforcement in the United States especially. *Cheerio* also aimed its appeal at women, with suggestions that under prohibition their homes would be subject to inspection by government agents looking for liquor, household essentials would increase in price because of lost tax revenue, and civil liberties would be eroded as prohibition was the thin end of the wedge to other moral 'reforms'.[57]

The most significant factor in shaping public opinion at this time was the Trade Defence Fund. As noted in the previous chapter, the trade had been accumulating a fighting fund to pay for advertising and other anti-prohibition literature since the 1890s – primarily through a levy on brewery output but also a percentage of wine and spirit sales and contributions from publicans. This was well enough known. However, the fund increased significantly with the formation of NZB in 1923. Indeed a clause in the memorandum of association for the new venture stated that it would 'contribute funds to or otherwise assist or encourage or carry out any propaganda or take other steps to further the interests of the company or to prevent the company's interest being prejudiced or affected'.[58] Calculations made during the 1945–46 Royal Commission on Licensing estimated contributions to the fund in the vicinity of £20,000 ($2.2 million in 2022) per annum during the mid-1920s and £40,000 ($3.85 million) by 1945. Obviously, a large portion of this was devoted to campaigning at the licensing polls, and the National Council later acknowledged spending £111,000 ($12.2 million) in 1925 and £87,000 ($9.6 million) in 1928. Beyond that, arrangements were shrouded in mystery. Eliot Davis, a son of Moss Davis of Captain Cook Brewery fame and a prominent figure in the industry in his own right, told the Royal Commission that no formal accounts were kept for the fund and expenditure was made as the controlling officials saw fit and without any obligation to report to those who had contributed to it. At the same time, the trade also won a judgment in the Court of Appeal preventing the Commission from investigating any trade contributions to political parties – on the basis that this was beyond its brief.[59]

It is clear that the trade and certain politicians had much that they wished to keep hidden. In 1975, former Labour politician, controversialist and writer John A. Lee published *For Mine is the Kingdom*, an account of his relationship with Sir Ernest Davis, Eliot Davis's brother. Notwithstanding Lee's tendency to exaggeration, self-aggrandisement and a blending of fact and

Continuance campaigns during the 1920s found much supporting evidence in the problems of prohibition enforcement in the dry United States. Posters such as this were paid for by the extensive Trade Defence Fund administered by the National Council of the Licensed Trade. Alexander Turnbull Library, Eph-D-ALCOHOL-Continuance-1920s-01 fiction that approached fantasy, there was undoubtedly some truth in his portrayal of dealings between the trade and the Labour Party. In the tradition of public-spirited and philanthropic brewers noted in the previous chapter, Ernest Davis was a leader in many sporting and civic bodies, a prominent racehorse owner and yachtsman who served as mayor of Auckland from 1935 to 1941.[60] His sympathies for Labour can be traced in part to a friendship with its first prime minister, Michael Joseph Savage, while the latter worked as a cellarman at the Captain Cook Brewery. As mentioned earlier, Davis also posted bail for Waihī strikers in 1912. By the 1920s, according to Lee, he was overseeing payments and gifts to Labour politicians, MPs of other persuasions, customs officers, and prominent and corrupt union officials. Donations to Labour candidates were crucial during the 1920s and early 1930s when party finances were fragile, although it is not certain how much came directly from Davis or from the Trade Defence Fund. Indeed, Lee's focus on Davis alone rather neglects the influence of a broader group of Auckland brewers, wine and spirit merchants and other business interests. As Lee's Auckland East electorate was said to have the greatest concentration of hotels in the country, NZB naturally wished to protect its interests. After Lee lost the seat at the 1928 election, Davis installed him as manager of the Palace Hotel, Rotorua. In this position Lee was repeatedly asked to provide cheap or free accommodation to anyone of prominence supporting the trade. When Lee was back in Parliament, he kept Davis informed about events, views of politicians and proposed legislation.[61]

Even within the Labour Party not all were entirely compliant. In August 1930, James McCombs, always the most trenchant of the Labour prohibitionist element, used the debate on the Customs Amendment Bill to roundly condemn the Trade Defence Fund.

> So far as that industry and that fund was concerned, they constituted the greatest corrupting influence there was in the public life of the Dominion. That organisation was a positive menace to the free exercise of the people's will, so powerful was its influence, with the huge funds which it possessed. How was it that, when the government was looking around for more taxation, it treated that particular industry so tenderly?

McCombs was particularly incensed by a proposal to increase beer duty by only a halfpenny per gallon, compared to 3 pence for petrol.[62] In February 1932, Lee, in an effort to assert his independent streak, published two anonymous attacks on 'Beer Barons in Press and Politics' in the *New Zealand Worker*, pieces that described the power of the Davis brothers as 'menacing the interests of the working class', and claimed that the populist *NZ Truth* newspaper was controlled by the brewers, that Alfred Bankart was or had been a director of *Truth* and that many of its articles had been written or inspired within brewery offices. Several months later, the *Worker*, presumably under threat of legal action, was obliged to use other major newspapers to issue an unreserved apology and withdrawal of all statements linking breweries to *NZ Truth*.[63] Despite Lee's efforts, Ernest Davis remained close to Labour, including paying the legal expenses of Jim Edwards and others charged following unemployed worker protests in April 1932. Erik Olssen suggests that he may have intervened with Savage to prevent Lee's elevation to Cabinet when Labour took office in 1935.[64] Davis was knighted on the recommendation of the Labour Government in 1937. Shortly afterwards, among a pattern of gifts of racehorses, Eliot Davis gave a stallion to prominent Cabinet minister Bob Semple.[65] Yet Eliot Davis, and another NZB director, Sir William Perry, had strong connections on the other side of politics with the Reform and then National parties. As Conrad Bollinger observed, 'The trade as a whole is politically quite impartial, which seems to ensure that, whatever party holds power, the trades interests will not lightly be infringed.'[66]

BELOW & RIGHT: Thomas Macarthy was a prominent brewer in Wellington from the late 1870s until his death in 1912. Thereafter the T. G. Macarthy Trust generated large sums for charity while antagonising prohibitionists who objected to breweries being run by the Public Trustee. *Dominion*, 5 June 1937, 24 (below); Public Trustee. *NZ Truth*, 6 June, 1925, 12 (right)

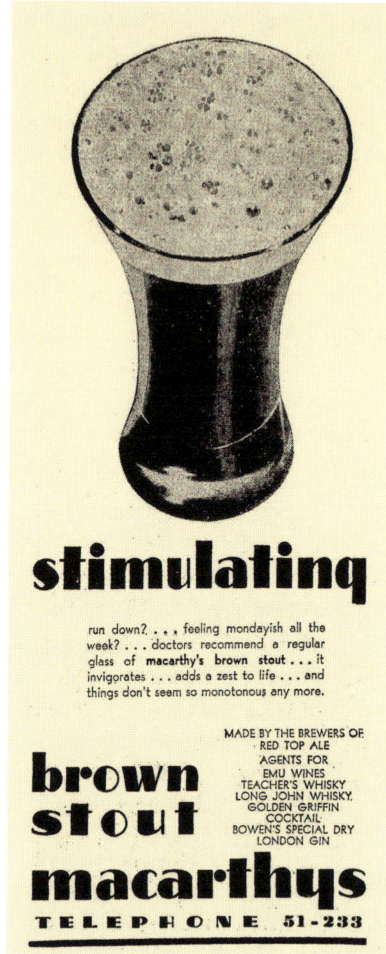

In the midst of these dealings, one brewer antagonised prohibitionists over money by an entirely different means. When Thomas Macarthy died in Wellington in August 1912, he provided for various bequests, and left half of his residual estate to his wife, and half to establish a trust for 'charitable and educational purposes or institutions in the Provincial District of Wellington'.[67] When his wife died without heirs in July 1934, the total estate was estimated at £369,689 ($51 million in 2022), easily within the top 0.1 per cent of all New Zealand estates. The Thomas George Macarthy Trust, augmented by continuing profit from the City and Phoenix breweries and various hotels, was administered by the Public Trustee and distributed an average of around £10,000 per annum during its first two decades through a board of governors including successive prime ministers, the Catholic archbishop of Wellington, the governor-general, and the mayor of Wellington.[68] However, this arrangement was not appreciated by some. In August 1928, Sir Robert Stout moved a motion in the Legislative Council condemning alcohol and recommending the passing of a law to prevent the Public Trustee from 'being a brewer'. Others pointed out that the Public Trustee was merely acting as a trustee for the estate of Macarthy and was bound to serve its interests. The proposal was defeated 22–6. It perhaps did not help that the breweries antagonised matters by advertising under the slogan 'The People's Ale – the more you drink, the more there is for charity' and that their bottle labels noted that the contents had been brewed and bottled by the Public Trustee, although the Licensing Act did require the name of the brewer and bottler to appear on any labels used. The City and Phoenix breweries continued to prosper until closing during the late 1940s in the face of NZB dominance. The Public Trust still controlled the hotel interests and by 1989 over $10 million had been distributed by the Trust. It was still operating in 2023.[69]

*

In both 1922 and 1925 prohibition secured more votes than continuance but failed to reach the 50 per cent required for change. Much as prohibitionists complained bitterly that they had been thwarted by the contentious third option of state purchase and control, Paul Christoffel demonstrates that this vote was always at the expense of continuance. In 1925, the percentage of the vote fell for both prohibition and continuance as state control benefited from reports of its increasing popularity in Canada, where some provinces had rejected prohibition in favour of partial state control. The other feature of the 1925 campaign was that Ōhinemuri became the first electorate to vote itself wet again, after seventeen years as a no-licence district.[70]

On behalf of the WCTU, the *White Ribbon* offered a somewhat bitter explanation of successive election defeats.

> THE TWO BALLOTS.
> Along in the winter, when cold was the weather,
> Two ballots were cast in a box close together;
> Two ballots were cast in together.
> They nestled up close, like a brother to brother,
> You couldn't tell one of the votes from the other;
> You couldn't tell one from the other.
> Both votes were quite equally pro-liquor votes,
> And sanctioned the licensing plan;
> One was cast in by a jolly old Brewer, And one by a Christian Man.
> The Christian man – no one could be truer –
> Kept busy all summer denouncing the brewer;
> All summer denouncing the brewer,
> But his fervour cooled off with the change of the weather.
> And late in the winter they voted together;
> In the winter they voted together.
> The Christian man had always been noted
> For fighting the drink, except when he voted;
> For fighting the drink, except when he voted.
> He piled up his prayers with holy perfection,
> But he knocked them all down on the day of election;
> But knocked them all down at election.
> The foxy old Brewer was cheerful and mellow,
> Said he, 'I admire that Christian fellow;
> I admire that Christian fellow.
> He's true to his Church – to his party he's truer;
> He talks to the Lord, but he votes for the Brewer;
> He votes every time for the Brewer.'[71]

In 1928 the vote for prohibition plummeted, never to regain its former support. At its annual general meeting on 23 May, the Alliance expressed concern regarding its budget and stressed the need to keep expenditure within prescribed limits. Particular emphasis was placed on securing cooperation from churches, volunteers and strong neighbourhood canvassing to compensate for the funding shortage.[72] Consequently, the Alliance decided not to commit any funds to advertising at the 1928 election, thus effectively handing the undecided vote to its opponents and experiencing a 15 per cent decline in support by polling only 40 per cent. It is clear that the Alliance staffing levels and head office had overstretched its funds. It was now possibly preserving its limited resources for a major campaign in 1931, by which time it imagined that it could point to the success of prohibition in the United States. Indeed, the *White Ribbon* greeted 1931 by insisting, 'The facts are all with us. Unemployment is rampant, yet we waste over eight million pounds in strong drink. Were this money spent on food, clothing, or luxuries, how many more it would employ. The brewer pays less in wages by a long way than any other trade.'[73] But the plan was thwarted when Depression austerity took hold. The Alliance and trade eventually agreed to postpone the 1931 poll.[74] The Alliance's own resource base and public backing evaporated still further amid widespread cost-cutting. The 1932 annual report lamented a very difficult year and regretted that hard economic times prevented the Alliance placing wall charts in schools to assist teachers.[75]

By the time the next licensing poll was held in 1935, Canada, Finland, Iceland, Norway and crucially the United States had all abandoned prohibition. The Depression, with its origins in the United States, had starkly exposed the superficiality of the frequent claim that prohibition would lead to a more efficient and economically buoyant society. The National Council, which spent £79,000 ($10.5 million in 2022) on the 1935 campaign, also made much of the flouting of the law, the rise of organised crime and abject failure of enforcement that characterised the United States during prohibition. On the other hand, although volunteers delivered 750,000 pieces of literature to households during 1935, many of them describing repeal of prohibition as 'America's Greatest Mistake' that had already led to an increase in alcohol-related crime and insanity, the Alliance had no money for press advertising beyond Auckland. It spent only £1350 during the entire campaign. Prohibition secured only 29 per cent of the vote in the 1935 and 1938 polls.[76] The Alliance noted that there was now a negligible response to questions it had issued to political candidates, and finance was inadequate to run the central office in Wellington.[77]

Paul Christoffel also suggests that the decline in support for prohibition reflected a broader social change and liberalisation of society during the 1920s. While New Zealand was perhaps not a bastion of the 'roaring twenties', there was certainly more acceptance of new fashions and opportunities for women in employment and sport, while new technologies such as radio, the gramophone and cinema introduced Hollywood and other more liberating influences. Indeed, Edith Thornton, who wrote a sympathetic history of the prohibition movement as an MA thesis in 1936, claimed that regular drinking in modern films created a tolerance for it. She also suggested that cinema-going provided people with entertainment in the evenings, where once they may have found it at a lively prohibition rally.[78]

*

No sooner had prohibition started its retreat in 1928 than the Great Depression arrived to buffet brewers and drinkers. Following from the October 1929 New York stock market crash and its reverberation across the Atlantic, New Zealand was further exposed because of its reliance on British purchase of agricultural exports. As revenue declined by perhaps 40 per cent and spending contracted, the official unemployment rate rose beyond 15 per cent, but was probably nearer 30 per cent, and wages were cut. Government support, such as relief work schemes, was inadequate and private charity had its limits.[79] In search of revenue, the government increased beer duty by a halfpenny to an even 1 shilling in the 1930 Budget. However, this was not enough to stem the deepening crisis. In March 1931, James McCombs suggested that rather than reducing wages and salaries to meet the fall in revenue, an increase in duty on beer and spirits could raise £1.5 million. But later in the year his Labour colleague Tim Armstrong criticised increased taxes on essentials, within which he evidently included beer and tobacco, rather than on wealth. He predicted that an increased tax on beer, which the brewers would inevitably pass to the consumer via the publican, would reduce consumption and hence reduce revenue. In due course the 1931 Budget increased beer duty by 50 per cent to 1 shilling 6 pence and imported beer to 3 shillings per gallon. Both were quickly passed on to drinkers.[80]

Notwithstanding the overall increase, Armstrong was to some extent vindicated. Overall beer production fell from 64.5 million litres in 1926 to 58 million in 1929 and only 44 million in 1931, before gradually recovering to 52 million litres by 1936. Per-capita consumption fell from 40 litres in 1929 to 30 in 1931 and only 25 the following year. There was a gradual increase thereafter, but not until 1937 did consumption again exceed the 1926 level of 41 litres. At the 1932 NZB annual general meeting, Alfred Bankart pointed out that beer duty was now six times the pre-war level and 'acted as an artificial burden on an industry which had already borne more than a fair share of taxation'. He argued that reduced consumption led to a loss of employment including in agriculture with the decline in demand for barley and hops. But perhaps a reported net profit of £88,000 ($12 million in 2022) undermined public sympathy for this position. On the other hand DB reported a £14,002 loss for the 1932/33 financial year due to 'crippling taxation'.[81]

With wages and salaries falling and beer prices rising, drinkers soon found alternatives. In July 1931 Alfred Bankart complained that increased beer duty had 'given a great impetus to the home brew industry, which notoriously is not confined to the legitimate brewing of a beer for consumption in the home' and now seemed to be in the hands of those with commercial intentions.[82] Shortly afterwards, NZB wrote to William Downie Stewart, minister of customs, highlighting the significant increase in home brewing and public demand for brewing materials and equipment. While most home brew was of standard strength, investigations and sample analysis by breweries was revealing increasing cases of beer brewed well beyond commercial strength. Based on sales of hops, NZB estimated that about 11 million litres were brewed privately during the 1930/31 financial year – roughly 25 per cent of the amount produced commercially. The minister was urged to follow the example of Australia and Britain where any beer over 2 per cent proof required a licence to brew and was dutiable. A few weeks later an 'NZ Free Trade Brewers' Association', apparently representing twenty-five breweries outside the NZB network, urged that all brewers should be licensed. In January 1932 NZB wrote to the acting comptroller of customs enclosing a home brew recipe and advertisement, and protesting at the unfairness of people being allowed to brew beer free of duty while breweries paid 1 shilling 6 pence a gallon.[83] For its part, the Alliance expressed regret that the Depression was bringing brewing into the home with a detrimental impact on children. 'The only true remedy is to educate the public to realise that home-brewed beer is just as noxious as that sold over bars.'[84]

For some home brewers it seems that the stakes were about more than cheap beer. By October 1931 a correspondent to the *Australian Brewing and Wine Journal* described it as 'a

fashion' in Whanganui with recipes, 'many of them involved to a degree', being eagerly sought and exchanged, and their attributes debated from the perspective of economy and potency. A month later the same source noted, 'The degree of success achieved varies greatly. Some produce a clear, sparkling and palatable beverage, while others are rewarded with a muddy-looking mixture whose only virtue lies in its hefty "kick". The fortunate ones are those who possess a suitable barrel, the brew matured "in the wood" being superior to all others.'[85] Yet, if the apparent spike in advertising of malt extract during 1931–32 is any indication, most brewers were not especially sophisticated and certainly not doing their own malting. One recipe from Central Otago is perhaps typical.

> A POPULAR RECIPE. HOME BREW. Twelve gallons cold water; 14 ounces hops; one cup rice; 7lbs sugar; 7lbs amber maltexo; 2 tablespoons salt. Put 12 gallons water in copper and immerse loose bag containing hops and rice, bring to boil and boil for at least half an hour. Remove to open barrel adding sugar, amber maltexo and salt. When at blood heat (80 deg. F) add sixpence worth of brewer's yeast. Skim daily for nine days then add two packets of Isinglass, skimming and straining after 48 hours, when the brew is ready for bottling. Fresh yeast will not be required for the next brew as the sediment left in the barrel will work the next lot.[86]

1930s advertisements frequently linked beer to active sportsmen, physical labour and other strenuous outdoor activities.
New Zealand Herald, 21 September 1936, 26 (far left);
The Press, 17 November 1939, 12 (left);
New Zealand Herald, 6 April 1938, 2 (below);
The Press, 1936, 18 (top right);
Poverty Bay Herald, 2 July 1932, 3 (bottom right)

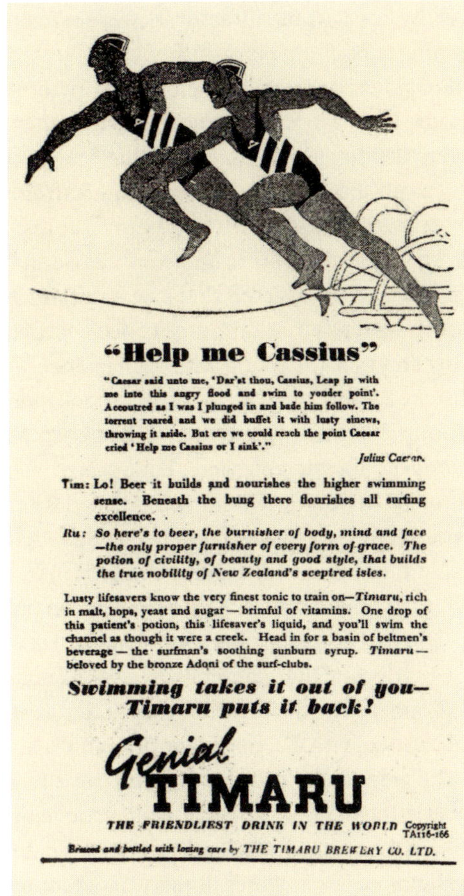

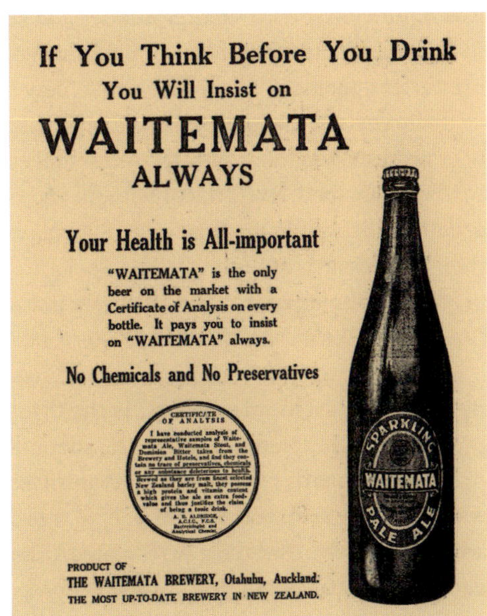

However, by late 1931 manufacturers were refusing to sell barrels to home brewers and bars were complaining at declining patronage.[87]

From 1932 enforcement closed in on home brewers. In April a magistrate noted that much home brew was considerably stronger than normal beer, as evidenced by a drink-driving case where the defendant claimed to have had 'four small glasses of home brew'.[88] In December an Invercargill shopkeeper was prosecuted after being found with 123 bottles of beer and 68 empties on his premises, with another 120 at his home. According to police, the shop had become known as the 'Dee St Brewery'. The defendant claimed that he had started brewing for friends who then insisted that it was unfair for him to supply them with beer at his own expense. He said that his large stock was in anticipation of a Christmas rush. He was fined £10 for selling and £5 for brewing without a licence. Early in 1933 the Temuka police offered 450 litres of confiscated home brew at auction on condition that any purchaser paid duty on the beer.[89] By this point, home brewer detection had become a component of the Police Examinations for detectives. Question 6 asked: 'You receive a complaint that a man is brewing and selling home-brewed beer. How would you proceed when making necessary inquiries? What evidence would you require to prove an offence? What charges would you prefer?'[90] In the inaugural editorial of the *New Zealand Licensee* in July 1933, home brew was condemned as being without any guarantee of quality, therefore a danger to public health and likely to produce an alarming increase in illicit consumption. But by mid-1934 evidence based on sales of ingredients suggested a significant decline in home brewing as the economy slowly improved. All commercial brewers were no doubt pleased when the government responded by reducing beer duty by 3 pence in September 1934, with a consequent drop in price for consumers.[91]

The gradual return to prosperity can also be seen in more voluminous and sophisticated beer advertising in newspapers. Although developments in lithography from the 1880s improved on time-consuming engraving and allowed greater use of illustrations and cartoons, the New Zealand press was generally slow to adapt to the international trend. Advertising was still very much dominated by words during the early twentieth century. However, by the 1920s there was significant growth in the number and sophistication of advertising agencies, formal training for commercial artists and the beginnings of market research and campaign planning. For the breweries specifically, there were also new opportunities.[92] In the absence of precise figures, it is virtually certain that six o'clock closing triggered a marked increase in the proportion of bottled beer produced to meet the after-hours takeaway market. The improved bottle supply from the 1920s also helped. In turn, consumers had choice beyond what was available on tap at their local, usually tied, establishment. Indeed, it was noted earlier that DB used bottled beer to overcome their early dearth of hotel outlets. Speight's was another to specifically focus on its bottled product, claiming of its Three Star Ale that 'From the Wood Speight's Ale is great. From the Bottle – SUPERB'.[93] In sum, there was renewed incentive to advertise.

Advertising drew on familiar themes from the nineteenth century, but now augmented by scientific developments such as the discovery of vitamins in 1912 and the consequent popularity of supplements from the 1920s. Although it was less common to emphasise medicinal or curative qualities in the way that spirit-based beverages were frequently promoted, much was made of the importance of beer as a provider of health and vitality. A 1932 Waitemata Ale advertisement presented a certificate of analysis by a bacteriologist and analytical chemist declaring that all beers from the brewery 'contain no trace of preservatives, chemicals or any substance deleterious to health. Brewed as they are from finest selected New Zealand barley malt, they possess a high protein and vitamin content which gives the ale an extra food value and thus justifies the claim of being a tonic drink.'[94] In 1936 Timaru Ale was advertised as ideal for lifesavers and other vigorous athletes.

> Lusty lifesavers know the very finest tonic to train on – Timaru, rich in malt, hops, yeast and sugar, brimful of vitamins. One drop of this patient's potion, this lifesaver's liquid, and you'll swim the channel as though it were a creek. Head in for a basin of beltmen's beverage the surfman's soothing sunburn syrup. Timaru – beloved by the bronze Adoni of the surf-clubs. Swimming takes it out of you, Timaru puts it back.[95]

In similar vein, Macarthy's Red Band Ale was presented as 'MALT for digestion, HOPS for appetite, SUGAR for energy, YEAST for vitality, RED BAND ALE for perfect refreshment. . . . It is the drink that will put health and heart into you – the drink needed by active workers and keen players.' Advertising for Macarthy's brown stout asked 'Run down? . . . feeling Mondayish all the week? . . . doctors recommend a regular dose of Macarthy's brown stout . . . it invigorates . . . adds a zest to life . . . and things don't seem so monotonous any more.'[96]

In her detailed examination of New Zealand alcohol advertising, Marie Duncan identifies an important contrast between beer and spirits. Whereas the latter tended towards images of middle-class men sitting in clubs, walking on golf courses or engaging in sport, a strong theme of beer advertisements portrayed an ideal, healthy, masculine type engaged in hard physical labour. Typical is a September 1936 advertisement for Lion Pale Ale under the title 'good honest beer for bodily health', featuring a shirtless male with pick in one hand and foaming pint of beer in the other. The text also states, 'The Khyber Pass Breweries not only constitute an important Auckland industry in themselves, but their products also give strength and energy to thousands of workers in other industries throughout the Province.'[97] As Duncan argues, this can be interpreted as a symbol of productivity, post-Depression recovery and expansion and the hope for a prosperous future.[98] Other breweries, especially the likes of Ballins and Timaru, regularly associated their beer with active sportsmen or those at sporting events. From 1928 the laconic beer-drinking twins Tim and Ru offered comment on sport, social issues and politics in advertising campaigns that gained national prominence and resulted in a variety of local sports being contested for a Tim and Ru Cup or Trophy.[99]

The Alliance naturally took exception to claims of a link between alcohol and health. Their particular target was 1934 ads for Mello Stout from NZB and Vita-Stout from DB.

> Vita-Stout gives you vim, vitality, zest with a z-i-p! Brisks you up . . . puts a sparkle in your eye . . . a tingle in your veins. . . . Actual vitamins are added to enrich the already tonic virtues of stout. . . . To be specific, there's as much sunshine vitamin in one bottle of Vita-Stout as in 2 pints of milk or 4 eggs. And as much life-giving vitamin B as in a whole meat loaf.[100]

Shortly after this appeared, H. S. Adams for the Alliance contacted John Malcolm, professor of physiology at the University of Otago Medical School, describing the comparison between alcohol and food as 'grotesque' and asking for a written statement about vitamins. Malcolm, no doubt weary of becoming involved in prohibition politics, informed Adams that he was not free to take commissions for private work outside his university commitments. He would therefore not make a formal statement, but pointed out that experiments had been done to prove that vitamins were present in beer, yeast especially was rich in vitamin B, and his advice to the Alliance was to leave

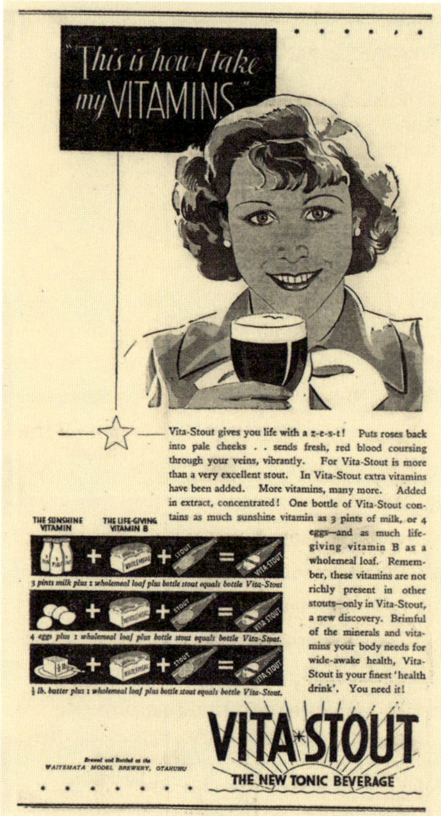

The health claims made for Vita-Stout and Mello Stout during the mid-1930s were strongly opposed by the New Zealand Alliance. New Zealand Alliance Records, Alexander Turnbull Library

the ad alone. He did however concede to Adams that the matter of vitamins in beer was open to misrepresentation. Undeterred, the Alliance then wrote to Sir Alexander Young, minister of health, complaining that from the perspective of food values the claims of the ads were 'fantastically untrue', deliberately intended to mislead, and therefore breweries should be compelled to prove their claims. An analysis of stout, indicating it had no vitamin value, was also enclosed with the letter, although this was an older Yorkshire analysis and not Vita-Stout. Further, 'The Mello stout advertisement, while not so gross in its perversion of facts, is deplorable as an incitement to young women to acquire the alcohol habit.' A month later the minister replied that the ads had been carefully considered and the breweries could not be prosecuted with any hope of success as they were not required to prove their statements. Rather, health authorities were required to disprove them, which was difficult. Somewhat exasperated, the Alliance then misrepresented Malcolm in arguing that as it was impossible to assess the vitamin content in food, the ads must therefore be untrue and consequently subject to heavy penalties under the Food and Drugs Act and the Quackery Act for making misleading statements. Despite being in a beer bottle with a stout label, the ads also apparently failed to notify the public that the product contained an appreciable amount of alcohol. Finally, 'to describe it as a tonic when the chief ingredient is recognised in the medical profession as a narcotic is also misleading'. The minister did not reply.[101]

When attention was turned to the growing issue of alcohol and cars during the 1930s, the Alliance had more success in harnessing popular and official support. Replicating Britain's Licensing Act 1872, the Police Offences Act 1884 outlined penalties for 'every person who is drunk while in charge, on any public place, of any carriage, horse, cattle, steam-engine . . .'. The Motor Regulation Act 1906 outlined more specific provisions against reckless driving in the same year that New Zealand recorded its first motor vehicle fatality.[102] By 1909 concern was being expressed about the risks posed by 'intoxicated motorists' and the lax measures against them. There was a national outcry when William Pavitt was killed in Christchurch following a crash in which the driver, W. A. Humphreys, son of a prominent wine and spirit merchant, admitted to having seven or eight beers at various establishments while travelling to and from town but still insisted he was quite sober. After a coronial jury took only fifteen minutes to return a verdict of accidental death, *NZ Truth* fumed that 'Christchurch has probably got a more appalling reputation for motor accidents than any place of the same size in the world, and no amount of killing seems to be able to stop it. The motor-hog is a dangerous person, who is seen so frequently in the S.M.'s Court that he is an established disregarder of law and order.'[103]

The drink-driving problem escalated as the number of cars in New Zealand increased from two in 1898 to 17,000 by 1915, 50,000 by 1920, 106,000 by 1925 and 205,000 by 1935, even allowing for a decline in vehicle use during the Depression. Road deaths, which were first recorded in detail from 1921, accounted for 1273 people during the decade to 1930, and another 1939 during the next decade, in addition to perhaps 4000–5000 injuries per annum by the late 1930s. Over the same period, vehicles were becoming more powerful but the quality of roads, especially in rural areas, was not keeping pace with the change.[104] While there is no detail as to how many of the casualties were caused by alcohol or by beer specifically, there was certainly a strong perception that this was the major contributing factor. It can also be inferred from a survey in 1939 which concluded that 40 per cent of all accidents occurred between 4 p.m. and 8 p.m., and over 20 per cent occurred on Saturday, thus suggesting some correlation with the most popular hotel drinking hours.[105] However, although the Motor Vehicles Act 1924 imposed severe penalties ranging from imprisonment to a fine of £500 ($52,000 in 2022) for driving while intoxicated, the definition of intoxication was vague and enforcement was haphazard. In 1926 a driver who was to be charged with being intoxicated in charge of a car was told by a constable to drive to the police station to be arrested. Three constables gave evidence that the man was drunk, but the case was dismissed as 'the defendant's senses were not sufficiently impaired to render him incapable of driving the car'.[106]

In response to the increasing number of convictions for people who claimed to have had only 'a few beers', the Alliance highlighted what it termed the 'Beer Bluffer'. 'Many people start the drink habit by way of the beer route. Beer is introduced to the beginner as "the friendly drink of moderation", but is really the spring-board into the strong drink habit . . . Its mildness in comparison with whisky bluffs people into thinking it is not intoxicating.'[107] In a later pamphlet entitled *The Menace of the Motor*, the Alliance noted the 'significant' number of accidents caused by those under the influence of alcohol, yet only 6.4 per cent of New Zealand driving offences during 1935 related to alcohol – pointing to an evident lack of enforcement. In view of medical evidence regarding the lingering effects of alcohol on drivers, it was necessary to amend the legislation to cover not merely those 'in a state of intoxication', but those 'under the influence of intoxicating liquor'.[108]

Following the election of the Labour Government, Minister of Transport Bob Semple was determined to bring change. Navigating a careful path through the broader alcohol issue, he observed in August 1936 that 'A man is entitled to get drunk in New Zealand as long as he goes to bed or plants himself somewhere where he is not a danger to anybody, but he is not entitled to take liquor and become a menace to the life of his fellow citizens, and as far as is humanly possible to stop him from doing that, I will.'[109] The following year traffic inspectors were appointed and the first national road safety campaign was launched. The Transport Law Amendment Act 1939 prohibited driving 'while in a state of intoxication'.[110] However, later chapters will show that the problem would get much worse before it showed any sign of improvement towards the end of the twentieth century.

*

As another world war loomed in 1939, the brewing industry witnessed its final consolidation of the period as previously rather fragmented local unions mostly came together in a single national body. With less than a thousand brewery employees nationwide, and variations in conditions of employment still preferred between different areas, such as the forty-six-hour week for brewery workers in Auckland but a forty-four-hour week in Wellington, earlier attempts at unity had come to nothing. But with the advent of compulsory union membership under the Labour Government and the formation of the Federation of Labour in 1937 to consolidate and coordinate union activity, there was renewed incentive. A national conference of the regional brewing and related cordial and aerated water worker unions was held in January 1938 to work towards a national union and to draft claims for a national award. Finally in September 1939 the New Zealand Brewers', Bottlers', Bottlewashers' and Aerated Water Employees' Industrial Union of Workers was registered, but without Nelson or Otago–Southland representation as they maintained regional objections to unity. Even then, local branches maintained considerable autonomy.[111]

At almost the same time, brewers and publicans felt the wrath of an increasingly powerful union movement in another context. In the 1939 Budget, beer duty was increased by 40 per cent to 1 shilling 9 pence, which represented a sevenfold increase since 1915.[112] Within a fortnight, in response to a price increase from the breweries, Wellington publicans decided to increase the price of beer from 5 pence to 6 pence per handle and to no longer provide any free drinks to regular customers, as had been tradition for many years. This triggered an impromptu meeting on 16 August of 700 Wellington watersiders, who unanimously agreed to boycott the hotels until the price of beer in Wellington was reduced from 5 pence to 4 pence per handle.[113] While publicans labelled the boycott a 'nine days' wonder' and insisted their increase was supported by the recently established Price Investigation Tribunal, they did suggest they would consider increasing the size of a handle from 10 to 12 ounces.[114] However, the boycott quickly spread beyond Wellington. The *Otago Daily Times* captured the complexity of the situation:

> Whether this action, which appears to have its sympathisers from Auckland to the Bluff, must be interpreted as a strike against the publicans, who retail the beer to the public, or against the brewers, who make the beer, or against the Government, which taxes it, it may be difficult to determine. That the protest is not directed against the beer itself is a safe, if intriguing, conclusion. Herein, it may be thought, is the weakness of the watersiders' revolt. Strikes are commonly directed against something that does not involve the worker in a privation. Distaste has lent a grim stubbornness to his action. But for beer he has, as a class, no distaste; rather the contrary. Grimness is lent to his present determination by the fact that, if he is loyal to his professions, he must repudiate that which is dear to him. Can his sober resolve endure against the clamorous promptings of a thirst unslaked?[115]

The boycott moved very quickly from watersiders to other unions and new demands. With only one voice opposed, the Addington Railway Workshops voted to declare Christchurch hotels 'black' and demanded that the standard beer glass measure should be the imperial pint in accordance with the requirements of the Weights and Measures Act.[116] Auckland Harbour Board employees added that 'in view of the enormous profits shown by breweries last year there is no justification for the increase' and that 'we also find that brewers are not only passing on to the publican the full 6 pence a gallon tax, but are also making an additional profit out of the same tax'.[117] In a lengthy analysis of the boycott at the end of August, Dunedin's *Evening Star* concluded that direct action by the unions was tantamount to a vote of no-confidence in the government in that it must have known that the increase in duty would be passed on to the drinker. Had it intended otherwise, the government could have imposed a specific brewers' or publicans' tax.[118] By the end of August some publicans were reported as being 'restive' at the decline in trade. In New Plymouth and apparently elsewhere the price increase on bottled beer had already been reversed. Some hotels had also moved from 10-ounce to 12-ounce handles as promised. But the watersiders were unimpressed and now insisted on 6 pence for a 20-ounce pint.[119]

The momentum of the boycott faded with the outbreak of war on 2 September. There also seems to have been intervention by the minister of industries and commerce, Dan Sullivan, who stated a determination to ensure that any price increase was no more than the level of the increased duty required. But as beer pricing was a very complex matter, the Price Investigation Tribunal fixed the price of handles at 6 pence for 12 ounces for the main centres only. Unconvinced, the Auckland watersiders in particular held firm until 20 September, when they conceded that measures of beer in relation to price had now been standardised and nothing was being gained by continued action. A week later beer duty was increased again to 2 shillings a gallon to help finance the war effort.[120]

The boycott was symbolic of a pronounced shift during the inter-war decades. While brewery advertising and elements of popular rhetoric still liked to emphasise the affinity between beer and the working man, the reality was that the brewing industry was now well on the way to domination by large corporate entities for which efficiency and profit exerted greater importance than the local loyalties and traditions of the nineteenth century. Much as NZB presented their 1923 merger as a refreshing departure from murky tied-house arrangements, the gradual rise of DB during the 1930s prompted an even more intense scramble for control that would be accentuated by a new set of challenges confronting brewers during the Second World War, but challenges that the industry adapted to rather quickly and comfortably.

Packaging beer at New Zealand Breweries, 1954. Lion Technical Library

EIGHT

1036 and All That: War, Swill and Technology, 1942–1957

In 1949, DB waxed lyrical about the products of its recently upgraded Waitemata Brewery.

> Our New Zealand taste – in food, cigarettes and in beverages – has tended away from the artificial, the de-natured and the doctored articles. We are a sturdy race, close to the earth, with a marked 'pure food and drink' preference. This brewery, therefore, prepares its products to satisfy the New Zealand taste. Its beer is a rich, golden colour, distinctive in body and flavour. . . . This drink has something of the sorcery of a rare wine. It is made to be sipped rather than consumed. The epicure finds it a fit and proper accompaniment to fine food, savoured appreciatively.[1]

Yet the late 1940s and 1950s are crucial to a rather different and enduring image of New Zealand beer, that of the six o'clock swill and its legacy of binge drinking. While DB painted a picture of refinement, drinking in hotel bars was described by one returning soldier in 1945 as a 'barbarous custom', a 'five to six o'clock scrum' that few could really enjoy 'with one elbow in a puddle of beer on the bar, with one's neighbour's elbow in your ribs and another one with his foot on your pet corn, to say nothing of the occasional splash of beer down the neck as someone reaches over you for a refill'.[2] Others noted that between 5 p.m. and 6 p.m. pubs were a 'lunatic buzz' with half of New Zealand 'noisily engaged in hoisting beer against time' and people 'lined three or four deep' at the bar.[3]

While there is a common association between six o'clock closing and the swill, references to swilling in relation to beer do not appear anywhere in New Zealand or Australia until 1945 – nearly three decades after early closing was introduced. Likewise, terms such as 'perpendicular' and 'vertical' drinking, describing the common practice of standing rather than sitting while drinking in crowded pubs, were also 1940s additions to the lexicon. The origins of the swill are found instead in crucial decisions about beer strength taken in 1942 and in a rapidly emerging economic model that suited the major breweries so well that they did not depart from it when they legally could after 1948. The breweries, having deflected some of the more radical suggestions of the 1946 Royal Commission on Licensing, preferred the new status quo and during the 1950s rationalised the production of beer into a very efficient and lucrative industrial process. While consumers periodically complained about the results, there were only occasional displays of concerted resistance.

The Second World War on the New Zealand home front was not initially accompanied by the level of angst around alcohol that was witnessed after August 1914. With early closing in place and prohibition no longer as dominant a political force, there was less pressure to restrict alcohol. The first government action on drink was to increase beer duty to 2 shillings per gallon in the 1939 Supplementary Budget and establish the Price Investigation Tribunal, to guard against profiteering. This set a maximum price of 6 pence for a 12-ounce glass of beer, a generally lower price than had prevailed just before the war. These moves passed without comment in the face of the growing cost of the war.[4] More positively for the vast majority of enlisted men who enjoyed a beer, and contrary to the unjustified hysteria about drunken soldiers that had prevailed during 1915–16, the minister of defence, Fred Jones, announced in November 1939 that Cabinet had approved the establishment of 'wet canteens' in all military camps. The impetus for this measure had come from both military officers and returned soldiers' representatives and a review of the practice in other countries. Sales would be confined to lighter ales and no spirits. Proposals were also afoot to establish milk bars and cafeteria services, and nothing, according to the minister, 'would be done to encourage the consumption of ale in preference to milk shakes, soft drinks, tea and coffee'.[5] Profits from the ventures would be reinvested for the benefit of the soldiers. By early 1941 there were twenty-three canteens at various defence bases, including one on HMS *Monowai*, with a total profit to that point of £16,250 ($1.7 million in 2022).[6]

Breweries moved quickly to support the war effort. By November 1939, all had agreed to subsidise the military pay of employees, initially for one year, so that none would be paid less when enlisting.[7] Early in the war a large portion of NZB's truck fleet was commandeered by the

BURNHAM'S BEER GARDEN: FIRST GATHERING AS WET CANTEEN WAS OPENED AT THE CHRISTCHURCH MILITARY CAMP

In contrast to the hysteria around alcohol that prevailed during the First World War, the government quickly agreed in 1939 that wet canteens could operate in military camps – a move celebrated in Burnham's 'beer garden' near Christchurch. Auckland Libraries Heritage Collections, AWNS_19391206_40_2 army and shipped to the Middle East. The outcome, according to a later recollection, was that 'many an astonished Kiwi thought he had gone "troppo" when he saw a N.Z. Breweries truck come trundling across the desert'.[8] The greater task was supplying beer to troops overseas. In February 1941, as the result of private negotiations between the brewery and military canteen authorities, DB filled an order for 200,000 litres of Waitemata Ale in quart bottles – requiring an entire goods train of eighty trucks to move it from Auckland to Wellington for shipment to Egypt. When the consignment went on sale in April, it was so popular that rationing was necessary to ensure wider distribution. In response to criticism of sending beer, a Dunedin officer wrote to New Zealand:

> Some folks have evidently been complaining about New Zealand beer being supplied to the troops, saying that it might make them inefficient. Some people haven't much to occupy their minds if that is their idea. The lads will have their beer, and if they can't get decent New Zealand stuff they will just continue to drink the local stuff and I think most of them have enough sense not to overdo it, especially in the hot weather. In any case, I wish some of the armchair critics could have a day or two out here – they would sell their souls for a drink of beer after experiencing the heat.[9]

Nevertheless, official assurances had to be provided in June 1941 that the export of beer was not at the expense of food for the troops and was not being paid for from the National Patriotic

As well as supplying beer to the forces, New Zealand Breweries trucks were also commandeered for overseas service and sent to Egypt in 1940. Lion Technical Library

Fund. There were also practical difficulties in satisfying this thirsty market. By January 1942, at the beginning of a year in which 1.3 million litres were exported to Egypt, the perennial shortage of beer bottles in New Zealand was being severely accentuated by the fact that bottles sent overseas were not coming back for further use. By the middle of the year, there was also a shortage of cork and tin for crown bottle caps.[10] At least New Zealand became self-sufficient in barley by late 1942, after the government directed an increase in the price to encourage local production and reduce any need for imports.[11]

Inevitably, there was an undercurrent of prohibitionist protest at all of these developments. In its 1941 annual report the Alliance raged that:

> In the face of this mighty challenge, the enemy in our midst, the liquor traffic, continues its fifth column activities, blatantly breaking the law, inciting crimes, bespoiling youth, undermining national health and efficiency, and wantonly wasting the Nation's wealth, manpower and resources, in spite of the need for an all-in effort for an early peace, and complete victory.[12]

There were also complaints against the apparent provision of beer in camps to those men under twenty-one, the legal drinking age, and to the provision of a wet canteen for the Women's Auxiliary Air Force based at Levin. In the latter case, the minister of defence, echoing a common moral refrain that drinking was a male preserve, stated that permission had never been given and would not be.[13] Meanwhile, in October 1941 a remit to the Dominion Conference of the National Council of Women deplored the wastage of shipping space on beer for troops overseas when such space could have been used for valuable foodstuffs produced in New Zealand, to be sent to Britain to assist the war effort. At the beginning of 1942 the WCTU noted that conscription of young men was essential, but asked 'must they also be handed over to that bloated capitalist, the Brewer, to increase his dividends? . . . They do not get drunk in wet canteens, but they learn to drink. And so the canteen obeys the command of the "Brewer's Digest" and creates appetite. Many mothers fear death for their boys less than they dread the physical and moral ruin wrought by alcohol.'[14] Three months later it was claimed that pigs and poultry were either starving or being killed off for lack of food while enough barley and maize to feed millions of them, and enough sugar to give every consumer an extra ration, was being wasted on the production of beer.[15]

These dramatic claims did not gain traction during the first two years of the war, but that changed in early 1942. While it was never explicitly stated, it is evident that attitudes to alcohol intensified once Japan entered the war, especially following the fall of Singapore in February and the emergence of a more direct invasion threat to New Zealand. In early March, Canon R. G. Coats claimed that 'If enemy aircraft had come over Auckland between 5 and 7 p.m. on Saturday, there would have been some terrible casualties in the city because of the drunkenness that was on every hand. Many, perhaps hundreds, of valuable lives would have been lost in rescuing those who were utterly incapable of looking after themselves.' Although Wellington Police Superintendent Charles Lopdell insisted that 'In proportion to the facilities for getting drunk, I should say the evidence of drunkenness is not large', action was soon taken.[16] Following the 1941 move to reduce beer strength in Britain by 20 per cent and the recent decision of the Australian government to direct the reduction of brewery output by one-third,[17] Prime Minister Peter Fraser announced in April 1942 that the liquor trade would be controlled and said that even the trade acknowledged the need for reform. After meeting with a large non-denominational deputation of Whangārei women a few days later, Fraser observed that

Despite the suggestion that the soldier would miss his beer, DB was a major supplier of bottled beer to New Zealand forces overseas during the Second World War. DB Breweries

'Allegations of the effect of over-indulgence on the national morale and efficiency raise a most important issue. Anything inimical is as much an enemy to New Zealand as is an enemy in the field.'[18] While prohibition was not an option, as this had been consistently rejected at the polls, the government certainly needed to investigate an issue around efficiency and the abuse of alcohol and licensing laws.[19]

When Fraser presented the government's financial statement to the House of Representatives on 30 April 1942, he declared that New Zealand, and the rest of the British Commonwealth, was facing 'the most ominous threat in the history of our civilisation'.

> Crisis has succeeded crisis throughout this world-wide conflict. . . . To-day there is danger on every side, but the deterioration of the situation can have but one result – the calling forth of the best and most united exertions of every freedom-loving man and woman. With the successive advances of the Japanese the threat to the security of this country has never been more critical or near, nor have we ever before set ourselves a task of such magnitude as that which is now engaging our attention. Our entire national effort and, indeed, our whole economy must become diverted towards placing New Zealand on a complete war footing. There is no need to stress the fact that this war will not be won by the fighting Forces alone. It is in the workshops and in the industrial efforts of our people, in savings and sacrifices in the home – in short, by the wholehearted effort of each and every one of us – that we can hope to survive and emerge triumphant from the dangers that now beset us.[20]

As the circumstances justified obtaining additional revenue, the decision was made to increase taxation on beer, wine, spirits and tobacco, although the tax on cigarettes and tobacco would not apply to official military canteens as 'a well deserved concession to those in the fighting forces'. Beer, seemingly, did not enjoy this status. By these means the government estimated an additional £1 million in revenue for the year.[21]

The beer tax was an increase of 1 shilling per gallon on all worts that did not exceed a specific gravity of 1036 (perhaps 3.8 per cent alcohol by volume) and a further 1 penny per gallon for each unit of gravity above 1036. Conversely, there would be a deduction of 1 penny for each unit of gravity below 1036, down to and including 1027, where the duty would be 2 shillings 3 pence per gallon. The acting minister of customs, Arnold Nordmeyer, stated that brewers now had every incentive to brew at a lower gravity and these measures would likely reduce the alcoholic strength of beer by 25 per cent, as it had tended to be in the vicinity of 5 per cent alcohol by volume following an earlier reduction in strength in response to the Finance Act 1915. 'A specific gravity of 1,036 is the point at which the Government would desire beer to be brewed in New Zealand, particularly during the war period. If the brewers wish to brew a few points higher than 1,036, there is nothing to stop them doing so, and they may brew at 1,040, but only on payment of an additional 4d a gallon.' Given that output had increased from 50 to 86 million litres since 1935, Nordmeyer seemed to imply that breweries had sufficient profit to absorb the change. While the price of beer would now increase by perhaps 1 penny per glass, this would be offset by a decrease in the cost of raw materials. Therefore, any price increase would be subject to investigation by the Price Investigation Tribunal. But Nordmeyer also observed that if the desired beer strength was not obtained by taxation, consideration would be given to achieving it by other means, as the government had two objectives in proposing the measure. 'The first was to bring in a very considerable increase in revenue, and the second was at the same time to ensure a diminution in the alcoholic content of the beer brewed, and in consequence an increased sobriety.'[22]

In an extremely prescient assessment of things to come, former Prime Minister Gordon Coates argued that the changes in beer strength were in favour of the larger breweries that already made lower-strength lager or had the equipment to adapt, but 'the little struggling brewer – the small man – has to adjust the whole of his plant in order to take advantage of this "most favoured nation" clause. It is a fact that some brewers have all the equipment necessary to enable them to take advantage of the tax adjustment, while other well-established breweries may not be able to comply readily with the conditions.' Coates then asked whether the minister had investigated the matter and added that he understood that Nordmeyer was capable of judging the excellence of beer. 'There is nothing like having an honourable gentleman in charge of this matter who is able to assure patrons that they will not notice any difference, and that his proposal will work smoothly. If he can show that, he will make the case more acceptable.' As Nordmeyer was a former Presbyterian minister and well-known teetotaller, the point that he was not qualified to assure drinkers will have been lost on no one. He replied that while the changes may be of greater benefit to some breweries than others, the government was not in a position to differentiate.[23] Thereafter, aside from the musing of Harold Dickie, MP for Pātea, that there was no equity in levying £1 million on beer but nearly £2 million on tobacco merely because the trade was well organised and smokers were not, the debate played out along familiar prohibition and continuance lines and the proposal passed comfortably on 9 May.[24] The next day the *New Zealand Licensee* scoffed that prohibitionists had claimed excessive drunkenness despite counter arguments from the military and police.

> The average man has laughed and thought nothing would come of it. However, Cabinet, with its number of dry members, has listened to the loudmouthed ones, and now New Zealand is to be made a laughing stock, is to have the weakest beer in the world; and those who are the mainstay of this country whether in uniform or at the work bench are to be deprived of their favourite brews and are to be invited to drink a concoction at which they must look ashamed every time they raise their glass.

The risk now was that there would be a decline in worker productivity due to weak beer, and a decline in consumption would also lead to a decline in beer duty.[25] This feeling was exacerbated as the Price Investigation Tribunal also increased the maximum price of a 12-ounce glass to 7 pence – although not on the West Coast, where the Licensed Victuallers' Association, for reasons not made apparent, unanimously agreed to retain 6 pence, but for a 10-ounce glass.[26]

This was not the end of government intervention. Changes were made to hotel hours in mid-June. Opening time was moved from 9 a.m. to 10 a.m., and all had to close between 2 p.m. and 4 p.m. on Saturdays, presumably to stop all-day drinking.[27] On 14 August, in a clear contradiction of his earlier suggestion that brewers could brew above 1036 if they were willing to pay for it, Nordmeyer followed through on his earlier warning when the War Cabinet decreed that no beer with a specific gravity above 1036 was to be brewed without the permission of the minister of customs. While most had 'loyally adhered' to the earlier measure, one or two, especially the Paeroa Brewery, had openly advertised that they would supply beer at the old strength. Therefore, Nordmeyer stated that the new measure was taken so no brewer could have an advantage over others. It was acceptable for the purposes of maintaining yeast to brew above 1036 and then blend with beer of a lower specific gravity, as long as the final product was below 1036. It was also permissible to brew stout, which was usually above 1036, as long as the average production per month in each brewery did not exceed that brewed between January and June 1942.[28] As available 1936 figures had stout at 7.8 per cent of total beer manufactured, and this was likely reduced amid

In spite of the need for wartime efficiency, the lowering of the beer strength in 1942 caused some concern for drinkers. *New Zealand Herald*, 20 May 1942, 6

wartime constraints, this would not be a significant component of the total beer supply. Strangely, none of the wartime regulations made reference to home brew, or the two companies offering 'beer making concentrate' for full-strength ale and stout via mail order to anywhere in the country.[29]

Nor was there any sign of vocal brewers in these debates. Indeed, Conrad Bollinger, author of the most influential critique of New Zealand liquor laws, suggests that brewers actively promoted reduction of beer strength during 1942 as a strategy to deflect prohibition.[30] This seems doubtful given that prohibition, with 29 per cent support in 1938, was hardly an immediate threat and many brewers struggled to adapt to the new regime. On 1 September 1942 permission was given to the Captain Cook Brewery to continue brewing lager above 1036 until 31 October, subsequently extended to 31 December, as NZB contended that due to the special nature of its lager yeast they had had considerable difficulty in reducing the gravity from 1047 to 1040, let alone 1036. Permission was not given to any other brewer to brew above 1036 except for the stout provision.[31] Evidence soon mounted, however, that some brewers were using significant amounts of sugar, a scarce wartime commodity, in the brewing process, adding it after brews had been recorded in the brewing book to increase alcoholic strength. There was also evidence of publicans adding sugar for the same reason. Consequently, a further addition to the Emergency Regulation in May 1943 restricted the use of sugar to 3 pounds per 40 gallons, stipulated that brewers were not to add any further sugar after the brew had been recorded, and confirmed that nobody else was to add sugar after beer had been delivered from the brewery.[32] Customs also displayed some rare sympathy for brewers. A technical report following a thorough

In the face of prohibitionist hostility, New Zealand Breweries took every opportunity to emphasise its contribution to the war effort. *Timaru Herald*, 27 March 1942, 6

check of breweries during 1943 highlighted difficulties in accurate and consistent recording of materials used and beer produced.

> While science is applied to a considerable extent by some brewers, the fact remains that brewing is still an art. The results achieved depend to a great degree on the judgement of the individual brewer. His aim in general is to obtain in as economical a manner as possible, and to the extent that legislative restrictions will permit, a product that will please the consumer, and he operates on his materials in the way which he thinks will best achieve this end.

It was impossible to state that a particular amount of malt would make a particular amount of wort as this could vary depending on the type and quality of the malt or the season.[33] To this end, during 1945 a Dunedin brewer stressed to Customs that irregular yields of worts in the brewery were caused by such things as differences in types of malt and possible errors in the weighing of malt. 'Further, our materials and practices are adapted to produce a satisfactory product, not with the object of complying with a set of figures laid down by the Customs Department. . . . Hence we feel that we must protest against any growth of the idea that the Customs Department can control the action of the breweries beyond the legitimate control necessary to ensure the honest payment of Customs Duty.' Customs replied that they had no intention to control breweries – merely to uphold the law and collect correct amounts of duty.[34] A 1947 Customs report indicated that the main concern during the war was with supervision of country breweries, where it was easier to conceal brewing practices, and inspections were frequently done by postmasters rather than regular customs inspectors. A former English brewer informed the Royal Commission on Licensing that if the Customs Department had employed a competent brewer they would have saved themselves a lot of difficulty.[35]

Even for those who could easily brew to the lower strength required by government, questions remained as to whether the result was especially drinkable. Speight's, which had long enjoyed a reputation for producing strong beer and had invested a considerable sum completely rebuilding the City Brewery in 1940, immediately struggled to bring its standard offerings down from 1047 (about 5 per cent alcohol by volume) to 1036. The brewers had to proportionally reduce the hop rate to maintain a balanced flavour, but the reduction of hops and alcohol reduced the keeping quality of the beer, which had enabled widespread distribution throughout New Zealand. Speight's effectively became a local Otago and Southland brewery again, with output declining from 957 brews in 1941 to 236 in 1943. Insofar as Speight's did reach Auckland again by the end of the war, it was claimed that local bottlers had to add sugar as the beer had gone flat during its long journey from Dunedin.[36] In 1945 Government Analyst K. M. Griffin told the Royal Commission that it was 'a very difficult technical proposition to make a decent beer at that strength'. While he was evasive on whether beer at 1036 deteriorated more rapidly, he conceded that such beer could not have much flavour.[37] Others stated that beer of lower gravity needed to be kept cool in order to last, while Gilbert Lawrence, a consultant industrial chemist, was of opinion that a gravity of at least 1044 was necessary for good beer. At 1036, the alcohol content was too low for all beer needing to be transported some distance or for those hotels with a slower trade where it was required to last longer. Moreover, the average nitrogen content of New Zealand barley was not entirely suitable for all-malt beers at lower strength, and the relatively small scale of malting operations in New Zealand did not lend itself to the selection and blending of malts in a manner that could address the problem.[38] The solutions to these problems ultimately resided with the larger breweries during the brewing innovations of the 1950s, with many smaller breweries unable to adapt, in exactly the manner Gordon Coates had predicted. One exception, however, was Morley's in Westport, which found a renewed market for its stout on the West Coast and in the upper South Island without competition from Speight's.[39]

Irrespective of quality, any notion that weak beer would reduce wartime consumption very quickly disappeared. Production of 85 million litres for the 1940/41 financial year increased to 95 million during 1942/43 and 111 million by mid-1945. Per-capita consumption, which had reached 51 litres by 1940 after the lull of the Depression, had reached 76 litres by 1946 when figures were available again – a level that had not been attained since the 1860s. The only possible explanations are that drinkers were consuming significantly more but significantly weaker beer to achieve the desired alcoholic effect, or that the lower-strength beers were easier to drink than the previous heavier beers, and took longer to reach the threshold where one decided they had had enough. Either way, although much had been said over the years about the poor quality of New Zealand hotels, it is only from 1945 that public discussion emerges about a tendency to 'swill' as much as possible in the shortest possible time, especially in the hour between the end of work and closing time at six o'clock.[40] The full implications of this consumer response will be discussed shortly.

*

If the beer strength measures of 1942 had made a difficult year for the trade, 1943 offered promise. On 25 September the electors of Invercargill decided by a majority of forty-five votes to restore licences and end the dry district after thirty-eight years. Crucial to the outcome, but not delivered for five weeks after election day, were the votes of soldiers overseas, 81.5 per cent of whom favoured restoration. Following calls for the voting papers to be returned to New Zealand for recounting, it was discovered that they had been destroyed in the Middle East. While this prompted accusations of skulduggery, the vote of 1919 indicates that it was highly likely soldiers would vote as they did.[41] The immediate problem was that Invercargill needed to elect new licensing committees, with further potential for contention, and to establish at least fifteen new hotels, given a minimum ratio of one hotel per 1000 people, but with few suitable buildings for them and little prospect of building new ones amid wartime restrictions. The government had no desire to become involved in such a prickly licensing issue, but also understood the special circumstances of restoration. In this setting various proposals for alternative licensing arrangements began circulating, including reference to the 'Carlisle experiment' for state management that had operated in parts of Britain since 1916. Here the ethos was 'disinterested management', where the managers of hotels had no personal incentive to sell alcohol and would therefore not be drawn into the 'abuses' that prevailed elsewhere. Under the influence of Minister of Justice Rex Mason, the government developed a proposal for control via a community trust that would hold a monopoly on the operation of licensed premises, with all profits from these being reinvested in community activities. The Invercargill Licensing Trust Act was passed on 29 March 1944 and came into force on 17 April. Alcohol was available for sale in Invercargill from 1 July, although initially in temporary establishments.[42]

The Trust was seen by many as a fresh opportunity to overhaul the customary hotel drinking environment. Following a visit to Invercargill in July 1944, Minister of Works Bob Semple declared that he was very impressed with the new arrangements and particularly the determination of the Trust to discourage 'shouting'. 'If I want to buy a pair of socks I don't have to buy half a dozen other fellows a pair of socks, and they feel under no obligation to buy me any either.' In other words, a man who wanted a glass of ale should be able to buy it for himself. The minister was also very interested in the Brown Owl, a local café that had been redeveloped as a licensed restaurant where a man could take his wife and have a glass of ale while she had a milkshake or a cup of tea. 'That is a big improvement on standing like a lot of calves round a milk trough trying to elbow one another out. In this way you can cut out perpendicular drinking.' Indeed, the Trust was very determined to discourage 'perpendicular' in favour of 'sit-down' drinking in the belief that the latter produced less consumption. Returning to older arguments against tied houses, Semple added that there was no incentive for managers to sell liquor after hours or serve intoxicated patrons, whereas high rents and goodwill had caused many publicans elsewhere to ignore the law to make a living. Finally, he stressed that as voters had comprehensively rejected prohibition, the only sensible approach was to ensure that the sale of alcohol was conducted in a civilised fashion. That could only be done by cutting out the incentive to abuse the law.[43]

The corollary to the Invercargill experiment was to bring back into focus other arguments about licensing laws and drinking conditions in New Zealand that had lapsed during the war. In 1939 Bert Dallard, under-secretary for justice, questioned the logic of colonial licensing legislation still in force and indicated that a comprehensive review was required. However, the plan was thwarted by the outbreak of war. But by 1945 it was clear that significant population

TOP: Soldiers of the 28th Māori Battalion at El Amariya, Egypt, taking time off for a beer after the evacuation of Crete in May 1941. War History Collection, Alexander Turnbull Library, DA-01171

BOTTOM: A serviceman and his girlfriend share a bottle of beer during VJ Day celebrations in Queen Street, Auckland, August 1945. Auckland Libraries Heritage Collections, p20062coll1_38970_large

increase and other changes during the twentieth century, including northward drift and the growth of cities, was starkly exposing the effective freeze on new licences that had prevailed since the 1890s. The proportion of licensed hotels to population for New Zealand as a whole had declined from 1 to 357 in 1881 to 1 to 583 in 1905, 1 to 1182 in 1925, and 1 to 1542 in 1945. The distribution was now entirely skewed. Whereas greater Auckland had 87 hotels spread over 8 electorates with a combined population of 209,000 in 1945 (1 to 2406), the Westland and Buller electorates combined had 132 hotels for a population of just over 34,000 (1 to 260). Hokitika had 1 to 195 while Greymouth had a comparatively sparse 1 to 403. Hamilton was in the extreme at 1 to 4396. By comparison, out of the Australian states that had also pursued a mix of early closing and other licensing restrictions, only New South Wales had a similar ratio to New Zealand during the early 1930s, while Queensland was the lowest at 1 to 696.[44] Moreover, and despite the scramble for hotels between NZB and DB during the 1930s that certainly brought improvements to some, the overall hotel stock was widely regarded as being in a poor state due to lack of investment and lack of incentive to improve. Wartime emergency regulations had relaxed some of the standards required of licensees, especially in the provision of accommodation. Capital issues controls from 1940, designed to ensure that investment for growth was strategically managed, did not favour investment in hotels.[45]

Shortly before the 1943 election the Labour Government promised a Royal Commission on Licensing if re-elected. This was appointed from January 1945 under the chairmanship of Justice David Smith who, notwithstanding that he had served on the national executive of the Alliance during the early 1920s, was widely regarded as impartial and extremely competent. From March 1945 the Commission travelled throughout the country, sitting almost continuously for eighteen months, far beyond the original estimate of seven months. It heard witnesses and digested reports from all parts of the trade, its opponents and other interested parties. Ever present with the eight commissioners was a lawyer for the Crown, two King's Counsel representing the trade, and two lawyers for the Alliance. The proceedings eventually amounted to 7824 pages in 52 volumes and the final report was 450 pages.[46]

From the outset the trade was on the defensive. While there was much criticism of the licensing laws, there was equal condemnation of the way brewers and publicans had behaved within them and their failure to initiate improvement from their own considerable profits. On the opening day of proceedings, W. H. Woodward, SM, chairman of the New Plymouth, Stratford and Egmont licensing committees, declared:

> The present laws are the result of a battle between greed and fanaticism in which the interests of ordinary sensible citizens have been ignored. The outcome is a system in which citizens drink cheek by jowl like pigs at a trough what they are given instead of what they may want and, like pigs, gulp down more than they need of it while they can get it, and for the privilege of doing so, pay many times the cost of the hogwash they swallow. The gods must surely laugh at such incredible folly and the men returning from overseas, who have had their eyes opened, will not be patient under a law that threatens to make guzzling a national habit and furtive drinking a fashionable pastime of youth in New Zealand. Drinking, under our law and because of it, instead of being a pleasant and respectable aid to social intercourse, has become the subject of idiotic mirth and censorious reproach.[47]

Among the charges specifically directed at the trade was strong criticism of 'vertical drinking', as chairs and tables had gradually disappeared from bars to accommodate increasing demand, and also of the failure to provide snacks, let alone more substantial food, for drinkers.[48] The familiar attack on tied houses also surfaced. It was reported that 26 per cent of all hotels were in the hands of NZB or one of its associated companies as owner, mortgagee or lessee, and 58 per cent of all hotels were controlled by breweries or wine and spirit companies. It was also suggested that many others were controlled by means not publicly discoverable.[49] Moreover, the pressure to turn a profit from these establishments was apparently resulting in some extremely unpleasant practices. In October 1945 a barman gave evidence to the Commission of widespread recycling of beer slops whereby the spillage and unconsumed beer was retained for resale. Some apparently put back as much as 90 litres after closing time. Other barmen who were supposed to give evidence on the topic were reportedly intimidated into not appearing. Publicans naturally denied the claims, but a vigorous pamphlet on the matter, seemingly issued under the auspices of the New Zealand Communist Party, noted: 'The liquor trade is strong and they will stop at nothing that affects their interests, and they consider themselves so powerful that they are out of the reach of the law. And all for the sake of profits, and still more profits, by reselling the waste dregs.'[50] In his opening and closing addresses to the Commission, P. B. Cooke KC for the National Council criticised the lack of evidence for many claims made, and insisted on the need for a very clear distinction between matters that were caused by the trade and those due to defects in the law. Further, the trade collectively should not be saddled with the whole responsibility for those few, present in every calling, who fell short of the accepted standard. On the specific matter of brewery profits, Cooke argued that the public view was simply uninformed as they failed to realise that much of the price of a handle was taxation, not brewery profit. Since 1939 the price of beer to the public had increased by less than the amount of increased duties and taxes.[51]

When the Commission issued its final report in August 1946, some of its recommendations were considered decidedly radical. Among other things, it called for the extension of trust control if licensing districts favoured it; standard measures and prices for glasses of beer; bars to have more seating and also provide snacks and meals; a trial of licensed restaurants; the issuing of temporary licences for 'reputable' dance halls or cabarets as required; a system using Condy's Crystals in drip trays to prevent the reuse of slops; removal of all purchase restrictions against Māori – especially in the King Country; and hours of sale in hotels to be 10 a.m.–2 p.m., 4–6 p.m. and 8–10 p.m. to avoid the emerging 'drinking rush'. The question of distribution of licences and the apparently poor quality of many hotels was to be addressed through the establishment of a Licensing Control Commission with powers to redistribute licences by cancelling those deemed surplus to requirements and reissuing them elsewhere, and eventually to increase the number of licences if population trends justified. This body was established in 1949, although it did not gain full momentum until later in the 1950s.[52] There would be a mixed fate for others of the Commission's recommendations.

The most radical proposal to emerge from the Royal Commission was to nationalise the breweries into a public corporation, although it was stated that any such decision should be put to the voters. Paul Christoffel notes that it is not entirely clear why the Commission went down this path. Only a few witnesses advocated nationalisation and certainly none from the trade or even the Alliance. Support for state purchase and control at the national licensing polls had only once passed 10 per cent, although it was to spike to 20 per cent in 1946 following the Royal Commission recommendation. It seems that the accumulated evidence of acute competition for scarce hotel licences, and the related claim that publicans were

TOP: A DB Lager advertisement from the early 1950s. DB Breweries

MIDDLE & BOTTOM: Although prohibition was no longer a threat by 1950, New Zealand Breweries was determined to emphasise the obligations of publicans to ensure responsible drinking. *New Zealand Licensee*, August 1950 neglecting the accommodation element of hotels in favour of increasing liquor sales by means legal and otherwise, shaped a belief that the only way to end this unhealthy competition was via a state brewery monopoly.[53]

Not surprisingly, the proposal was condemned by brewers. Dominion Breweries' Henry Kelliher described it as 'impracticable, cumbersome, and extreme' and 'State bureaucracy on a grand scale'. The elimination of competition in the brewing and distribution of beer would lead to a decline in quality and service, to a level a long way below the standard set by private enterprise. Kelliher felt that while proposals for the redistribution of licences, the alteration of opening hours and the extension of sales to restaurants and nightclubs were all constructive measures in accord with overseas trends, in other respects the trade was still labouring under wartime regulations and restrictions a year after the conflict had ended.[54] Others were entirely bemused by the claim from the Commission that the cost of acquiring all breweries in New Zealand would be less than £2 million, based on evidence from balance sheets. Some suggested that brewery assets, conservatively valued on balance sheets, were worth at least double that figure. Moreover, previous state acquisitions had been based either on the market value of shares or on the value of the assets, and shareholders were paid out at the rate of Stock Exchange quotations. The shares of NZB alone were worth almost £5 million.[55] However, as Conrad Bollinger points out, the enduring connections between the likes of NZB's Ernest Davis and the Labour Government made it very unlikely that state ownership would ever be a viable option.[56]

*

If the might of the Royal Commission could be tamed with the help of friends in high places, the breweries initially had less success in containing a seemingly local dispute with interesting implications for the balance of power between producer and consumer. On 28 September 1947, a meeting of Wallsend State miners decided that if the price of beer was increased from 6 pence to 7 pence a glass as proposed for the West Coast, they would boycott hotels from 1 October. On 30 September a meeting of publicans in Greymouth unanimously affirmed the decision to increase the price. Having maintained it at 6 pence rather than 7 pence a glass since 1942, albeit with 10-ounce glasses, they could no longer continue to sell beer at prices below those elsewhere in New Zealand. At the same time, the West Coast Trades and Labour Council, representing thousands of workers, unanimously agreed to boycott. Given the profits being made by breweries, and the large sums paid for goodwill on the sale of West Coast hotels, they could see no justification for a price increase. T. Pratt, secretary of the Council, observed, 'We do not propose to be exploited. We consider that both the quality and quantity of the beer, as compared with that of 1940, represent a price increase to the worker of more than 100 per cent, while in that time the profits of the trade have increased.'[57]

During early October the boycott spread beyond its mining base to include watersiders, dredge and timber workers, railwaymen and apparently some non-unionists. Only the Westland Hotel Workers' Union passed a resolution condemning the boycott and asking the Trades and Labour Council to lift it and instead seek an increase in wages for hotel workers. The one Greymouth hotel that remained immune to the boycott by charging 6 pence a glass ran out of beer during the first weekend and was closed temporarily. A member of the public provided a barrel of beer, which sufficed until further supplies were received from Speight's.[58] While other publicans held firm, it was reported on 11 October that a representative was heading to

Wellington to seek support from 'a Dominion organisation which has control of a fund to which local publicans contribute' – presumably the Trade Defence Fund.[59] A few days later a Hokitika publican challenged Pratt to a public debate on the beer boycott, with the loser to contribute £100 to the Food for Britain campaign. The Trades Council was asked to prove that beer could be sold economically at 6 pence, taking into consideration the greatly increased cost of running hotels. The invitation was not accepted.[60]

By mid-October tensions were rising. A Dillmanstown sawmill was left idle for a day when men refused to work with a colleague who ignored the boycott. A fight broke out at a Kumara Junction hotel when a group of miners demanded 6-pence beer, and one was injured when struck with a mallet by the proprietor. A three-hour meeting on 19 October, including an offer from publicans to supply 12- rather than 10-ounce glasses for 7 pence, revealed no prospect of an immediate settlement.[61] At the end of the month passengers on an excursion train from Christchurch to Greymouth were warned by two men going from carriage to carriage that all but one hotel in the district had been declared 'black'. 'You are asked to support this boycott by the workers of the West Coast by refusing to drink beer during the holiday weekend for which a charge in excess of 6d is made.'[62] Greymouth and surrounds were also plastered with posters stating '7d beer is black'. But it seems that many tourists did not abide by the boycott and hotels reported good trade.[63]

By mid-November, there were signs that the boycott was fragmenting, with unionists not holding to collective decisions and few hotels complaining about slow trade.[64] On 24 November men at the Dobson and Wallsend mines refused to work in response to the presence of men who had breached the boycott. When interviewed, one of the 'offenders' explained:

> A meeting was called at the bathhouse, when I was asked whether I had broken the beer boycott, I replied yes. I was then asked for an explanation, and said I would not mind supporting the boycott if I thought it was genuine; but from what I had seen and had been told I considered the boycott had come to a hypocritical stage, with half the members of the union going into the country to do their drinking behind the officials' back while I was doing my drinking in front of them.[65]

In response, *The Press* editorialised that although the rights and wrongs of the boycott argument were of little interest to the general public, they should be concerned 'when the actions of union members, or anyone else, disrupt industry and hamper the production of much needed materials. . . . The unions are entitled to fight for cheaper beer, and for cheaper commodities in general; and they will have the sympathy and support of the consuming public in their efforts so long as they follow constitutional methods.'[66] A day later the Council of the New Zealand Federation of Labour meeting in Wellington adopted a resolution endorsing the boycott. The New Zealand Waterside Workers' Union then asked the New Zealand Transport Workers' Federation to take action to prevent the transport of beer supplies from an unnamed Dunedin brewery, presumably Speight's, to all parts of the country until West Coast hotels reverted to 6 pence. At some point this evolved into a demand for the brewery to send adequate supplies to the few West Coast 6-pence hotels or its beer would be declared 'black' to all districts – an interesting approach given the general consensus that Speight's did not travel well. 'Unless the demand is complied with it will be an unfortunate Christmas as far as beer supplies in New Zealand are concerned. Our desire is to see that enough beer reaches the West Coast sixpenny hotels from Dunedin. Transport workers undoubtedly should support West Coast unionists in their fight.'[67]

The brewery eventually agreed to guarantee supplies to 6-pence hotels. Westland Breweries had apparently agreed to boycott 7-pence establishments.[68]

By the end of the year, amid further workplace disruption as a number of boycott breakers were blacklisted, it was clear that the unity of the publicans was also faltering. Two Blackball hotels reverted to 6 pence in mid-December and a number of others, particularly in small mining towns where the consequences of the boycott had been especially severe, were allowed to follow suit in mid-January.[69] In late December their position was further eroded when groups in Brunner, Greymouth and Rūnanga secured charters to establish working men's clubs as alternative venues to drink and socialise.[70] Notwithstanding a statement on 2 January 1948 of 'definite and unbounded loyalty' to the decision to retain 7 pence, a special meeting of the West Coast Licensed Victuallers' Association the following day acknowledged that at least thirteen of twenty-two Greymouth hotels were selling beer at 6 pence. A week later it was agreed that all would return to 6 pence, although there would no longer be any shouting of free drinks by the publican nor free bar lunches as had prevailed before.[71]

No doubt the breweries were concerned at such a successful display of consumer resistance. Ultimately, though, this could only work on the West Coast with a small and highly unionised population. Even allowing for the leverage exerted by national unions, there was no chance of such a boycott succeeding in a large urban centre. Victory for West Coast drinkers was only a blip for the breweries, as almost immediately it became apparent that many of the recommendations of the Royal Commission were to be ignored in a decidedly watered-down Licensing Amendment Act. On similar principles to Invercargill, further licensing trusts were established in Masterton in 1947, Ashburton in 1948, and Geraldine in 1950. Five more followed during the 1950s with a peak of eighteen by 1975.[72] All purchase and alcohol restrictions against Māori were also removed. But little else survived of the reform proposals.

The recommendation to nationalise the breweries was the first casualty. Speaking to the Licensing Bill in Parliament, Minister of Justice Rex Mason suggested that popular sentiment would not be in favour of taking over breweries.

> As far as the government is concerned, it does not feel that the benefit to be gained would be commensurate with the expenditure of energy and effort that would be involved in making the change. . . . I feel that there would be almost unanimous opinion that the turmoil and disturbance due to the discussion of the matter would not be warranted by the slender chance there would be of any positive or other result eventuating from any such course.[73]

Nor was there concerted action to provide seating, food or standard measures in bars. To the contrary, the situation would become worse during the 1950s and the slops problem certainly did not disappear. Licences for restaurants and entertainment venues were also rejected, while six o'clock closing was to be the subject of a referendum in March 1949. It was evident, even allowing for the decline of prohibition sentiment, that alcohol remained such a sensitive subject that politicians, especially within a two-party system, were reluctant to innovate to any degree that could antagonise swing voters and potentially lose marginal seats. Hence the status quo and cycle of triennial licensing polls endured.[74]

The status quo also prevailed despite the final clause of the Licensing Act, which revoked all of the wartime emergency regulations including the control on the strength of beer. While it is doubtful whether most brewers supported the reduction in 1942, it did not take long for some of them to appreciate its virtues. Certainly the report of the Royal Commission recommended the strength be increased to 1040 and that all-malt beer should be encouraged rather than use of sugar, which had long generated complaints about the flavour it imparted.[75] The following year, one correspondent described New Zealand beer as 'the present watery apology' and the lawyer for a barman prosecuted in Christchurch for serving a person under twenty-one stated, 'The beer, which is not very strong today, was liberally supplied with lemonade, so not much harm was done to the youth.'[76]

In mid-1947 the government moved to counter an increasing tendency to brew below 1036, and therefore to pay less duty, by abolishing the 'discount' of 1 penny for each unit of gravity below that level. Consequently, there was no incentive to brew below 1036. NZB immediately advertised that their brewing books for the past twenty months had been subjected to formal audit which confirmed 'That the strength of beer on which Beer Duty has been paid over that period on each and every brew has at no time been less than the maximum Specific Gravity of 1036 permitted by law and for the whole of the above period Excise Duty at 3/- per gallon has been paid to H.M. Customs.'[77] This did not satisfy T. F. O'Byrne, who argued in the Legislative Council in late 1948 that the Licensing Bill did not go far enough in merely revoking the emergency regulations and that steps should be taken to return beer to pre-war strength. He contended that timber workers who came out of the bush did not drink beer for the sake of drinking. They drank two beers to quench their thirst but also wanted to get a 'kick' out of it. There was plenty of water in the bush and these men did not want to drink beer that was 99.5 per cent water.[78] On the other hand, a Wellington publican argued in early 1949 that young

While the trade did not campaign to end six o'clock closing in the 1949 referendum, they certainly maintained vigorous support for continuance in widespread newspaper campaigns. *The Press*, 9 November 1949, 9

ABOVE: In the March 1949 referendum, the New Zealand Alliance campaigned vigorously for the retention of six o'clock closing. It helped that the trade saw no need to oppose them. Alexander Turnbull Library, Eph-C-ALCOHOL-Hours-1948-01

RIGHT: The New Zealand Alliance campaign to retain six o'clock closing included a strong appeal to home and family amid the renewed post-war emphasis on domesticity. Alexander Turnbull Library, Eph-C-ALCOHOL-Hours-1948-03

men increasingly preferred lager, which had not been an option twenty years ago. 'Many drinkers today had never known anything else, and old drinkers had acquired a taste for the light beer, so it was possible that even if the strength of beer were increased, drinkers would not care for it.'[79]

For the breweries, the issue was straightforward. While they could now brew above 1036 if they wished, there was little incentive to do so, as the wartime stipulation of an extra 1 penny of duty for every unit above 1036 remained in place. Speaking at the annual general meeting of NZB in June 1949, Chairman of Directors Sir William Perry said that it was impossible to increase the strength of beer unless either the public bore the increased cost or the government reduced the duty. By his calculation, the purchase price of a 10-ounce glass of beer that sold for 6 pence comprised 3 pence in duty and sales tax, 2 pence to the retailer for freight, wages, rent, interest, overheads and taxation, and 1 penny to the brewer for materials, wages, overhead, taxation and shareholders' dividends. He claimed that aside from duty and other taxes, the wholesale price of beer had not been increased for thirty-five years. Moreover, the cost of hops and sugar had also increased significantly in recent years and brewers were also having to import barley to meet a shortfall.[80] Given that NZB's net profit for the year, its lowest since 1946, was £140,000 ($12.2 million in 2022) and would be £198,000 by 1952 ($13.4 million), consumers were inclined to muse darkly about brewery profits and the ability to absorb the cost of stronger beer. Brewing above 1036 became something of a novelty. In 1950 Waikato Brewery launched 'Waikato XXXX pre-war strength bitter ale, produced for enjoyment of those who prefer beverages of greater strength and fuller hop flavour'. However, this was something of an exception to the lower-strength norm.[81]

NZB's plea of poverty as a defence for weak beer must also be judged alongside the response of the trade during the March 1949 referendum on six o'clock closing. Although the National Council seemed to support altered trading hours during the Royal Commission, they scarcely pushed the point as the referendum approached. It was evident that the Council was not campaigning. As Percy Coyle explained, 'The licensed trade did not participate actively in the campaign for the extension of hours, preferring to leave it to the popular vote, which in a democratic country must be the only way to decide such an issue.'[82] But the energy with which the trade campaigned for continuance later in the year and against prohibition at previous national licensing polls indicates that motives other than respect for democracy were at play. Many publicans, and no doubt those who tied many of them, were keen to do most of their business in a short period, as this reduced the need to find staff and pay penalty rates in a tight labour market. There was also little incentive or need to invest in improvements to hotels beyond minimum standards required under the law. Given one hour to drink after work, most tended not to dwell on the aesthetics of their surroundings and those who were more likely to do so, women, had long been almost entirely excluded from public bars. Some also believed that such things as the shortage of licences and uneven distribution of hotels, rather than the closing hour specifically, were the cause of problems such as overcrowding. Longer hours would merely exacerbate already bad drinking habits.[83] Finally, as Paul Christoffel argues, renewed emphasis on the 'cult of domesticity' also played a part in the immediate post-war years, with an emphasis on home and family rather than evening drinking as a key to the return to normality and prosperity. There was only a 56 per cent turnout for the referendum, perhaps a sign in itself that the outcome was inevitable, and it was rejected in all eighty electorates with 76 per cent voting to retain six o'clock closing.[84]

*

Having confirmed an environment where weak beer was to be consumed in a confined period with a minimum of amenities, it was inevitable that those who could satisfy these requirements most efficiently would thrive. Information collected by the Royal Commission for 1944 indicates that the process of brewery concentration was already well under way, with powerful large breweries dominating in the main centres and smaller establishments only surviving in regional markets.

Table 8.1: Change in the Brewing Industry, 1936–1971

	Number of breweries	Beer produced per year (L)	Average beer production per brewery per year (L)	Employees per brewery	Population per brewery
1936	44	52,282,286	1,188,233	19.7	33,768
1946	38	124,445,728	3,274,887	38.6	44,709
1951	35	160,906,010	4,597,314	37.6	55,156
1956	27	211,256,365	7,824,309	50.2	80,107
1961	17	236,802,958	13,929,585	73.0	141,389
1966	18	284,729,618	15,818,312	80.3	147,739
1971	15	336,881,330	22,458,755	99.1	189,733

Source: New Zealand Census, various editions.

Of the 115 million litres produced in that year by thirty-nine active breweries, 67.7 million or 59.2 per cent was in the hands of NZB, with another 18.4 per cent from DB's single Waitemata Brewery, which also had the largest output at 21 million litres. For NZB the Speight's brewery produced over 18 million litres, with Lion and Captain Cook producing 14 million and 11.3 million, respectively. Of the 'independent' breweries, Ballins in Christchurch produced 6.3 million litres, while Tui was the only other beyond 2 million. Ten breweries produced less than half a million litres, with the smallest being the very part-time Stewart's Brewery in Greymouth, which reported 3000 litres for 1944. NZB stated that one-third of its output was bottled beer, with plans to increase this to half in coming years. Even then, more than 90 per cent of its product was sold through hotels. Tui on the other hand sold less than 5 per cent through hotels. While some of the smaller regional breweries had access to local hotels, most also did a good proportion of their trade direct from the brewery or through other retail outlets. Waikiwi sold only 500 of its 195,000 litres through hotels, while Gore and Stewart's sold none.[85]

That the average production per brewery in 1971, at a point close to peak per-capita consumption, was greater than that of the single largest producer in 1946 (Table 8.1) is testimony to a dramatic transformation of production in a fairly short time and to an almost total duopoly by NZB and DB as they took advantage of economies of scale and favourable legislation. By 1971, only Leopard and Waikato Brewery were in any sense separate firms, and both were half-owned by NZB.

Much as they complained about legislation and taxation, NZB and DB were better able than the smaller breweries to absorb price controls and duty increases. Indeed, the Price Investigation Tribunal, which endured until 1961, tended to shape its decisions around models of cost presented by the larger breweries. Moreover, the Tribunal did not necessarily allow price increases in line with inflation, but expected breweries to absorb some costs with greater efficiency. This was far easier for large breweries with more efficient production and the capital to pursue technical innovation. This point was very apparent when the Royal Commission had discussed 'wastage', the allowance made by Customs for the difference between wort prepared for fermentation, on which duty was paid, and the amount that was finally available as saleable beer. The allowance was 10 per cent to take into account evaporation, overflow, spillage, losses while bottling, other inefficiencies, and consumption

Open fermentation gyles at New Zealand Breweries in 1953. Within five years the bulk of beer was being produced in closed fermenters as the two major breweries adopted continuous fermentation. Lion Technical Library

at the brewery. By 1942 DB had apparently reduced their wastage to 2.5 per cent and Speight's to under 6 per cent, meaning that between them they were selling perhaps 1.8 million litres of beer per annum on which they did not pay duty. On the other hand, wastage for the Gore Brewery was estimated at 21.6 per cent, and the average for sixteen smaller breweries was 13 per cent, meaning that duty was being paid on wort that never became beer for sale.[86] Referring in 1958 to the gradual disappearance of the 'small family brewery', Sir William Perry observed, 'It is only, however, by centralisation and the use of modern equipment and methods that costs can be kept down. This is particularly important in an industry which has to bear on its back the colossal task of collecting such a large contribution to the Consolidated Fund.'[87] He also rightly pointed out that the trend to fewer but larger breweries was scarcely unique to New Zealand. The number in the United States had fallen from about 750 to 230 in twenty-five years, while the number in England had dropped from 492 in 1939 to 320 by the late 1940s.[88] The tax and price regime also provided no scope for small breweries to survive by such means as differentiating themselves with premium products. Nor, unlike NZB in particular, was their business reinforced by profits from a network of hotels. NZB owned

ABOVE: Keg racking at New Zealand Breweries in 1960. Metal kegs replaced wooden barrels during the 1950s. Lion Technical Library

LEFT: The New Zealand Breweries Central Laboratory, Wellington, 1960. While DB won renown for pioneering continuous fermentation during the 1950s, NZB also made significant technical refinements to brewing. Lion Technical Library or had a financial interest in 202 hotels, and many more were controlled by their associated companies Hancock & Co. and Campbell & Ehrenfried. Ballins had interests in sixty-one, DB fifty and Westland Breweries thirty-nine. Otherwise, only Macarthy's with two and Tui with one had any hotel connection by the 1940s.[89]

From NZB's perspective, reduction of breweries and centralisation were necessary as the building and material restrictions of the war years and beyond made it difficult to upgrade breweries to meet the technical challenges of brewing an ever-increasing volume of beer at 1036.[90] In the first instance, NZB consolidated its own operations and became much more of a unified business rather than a group of linked breweries. In Christchurch, Manning's was closed in 1940 and Ward's in 1955. In Dunedin, McGavin's and Strachan's were closed in 1951 and all brewing shifted to Speight's. The Gisborne Brewery was closed in 1966. Takeovers and subsequent closures of other breweries also proceeded apace during the 1950s. Among them were Southland, Taihape's Cascade and Whanganui's Western in 1955; the Dunedin Brewery in 1958; and Gore in 1959. While DB was generally not acquisitive until the late 1960s, it did purchase Napier's Sunshine Brewery around 1950. Elsewhere, Dodson's and Harley's had merged to form Nelson Breweries from February 1945; the Marlborough Brewery succumbed to the last of its many fires in 1946 and was not rebuilt; and Westland Breweries closed the Phoenix Brewery at Reefton in 1960. Others simply fell victim to competition. When the Paeroa Brewery closed in late 1956 it cited high railway transport costs making it impossible to compete with larger Auckland breweries. The closure apparently also removed 40 per cent of the stout brewed in New Zealand. The Masterton Brewery and the Kauri Brewery in Woodville had also been squeezed out during the previous two years.[91] The only counter to the trend was Newbigin's small Leopard Brewery in Hastings, which was purchased in 1956 by a New Zealand-registered company sponsored by Fraser & Neave and Malayan Breweries of Singapore. By the early 1960s Leopard would enjoy a solid presence in the New Zealand market.[92]

Before its brewing innovations gained full momentum, NZB moved to streamline beer delivery. Early in 1934 it had investigated an English system for transporting beer by tanker rather than barrels, and from 1938 one was used in Christchurch to move beer from brewery to bottling house. In 1946, the first stainless-steel beer storage tank was installed in the cellar of the Carlton Hotel in Christchurch. After various experiments with CO_2 gas in the tanks to maintain pressure and ensure that beer would withstand the handling from tanker to tank without loss of condition, NZB Auckland further refined the system from 1948. Thereafter delivery of beer via tanker became common – only constrained by various government controls on road transport and limits on the routes on which a full tanker could be safely driven. By 1957 NZB had forty-five tankers delivering to over three hundred hotels, and DB had another ten by the early 1960s and were looking to expand. Tanker capacity had also doubled to 40,000 litres in some cases. Consequently, hand pumps on bars gave way to pressure systems and later 'bowser' systems from which beer was rapidly hosed into glasses or more frequently jugs. While NZB emphasised hygiene, in that the direct transfer of beer minimised handling and the possibility of contamination, no doubt efficiency was also a key factor.[93] Where tanks were not used, stainless-steel or aluminium kegs were also replacing traditional wooden barrels by the mid-1950s. Suitable wood and trained coopers were in increasingly short supply. Russian and then Tasmanian oak were no longer available, and experiments with New Zealand wood were generally unsatisfactory as it influenced the taste of the beer. The new metal kegs were also more durable and easier to clean. The last bar in Auckland to serve beer from barrels changed in mid-1955.[94]

Not every tanker delivery went to plan. In October 1957 when a tanker overturned at a bridge near Kaikohe, and efforts to right it were unsuccessful, it was resolved to 'open the hatches and let the beer flow'. Immediately, 'glasses, gallon jars, milk billies, jugs, in fact, containers of every type and shape, appeared as if by magic' as locals took advantage of the calamity despite the best efforts of the Kaikohe Fire Brigade to keep the area clear and pump the beer away.[95] Further tanker capsizes late in 1959 did not benefit the public,[96] but one near Hāwera on New Year's Day 1960 certainly did.

> Soon there was hardly an empty container in the village. Basins, kegs, buckets, bottles, jugs, anything that didn't leak was used by the salvage crew. One man rushed off with two four-gallon jars under his arms, to reappear within minutes for a refill. 'I put the plug in the bath, and poured it in there in the meantime', he explained triumphantly, as he took his place again in the fill-up queue. The local publican felt it was one below the belt when two hopeful citizens appeared on his doorstep, asking for the loan of some empty kegs.

The publican was the only one who did not share the joy of the occasion as his establishment was apparently not required in the short term.[97]

New Zealand Breweries revolutionised its transport options for beer by the 1960s. Lion Technical Library

Tankers and kegs were not the most important innovations of the 1950s. Those were to be found inside the brewery as, following the likes of the Brewing Industry Research Foundation established in Britain in 1951, brewing science reached a new level of sophistication.[98] NZB centralised its technical laboratory in Wellington from 1953. Speight's, under the direction of Doug Cocks, who had worked at various NZB branches since the 1930s and studied brewing at the University of Birmingham, also pursued various innovations and modifications. The most important was the first production of lager in the South Island in conjunction with Danish brewer Tor Toft, who was brought to New Zealand for the purpose.[99] Nevertheless, NZB was cautious in emphasising its embrace of brewing science and technology. In December 1956, Sir William Perry was 'proud to state that New Zealand Breweries Ltd. employ many graduates to supervise quality control and carry out long-term research'. But he quickly added:

> Do not be confused. The aforegoing are not the brewers. The brewers are in charge of production, and produce from the finest malt, hops, yeast and sugar our quality products. We do not know the term Chemical Beers. Our beers are made from pure, wholesome ingredients, but chemists and scientists are essential to ensure that quality prevails.[100]

Morton Coutts, the pioneer of continuous fermentation and many other innovations at DB, at the Waitemata Brewery during the 1960s. DB Breweries

However, it was Morton Coutts at DB who entirely revolutionised brewing with the development of continuous fermentation. A grandson of nineteenth-century brewer Joseph Kühtze, Coutts was responsible for almost all technical innovation at DB from its beginning in 1929. Until the mid-1950s all beer was brewed on a particular day in individual batches in open vats. This process was labour-intensive, as all equipment needed to be cleaned and sterilised after each brew. There was also potential for each batch to taste slightly different, depending on even minor changes to proportions of ingredients or contact with the atmosphere during fermentation. With continuous fermentation, beer flowed through a series of closed tanks and was fermented under pressure with yeast added at each stage of fermentation and the yeast continually recycled from the end of the process back to the beginning. As the beer was not allowed any contact with the atmosphere even during bottling, the result was a product of uniform quality and taste with better keeping qualities. Further, a brew could be completed within a week, much faster than previously, and was significantly less labour-intensive as constant cleaning was not required.[101] NZB eventually completed a twenty-one-month continuous brewing run in Wellington that was only stopped when muddy water entered the local supply. Most DB Draught was produced in the continuous manner from 1956, and in August provisional patents were taken out. DB was soon applying for full patents in all of the main brewing countries.

Naturally, such developments quickly attracted the attention of NZB. An agreement was reached with DB to allow its products to be sold in NZB hotels in return for access to the new technology. Even though NZB capped the DB market share at 28 per cent, this was still an advantage as it had long struggled to secure widespread hotel access. In 1958 NZB rebuilt the Standard Brewery in Palmerston North, which it had acquired in 1947, as the first specialist continuous-fermentation brewery in the world, an event captured by its in-house magazine *Brewnews* under the headline 'Boffin Day at the Brewery',[102] and by 1960 the Auckland, Wellington, Christchurch and Dunedin breweries had been overhauled for the same purpose. Indeed, the timing was ideal for NZB in that continuous fermentation emerged at precisely the moment when it was due to embark on an overhaul of its breweries delayed by wartime and post-war building controls.[103] Not for the first time, NZB felt obliged to capture the moment with doggerel:

> Oh, Continuous Fermentation –
> Thing of utmost speculation –
> How is it possible, we ask,
> For yeast to do its wondrous task,
> With such astonishing rapidity.
> And yet, 'tis proved in all validity
> Indeed the process can't be wrong
> For Wort, like music in the song,
> After being fed in here
> Goes round and round and it comes out
>
> BEER[104]

By the end of 1960, 85 per cent of New Zealand beer was produced by the continuous method. Although representatives from the likes of Guinness and Courage visited the Standard Brewery during 1960 and the continuous fermentation process was sold to firms overseas in locations as diverse as Sweden and Singapore, the process generally struggled to adapt – a situation attributed to the absence of Morton Coutts' technical expertise in these breweries.

Not surprisingly, neither DB nor NZB in their respective corporate histories referred to the role of the other in these developments. Indeed, NZB described the process as 'controlled flow' fermentation.[105]

The legacy of continuous fermentation, for beer purists more than those happily consuming the products, was a question of the designated beer style. While the two major New Zealand breweries referred to various of their beers as 'draught' or 'ale', in reality continuous fermentation worked with lager yeast, which ferments more slowly and at cooler temperatures. The prevailing New Zealand styles were in fact amber lagers. In December 2005, *Consumer* magazine refused to include Tui East India Pale Ale in a review of ales because it was not one. It added that the labelling of Tui was misleading and should be changed. DB defended its labelling on the grounds that the New Zealand beer industry had evolved independently, and modern technology allowed it to make both ales and lagers with the same yeast strain. Lion, as NZB was to become, added that the overall taste, ingredients, colour and aroma of a beer were more relevant in determining its style than the type of yeast.[106]

Technical innovations did not end with continuous fermentation. The Australian section of the Institute of Brewing held its conference in Auckland in early 1966, while barley and hop conferences were held in alternate years. Although a bottled-beer-dispensing machine patented in Christchurch in 1960 and given a trial in a suburban hotel did not achieve success, the growing popularity of lighter, colder beer prompted various developments in refrigeration and dispensing. In 1969 NZB patented a mini tanker to supply cold beer to racecourses and other outdoor social functions. During the early 1970s, new laminar flow technology was created to serve colder and more carbonated beers such as Lion Super and Lion Brown.[107] But no New Zealand brewer was so sophisticated as to be able to respond positively to the South African company that wrote to the Department of Industries and Commerce in 1957 seeking the franchise for beer in tablet form.[108]

Hop growing also underwent major changes to align with modern brewing. In 1947, on the initiative of Auty Harley of Nelson Breweries, a New Zealand Brewers' Association was formed with the objective of better supporting the hop industry. Its first endeavour was a joint committee of hop growers, brewers, the Department of Scientific and Industrial Research and the Cawthron Institute to work on the eradication of black root rot, which remained a scourge. In 1949, partly funded by a 1-penny levy per pound on the price of hops, a Hop Research Station was established near Riwaka. During the early 1950s, it started a programme to breed disease-resistant plants by cross-pollinating male Fuggle plants, a low-yielding but resistant variety, with female Californian, a high-yielding but susceptible variety. Eventually, three new varieties of high-yielding hop were developed with resistance to black root rot, which soon ceased to be a threat. Thereafter the Research Station moved on to develop seedless hops, hops with a range of harvest dates to remove some of the pressure of the harvest, and hops with a high alpha acid content – being the resin in the hop cones that gives beer its bitter taste. In time this would produce hops that were highly desired in overseas markets. However, the focus remained very much on supplying local demand. In mid-February 1957 numerous members of the Brewers' Association, brewers and chemists, visited Nelson and Motueka to inspect hop gardens, the Hop Research Station and the Cawthron Institute farm and to hear lectures on various aspects of hop cultivation and disease prevention. As *Brewnews* reported, 'The attentive attitude of the audience, which included growers and their wives, demonstrated the acute interest taken today in the specialised field of hop production and usage.'[109]

In 1962, as part of a strategy to counter the boom and bust cycle that had periodically destabilised the industry, a ten-year supply agreement was negotiated between the Hop Marketing Committee and the New Zealand Brewers' Association in which all hops were purchased at a set price.[110] The 1940s and 1950s also saw the introduction of automated stoking in drying kilns, thus removing the need for manual stoking and constant temperature monitoring. Gas and oil kilns and fans to circulate dry air were also introduced by the early 1960s.[111]

But labour shortages were a continuing problem after the war, and hop growers were making the same appeals to use schoolchildren as they had during the 1880s. To counter the shortage, hop-picking machines, both locally built and imported, were trialled throughout the 1950s. These proved unsuccessful because they either shredded the hop cones or failed to pick them. Finally, in 1962 with the technology refined, the Brewers' Association imported picking machines from England and helped to finance growers to purchase them. Within five years, the entire Nelson crop was being picked by machines. Those growers without the means to purchase one could contract their picking. Although the communal and social aspect of the harvest had ended after more than a century, there was no longer a labour problem and growers felt confident to increase acreage.[112]

*

During the 1950s NZB and DB clearly adapted to weaker beer and six o'clock closing with industrial efficiency and without having to invest a great deal in the drinking environment. As Conrad Bollinger explained:

> Liquor was available only alone, without any concomitant amenities, without games, without music, without entertainment, without food. . . . served in a bare room without even seats, under conditions where the only thing to do is booze and the vertical position and pressure of time make the customers drink as much as is physically possible in a short time. And since drink was prohibited everywhere else, and certainly couldn't be taken publicly in connection with eating, dancing, or any sort of entertainment, the publican had a guaranteed market which could be supplied with the very minimum of trouble and expense.[113]

But remarkably, the breweries still struggled to meet demand and there were periodic reports of beer shortages and rationing schemes during the 1950s.[114] As Table 8.2 confirms, weaker beer under shorter hours produced faster drinking and in much greater volumes than ever before. From 1946 to 1975, eight years after the end of six o'clock closing, pure alcohol consumption increased by 56 per cent to a level not seen since the early 1880s. While there was an increase in consumption of spirits, especially of wine from the late 1960s, pure alcohol consumption of beer doubled. In 1955 and 1961 statistics compiled by the British Brewers' Society ranked New Zealand per-capita beer consumption fourth in the world behind Belgium, Luxembourg and Australia. By 1968 New Zealand was fifth behind Czechoslovakia, West Germany, Belgium and Australia.[115]

Unsurprisingly, this drinking pattern attracted much animated criticism. Writing in 1944, just as it was becoming apparent that weaker beer was prompting significantly increased consumption, poet and critic Rex Fairburn suggested that New Zealanders were a passive and complacent people caught between English traditions of sociability and American puritanism without having the initiative to chart their own course. 'We have made no attempt to absorb

A DB Lager advertisement from the early 1950s. DB Breweries

Table 8.2: Pure Alcohol Consumption Per Capita, 1940–1975 (litres)

	Liquid volume	Pure alcohol			
	Beer	Beer	Spirits	Wine	Total
1940	50.9	2.54	0.49	0.10	3.13
1946	76.4	2.90	0.23	0.17	3.30
1950	84.1	3.19	0.58	0.26	4.03
1955	99.1	3.76	0.89	0.21	4.86
1960	100.0	3.80	0.67	0.26	4.73
1965	103.2	3.92	0.85	0.34	5.11
1970	116.8	4.44	0.72	0.68	5.84
1975	133.9	5.08	1.24	1.07	7.39

Spirits are assumed as 40% and wine 15% abv. Beer is assumed as 5% in 1940 and 3.8% thereafter.

Source: Derived from *Statistics of New Zealand*, 1940–75.

these viewpoints and make something of our own. Our liquor laws and our drinking habits are extremely barbarous. On one side are the militant ranks of the prohibitionists, and on the other the wealthy and powerful "trade". In the no-man's land between them is the great mass of the public, flabby and inert, powerless to impose a civilized pattern on the drink situation.' For Fairburn, New Zealand was gripped by deep conservatism and a tolerance for dull mediocrity in which a minority was able to assert its 'negative rights'. As he explained: 'The right to drink, gamble, dance, sun-bathe, and play games on Sunday is a positive and legitimate one. The right to prevent other people doing these things because of pseudo-moral or pseudo-theological prejudices is negative and is not justifiable even in a majority.'[116]

In a similar vein, although produced as a parody, Rod Derrett's 1965 song 'Rugby, Racing and Beer' also epitomised for some the narrow, working-class interests of a philistine society. Here a 'little shaver' sitting on the knee of his 'old man' is advised that success and social standing depends on such skills as 'how to pick the second leg / And how to tap a brewery keg'. Whatever the problems of the world, salvation was apparently to be found in a 'national heritage of rugby, racing and beer'.[117] Indeed, as the next chapter will explain, the breweries reinforced this heritage throughout the 1960s with increasing sponsorship of both racing and rugby.

Yet some of the most graphic commentary on New Zealand drinking conditions came from overseas visitors. In June 1952 a Dutch immigrant announced that he was leaving for home after nine months. He had worked for a period as a barman and bottle-store hand in Taihape and was disgusted by local drinking habits. 'I don't think New Zealanders know the art of drinking. I have seen them drinking beer, whisky and gin alternatively at one drinking session. It really seems to me that the object of many drinkers is to get drunk as soon as possible, and your liquor laws, which in my personal opinion are really stupid, promote that quick drinking too.'[118] In 1954 American Fulbright scholar Robin Winks declared that New Zealanders had no idea of how to drink properly and seemed to drink to drown themselves. 'To the psychologist, the high drinking rate is a reflection on the same tendency which the high suicide rate shows – that the New Zealander is wrapped up in himself, turned inward, an introvert who exerts aggression against himself rather than aggression against others. New Zealanders not only do not appreciate good liquor when it is presented to them; they are unable to drink and hold it with any grace.'[119] British naturalist and writer Gerald Durrell, who arrived in 1962, captured the chaos and drama of bar life near closing time.

> I shall always attribute my uncertain start in New Zealand to the fact that I was introduced too early to what is known as 'the five o'clock swill'. The phrase has, when you consider it, a wonderful pastoral – one might almost say idyllic – ring to it. It conjures up a picture of fat but hungry porcines, all freshly scrubbed, eagerly and gratefully partaking of their warm mash from the horny but kindly hands of a jovial farmer, a twinkly-eyed son of the soil. Nothing could be further from the truth. The five o'clock swill is the direct result of New Zealand's imbecilic licensing laws. . . . Dozens of thirsty New Zealanders lined the bar some twenty deep, all talking at the tops of their voices and gulping beer as fast as they could. To facilitate the replenishing of their glasses with all possible speed, the beer was served through a long hosepipe with a tap on the end. As the empty glasses were slapped on the counter, the man behind the bar moved rapidly up and down squirting them full of beer. This was an operation fraught with difficulty, and more beer appeared to go on the counter than anywhere else. . . . Then suddenly an enormous brass bell let out an ear-splitting clanging, like the distraught cries of a fire engine thwarted in love.[120]

American psychologist David Ausubel was perturbed by the extent of drinking and public drunkenness compared to the United States, 'and the number of inebriated individuals who pour out of the pubs at closing time and weave through the streets is astonishing'. He further claimed that the prevalence of rowdy, drunken parties was more characteristic of socially demoralised lower-class groups in the United States than the middle class who sanctioned them in New Zealand. 'Inebriation commonly proceeds to the point of insensibility and is not infrequently accompanied by brawling, vulgar cavorting, the use of obscene language in mixed company, indecent propositioning and the smashing of windows and furniture.' Many Friday and Saturday night 'beer parties' in small towns were simply 'a desperate way of escaping from the oppressive boredom and emptiness of social life'. Alcohol facilitated the release of underlying hostility, and without it New Zealanders were apparently unable to shed their customary reserve and to 'escape from a brooding sense of failure to achieve ambition'.[121]

Such lurid depictions no doubt aided the sales for respective authors, but there are several qualifications to be made here. Firstly, some accounts, especially descriptions of drunken staggering in the streets, were eerily similar to prohibitionist flourishes of an earlier generation that owed more to rhetoric and magnifying the abnormal than hard evidence. Although convictions for drunkenness were always a measure of the enforcement whims of the police rather than the publicly visible consequences of drinking, it is nevertheless the case that convictions for drunkenness during the 1950s were scarcely a third of what they had been half a century earlier and little different to what they had been during the mid-1930s. Having been as low as 3 per 1000 adult males in 1947, they rose slowly to a peak above 5 in 1955 before declining to under 4 from 1959 onwards. Secondly, if after-hours drinking was as ubiquitous as evidence to the Royal Commission and a great deal of anecdote suggests, then many drinkers did not have an imperative to guzzle by six o'clock. Thirdly, the emergence post-war of the half-gallon flagon, the half-g, as an effective means for takeaway draught beer, to say nothing of the availability of bottled beer, may also have reduced the rush. Fourthly, the response to the publication in 1957 of James McNeish's *Tavern in the Town*, an idiosyncratic history of New Zealand hotels, dwelt on the enduring character of many rural establishments in particular, and the traditional role that many still played as community centres.[122]

Perhaps, then, the 'swill' ought to be narrowed down to a largely urban phenomenon, especially in those districts where the ratio of licences to population remained problematic. Indeed, in a summary of the 1954 licensing poll result, the *Dominion* referred to two New Zealands – one of urban indulgence and the other of rural evasion.[123] Perrin Rowland's argument that options for dining out in New Zealand during this period were not as basic and barbaric as historical accounts have frequently suggested is also a useful context for the swill. The popular image of New Zealand as an isolated outpost underestimates the range of international goods, ideas and services that were available. Even in the absence of licensed restaurants, it is safe to assume that those who wished to could find opportunities for convivial eating and drinking in private homes. Similarly, Tanja Luckins has shown how Australian drinkers found ways to circumvent six o'clock closing or found convivial drinking arrangements beyond the pub.[124]

There may also be an element of class prejudice in some reactions to the swill. While the ordinary drinker and their drinking environments were periodically maligned, the press seems to have treated the annual university student 'drinking horn' with a degree of jocularity. By some accounts the drinking contests that became a regular feature of university Easter and winter sports tournaments 'originated in the twenties with certain students, now Christchurch legal luminaries of some brightness'.[125] From 1936, when it was noted that Victoria's only success in the Easter tournament was when they 'romped home with the handy margin of two handles over Canterbury' in the unofficial drinking horn held at the Hotel St George, results were reported on a regular basis, although not in sufficient detail to satisfy all. In April 1939, 'Watchful' observed that while in the other contests at the tournament the *Otago Daily Times* had published the names and photographs of the winners, in the reporting of this contest even the names were suppressed.[126] By 1948, when Otago were regarded as 'infallible', the event evidently had very strict protocols. 'The constitution governing the contest insists that each of the five men following the leader of the team must remain with palms down on the bar till the previous drinker has returned his glass empty to the bar. This was adhered to, but all teams erred on the point of spilling.'[127] In response to some public criticism later in the year, a Canterbury student responded with mock indignation.

> As a pure, celestial, and holy student, who has been carefully secluded from the evil doings of this world, I should also like to register my abject horror of the shocking beer-drinking contest. Being of the same mind as previous correspondents, I should like to suggest to them that we form a small group and distribute soul-saving pamphlets to those who have committed this paramount sin. I am glad to know that there are other people as pure and celestial as myself, and I feel that they should try to register their complaints in the newspaper headlines, not necessarily because they have something to say, but because they have to say something.[128]

In more serious vein, the *White Ribbon* fumed in 1949 that 'It is deplorable that the University is developing students with no higher ideal than the "drinking horn contest". We look for something better from the young men who were once described as "the cream of our manhood". The fact that a young woman was pictured in the newspaper as one of the onlookers also shows a decline in the feminine taste.'[129] While there was periodic public complaint about such drinking, and periodic student complaint at the way the drinking was organised, such contests endured into the twenty-first century.[130]

By the late 1950s the brewing industry and the nature of beer drinking had undergone rapid changes in a relatively short time. The number of breweries had declined sharply and the brewing process had been fundamentally altered. It is surely no coincidence that the term 'brewing plant', with its more industrial connotation, was regularly used instead of 'brewery'. There is also no doubt that the larger breweries were comfortable with these developments and the profits to be made from them. However, there were limits. In June 1957 K. L. Usmar, Percy Coyle's successor as secretary of the National Council of the Licensed Trade, used an address to tourism representatives to condemn New Zealand's obsolescent licensing laws and to suggest that reform was long overdue. 'Whatever may be the drinking behaviour of New Zealanders, visitors want to tell people back home about it. All that most of them can speak about from experience is the 6 o'clock exodus.'[131] Within a decade, very substantial change would come and new settings and tastes would be presented to beer drinkers. However, reform would also bring new and sometimes unpalatable challenges.

As a counterpoint to popular images of the six o'clock swill, this 1951 DB advertisement presents a more civilised setting for beer. Alexander Turnbull Library, PUBL-0100-1951-05-back

FOR HAPPY MEMORIES . . .

D B LAGER

with its

Refreshing and Fragrant Flavour

A PRODUCT OF DOMINION BREWERIES LTD., AUCKLAND

R1

The New Zealand Breweries brew hall opened in Auckland in 1960 to accompany the disastrous Lucky branding. Lion Technical Library

NINE

Black Budget and False Dawn: Slow Liberalisation, 1958–1980

The years from the exaggerated depths of 'the swill' during the 1950s to the relatively more enlightened 1970s must be regarded as a period of liberalisation for beer drinkers against the backdrop of a rapidly changing post-war society. Licensed restaurants arrived in 1961, evening hotel hours in 1967, and the range of venues that were able to hold liquor licences expanded significantly thereafter. The variety of beer available also offered hope for the more adventurous consumer. On the other hand, these were the decades in which the NZB–DB duopoly became fully entrenched as the last few small or partially independent breweries were merged into the big two, which renewed their scramble for outlets and market share. The choice for beer drinkers was still relatively limited, and they periodically bore the consequences of rising prices, industrial unrest and drinking environments that did not deliver all that had been hoped, as the economy struggled from the mid-1970s.

If popular recollection is any measure, the so-called 'Black Budget' of 1958 is second only to the swill among the calamities of New Zealand beer history, with its tax increase on beer apparently depriving the common man of his simple pleasure. However, this view is somewhat embelished and owes more than a little to the self-interest of the two major breweries. Certainly, as Jim McAloon points out, the Labour Government elected at the end of 1957 inherited serious economic problems and did not handle them as well as it could. Walter Nash was an elderly and politically fading prime minister who did not provide the strategic domestic leadership that was needed. Hence, Finance Minister Arnold Nordmeyer had the major responsibility for delivering on Labour's expensive election promises to expand the welfare state while dealing with a serious balance of payments deficit triggered by declining prices for butter and wool exports. From 1 January 1958, emergency measures were introduced, including severe restrictions on import licences to prioritise food, machinery and raw materials. This effectively ended imports of beer and significantly reduced volumes of imported wine and spirits.[1] In mid-May, both NZB and DB announced that they would fill the import gap by producing stronger beer in pint bottles.[2] Given that total beer imports in 1957 had been less than half a million litres, around 0.2 per cent of all beer available for consumption, there was surely a degree of marketing opportunism rather than overwhelming consumer demand in these decisions.

As the Budget approached, Treasury was in favour of increased taxes on alcohol, tobacco, gambling and other 'luxuries'. Doubtful that it would get all that it wanted, it may have submitted an overbid to Nordmeyer and then been surprised when he agreed to all of it. Others have suggested that the government had little choice as it sought to reduce spending and raise revenue.[3] Consequently, the Budget announcement on 26 June 1958 included additional taxes on cars, alcohol and tobacco, with a doubling of beer duty from 3 to 6 shillings per gallon, quickly passed on to the retail price for drinkers. While the Budget did increase the family benefit by 50 per cent and offered other advantages to families, these measures were overshadowed by a sense that the government had betrayed its own worker constituency by eroding its simple pleasures. The term 'Black Budget', probably coined by Federation of Labour President Fintan Patrick Walsh, was soon adopted by the National Party and endures as a description of supposedly austere Budgets.[4] That both Nash and Nordmeyer were known to be non-drinkers and non-smokers also made it easy to construct the Budget as a wowser attack. Some may also have remembered that Nordmeyer had been the minister of customs when the strength of beer was lowered in 1942. Indeed, one correspondent, writing two days after the Budget, declared that Nash should have forced a reversion to pre-war beer strength if the price was to be increased. Having returned to New Zealand after twelve years, he found the beer 'like coloured water or cold tea . . . If the working man is good enough to vote for Labour (and I assume he does), it is up to Labour to give him a decent, palatable glass of beer.'[5]

While the breweries were understandably not pleased, there was also an element of pragmatism. As NZB's Deputy Chairman M. O. Barnett remarked at the 1958 annual general meeting, 'We must accept the conditions facing us and realise that "grousing" about them will do no one any good. The future of the country is not in doubt. It is facing a period of re-adjustment of expenditure to income and we must do likewise.'[6] A more tongue-in-cheek response came from Marilyn Dunn, described as 'Speight Branch's versatile and popular switchboard operator'.

> Budget Rhymes for Little Children
>
> I love little Nordy, his voice is like ice.
> And when he reads Budgets he's full of advice.
> The country is ruined, the party bereft. He promptly proceeds to take all we have left.
>
> Rock-a-bye Budget Taxes are tops,
> Petrol is soaring and beer made from Hops.
> Cigarette smoke is a thing of the past. Nordmeyer spoke. But how long can it last!
>
> Sing a song of Budgets Taxes to the fore,
> Everything is dearer than in days of yore.
> For the axe was sharpened with a dreadful skill.
> No one swallowed, happily, The Budget's bitter pill.
>
> Arnold Nordmeyer sat by the fire
> Scanning his Budget bold
> He raised every tax
> We straightened our backs
> And said 'There's a wolf in the fold!'[7]

Quite apart from a public outcry in response to the Budget, there were serious ructions within the labour movement. Early in 1959, Fred Young, the long-serving but highly volatile general secretary of the Hotel Workers' Federation, who had also expressed minority views as a member of the 1946 Royal Commission, informed the Federation's conference that while he wanted to keep Labour in power he had little faith in the running of the party, its fiscal policy and especially Nash as leader. Moreover, he had previously warned Nordmeyer of problems ahead if beer duty was increased in the Budget. Young called the Budget 'a vicious attempt to make the workers pay a disproportionate share of the country's increased taxation', adding that it was designed to support farmers who tended to demand support when things were going badly but insisted that government kept its nose out of their affairs when things were going well. Labour would lose votes if it did not reverse the decision.[8] In another letter to Tom Skinner, secretary of the Auckland Trades Council, Young defended the breweries against claims that they were well able to absorb duty increases without these being passed on to consumers. He argued that the real issue was the extent of duty as a brewery only made one-thirteenth of a penny on each 8-ounce glass of beer. Young was also critical of a 'fable' about beer barons, in that the companies had 7000 shareholders, including some of their own employees, who stood to benefit from the prosperity of the business. An increase in beer prices would also lead to more home brewing and hence a reduced tax take for the government. Declining hotel patronage would also jeopardise the employment of Young's members. Finally, he insisted that breweries were the best employer in the hotel industry.[9]

At the end of 1959, Young unleashed his fury directly to Walter Nash.

> [I]t is fair to assume that your Party now comes out openly as an anti-liquor Party and the actions of the Price Control Division and the Price Tribunal which the government controls indicate that the policy is designed to smash the hotel industry by putting those concerned in that business into the bankruptcy court. As an honest man why did you not indicate to the public during the election campaign the policy you have since adopted in this connection before the people were called upon to vote? You have acted in this matter despite the fact that during the election 883,000 people voted for liquor as against 260,000 who voted against it. It is well to remember that the government which you represent and which controls the Treasury Benches was elected with a vote of 559,000.[10]

The clear implication was that Labour could lose the 1960 election if it did not reverse its Budget decision. Young particularly defended hotel workers, who were tired of the criticism levelled at them for prices over which they had no control. He suggested that barmen would start wearing white coats emblazoned with 'Official Tax Collectors – Unpaid'. While his Federation had contributed £5000 to Labour's campaign fund because they believed this was in the best interest of the country, the government had shown how wrong this judgement could be.[11] Indeed, in 1960 the Hotel Workers' Federation did suspend its ties to the Labour Party.[12]

Naturally, brewers also had a gloomy assessment of the Budget's impact. In mid-September 1958, NZB claimed that the demand for bottled beer had fallen 40 per cent and draught beer 15 per cent. The trade was £4 million a year worse off. In its 1959 annual report, NZB argued that brewery profit per gallon was 1.65 pence while the government took fifty-five times that amount in duty and other taxes.[13] From Nelson, Auty Harley, president of the Brewers' Association, mused that consumer resistance to the price of beer was encouraging home brewing, and high duties meant that business for breweries would be far more attractive from the sale of raw materials for brewing rather than from actually brewing. On the other hand, there could be no control of the strength of beer made from ready-mixed ingredients unless sales were made through hotels. 'We will surely find that many people, including teenagers, will be able to brew beer of fantastic strength. This is completely undesirable, and would create grave social

Hops sold for private use, such as these from the 1960s, also came with recipes to assist the aspiring home brewer. Photograph: Neil Macbeth

problems.'[14] In similar vein, a representative of the Taranaki Brewery blamed the government for encouraging home brewing, but also wondered, 'Why should it be lawful for persons under the age of 21 to get drunk on home-brew and yet unlawful for them to be served on licensed premises?'[15]

The brewers had a point about home brew eroding taxable consumption, but also sought to capitalise on the opportunity it presented. Per-capita beer consumption, which had increased every year from 76 litres in 1947 to 103 litres in 1957, now declined to 96 in 1958 and 90 in 1959 before rebounding to 100 litres in 1960. Only in 1966 did it again exceed the 1957 level. Meanwhile, in early 1959 it was reported that packets of hops for home brewing had run out, and brewers who had become active after the Budget were having to stop until after the next harvest. However, later in the year the total volume of home brewing was estimated at 36 million litres.[16] By September, the major breweries had developed tins of ready-mixed dry home brew ingredients. The DB version was described as 'powdered beer' which just needed water and yeast added to produce beer after ten days. The government was apparently 'troubled' and Customs 'confounded' as to how duty could be levied on a dry wort that did not of itself contain alcohol. In the interests of due diligence, samples were dispatched to Wellington where Treasury staff apparently made a brew and pronounced it excellent. A demonstration was also provided to Nordmeyer's staff.[17] However, although firms such as the Brewers' Trading Co. of Christchurch continued to provide ingredients and equipment directly and by mail order, there is no sign that the major breweries continued very far down this path.[18]

For the beer drinker the silver lining from the Budget came in a reluctance by successive governments to again increase the tax for nearly two decades. In July 1962 when National MP Duncan McIntyre was asked why his government did not repeal a tax that was apparently so unpopular, he conceded that it was necessary to fund the family benefit, which had also been increased under Labour and which National had no inclination to reduce.[19] In May 1967 Minister of Finance Robert Muldoon increased the duty on spirits but not that on beer in a mini-Budget designed to curb economic instability following a crash in wool prices at the end of the previous year. The Labour leader, Norman Kirk, immediately attacked Prime Minister Keith Holyoake.

> [He should] abjectly apologise for the hypocrisy and humbug the National Party has employed to criticise and revile for nearly nine years the member for Island Bay Mr A. H. Nordmeyer. I remember well Budget night, 1958. How well I remember the gloating grin that came over the Prime Minister's face when certain provisions of that document caught his eye. He didn't really care about the people. He saw it as a stepping stone to office. How different it would have been if, instead of his extravagant abuse, he had announced then that he would continue beer tax unchanged, and add to it; that he would take the school milk away from children, even from the blind children.[20]

While *The Press* felt that there was 'no ground in equity or in economics for letting the beer-drinker off scot-free' when the spirit drinker was subjected to an increase, Muldoon continued to resist the temptation.[21] On the eve of the 1972 election, Kirk warned voters that National had a secret tax agenda, including beer, and reiterated that 'After what they said about Mr Nordmeyer, I don't know how they could even consider it.'[22] Labour's victory forestalled any such plan. But within three weeks of National's return to office at the end of 1975, with a faltering economy in the wake of the 1973 oil shock, there was finally an increase in tax on beer. Others would follow as the economy remained volatile into the 1980s.[23]

The sequel to the 1958 Budget came more than half a century later at the hands of DB. In an advertising campaign for DB Export, and without reference to the wider economic context, the non-drinking Nordmeyer was described as having 'taxed the price of beer out of the reach of Kiwi men' in addition to imposing import controls that restricted choice. Accompanying footage depicted angry demonstrators. Morton Coutts was apparently inspired to develop a new brewing process to create a new beer and thereby avoid the tax by brewing his export beer in New Zealand. Eventually the Advertising Standards Authority, acting on several complaints, ordered DB to withdraw the television and online versions of the ads. Aside from the fact that the new brewing process, continuous fermentation, was patented two years before the Budget, that few 'ordinary Kiwi blokes' were likely to have drunk the small quantities of more expensive imported beer reaching New Zealand in the post-war years, and that DB had announced its import-strength beer plans six weeks before the Budget, the footage of angry demonstrators in the ads was drawn from the 1951 waterfront lockout. In response, the executive creative director of the agency responsible for the ad remarked 'I don't think a beer brand wants to be the guardians of history, we want to tell great stories.'[24]

*

If the Budget was a challenge for the breweries to confront, the next obstacle, for NZB at least, was entirely self-inflicted. During a broadcast on New Zealand attitudes in December 1957, lawyer and politician Dr Martyn Finlay wondered, with transport costs high, why it was necessary or economic to transport beer around the country. 'Every regular drinker claims to have his favourite brew, but our brews are not so different that we should deny carriage space to more worthwhile things in order to ship Dunedin beer to Auckland, and vice versa.'[25] Whether or not NZB were listening, they were certainly of a similar view. Having rebuilt Palmerston

Lucky beer labels.
Lion Technical Library

North's Standard Brewery in 1958 to accommodate continuous fermentation, NZB then opened a new Lucky Brewery on Auckland's Khyber Pass Road site in 1960 to replace the old Captain Cook Brewery, and also refurbished breweries in the other main centres for the same purpose. With five brewing plants now able to work to the same formula, NZB announced on 22 July 1960 that all of its bottled brands throughout New Zealand were being withdrawn and replaced with standardised 'Lucky' brands – lager, draught, bitter and stout.[26] As a speaker at the opening of the Auckland brewery enthused:

> Having a tour of this brewery, you will know that the product is not synthetic. You have seen the malt, the hops and the sugar. You have seen the marvellous plant from which the extract of these ingredients is produced and which, in the form of wort, is placed in cold storage. . . . Now we have to admit that at this stage a mighty alchemist, the Minister of Finance, works upon it and extracts great wealth at rates from 12/- to 6/- per gallon – he's LUCKY.[27]

At each brewery there was a grand public ceremony. As *Brewnews* reported: 'LUCKY pennants flew high over the breweries for the official openings. Guests met in large numbers, hotel-keepers, other members of the Licensed Trade, shareholders, representatives of various trades and professions and leading citizens. At each place they were shown over a spic-and-span brewery, the visual effect of this being heightened by elaborate decoration schemes.' There were also special tours for scientists and for 'out of town' licensees who were brought to the main centres. Those attending in Wellington left with gifts: 'ladies presented with folding clothes brushes, and men with plastic wallets containing card packs – each gift appropriately inscribed "A Memento of a Lucky Occasion"'.[28]

By all accounts, these were very palatable beers. In particular, Lucky Stout with an original gravity of 1072 (8 per cent alcohol by volume) was twice the strength of standard beer, although also twice the price. But within months they were gone in a triumph of brand loyalty over substance. Precisely how consumer resistance unfolded is not clear. In its 1961 annual report NZB simply observed that the withdrawing of established brands had caused 'some disappointment' and they had been reinstated alongside Lucky.[29] In reality, only Lucky Stout endured to the mid-1970s.[30] It seems that consumer preferences were not moving as rapidly as the logic of brewing technology. Consumers liked the perception of choice even if the beer came from the same company.

There were other periodic exceptions to the quest for uniformity in the form of special brews to mark various occasions. In October 1959 a Red Band variation was produced to celebrate the official opening of Wellington Airport. This was followed in 1963 by a one-off Red Band Export Lager for the cruise of the SS *Rangitoto* to the Melbourne Cup. In the same year, Speight's created a special brew for the centennial of Otago Boys' High School, and in 1966 produced 'Taxes Bitter' for the Inland Revenue Department's national basketball tournament in Dunedin.[31] By the 1970s it was evidently easier to relabel an existing brand, such as the Commercial Bank of Australia Lion Brown label or that for the New Zealand Silver Plough Championship, both in 1973.[32]

*

After the Budget and branding challenges, matters began to move in favour of the breweries and the trade, and drinkers more generally, through a gradual liberalisation of the licensing laws. Some of the inspiration may have come from Australia. Although New South Wales retained six o'clock closing in a 1947 referendum, a wave of immigration from Britain and Europe brought people with more sophisticated tastes who were highly critical of the swill. At the same time, concerns about corruption in beer retailing led the New South Wales government to establish the Maxwell Royal Commission on the licensed trade in 1951. Its report in 1954 prompted fundamental changes to licensing law and drinker behaviour, and, following a positive referendum, six o'clock was replaced with ten o'clock closing from 1955. Tasmania had already taken this path in 1937 and Queensland in 1941, while Western Australia had never adopted six o'clock.[33] In New Zealand, the Licensing Control Commission had already been vocal in its opposition to six o'clock closing. In its 1952 report the Commission criticised rushed drinking and contrasted this with the 'leisurely and sociable' approach on the West Coast, where early closing was almost entirety ignored.[34] Following the 1954 licensing poll in which prohibition secured 23 per cent of the vote, the *Dominion* declared it a lost cause and said that the public clearly wanted to be presented with other options. 'The licensing laws in this country are flouted openly in many places and covertly in others. While such breaches cannot be confined, they do point to public dissatisfaction.'[35] The following year, in sharp contrast to the deliberate disengagement of 1949, the National Council's Percy Coyle responded to tourist industry criticism that the trade was comfortable with the status quo with a statement that the hotel industry now favoured ten o'clock closing and would enforce it properly.[36]

These developments clearly disconcerted the Alliance. It responded to the change in New South Wales by arguing that there was no less drinking under ten o'clock closing and questioned the rationale whereby people complained about crowded bars but not about crowded department stores or crowded trains. Early in 1956, after a poll in Victoria had comprehensively rejected later closing, they hoped this would erode the campaign for later closing in New Zealand, but warned there was no room for complacency. The trade would strike at any opportune moment, 'and no moment could be more opportune than when the Alliance is ill-fitted to oppose the trade's demands'. The Alliance was not in a financial position to mount a 'stiff campaign' and its membership was divided over its direction. Rather than prohibition, many now saw its best hope in a campaign of temperance education to shift the public and church groups from their apparent apathy in the face of declining support at the polls.[37] A subsequent meeting debated whether the Alliance should declare for or against trust control of the trade in response to clear evidence that drinking conditions were better in trust-controlled establishments. However, this could have a disastrous effect of making drinking more acceptable. 'Respectable drinking conditions will create a drinking fashion that can have a detrimental effect on the nation's health. If respect is engendered for drinking it is conceivable that increased drinking hours will ultimately result.' As the Alliance still contained 'sincere absolutists' for prohibition, it did not want to risk alienating them by anything amounting to endorsement of drink. 'To alter the law to allow evening drinking would multiply the difficulties of enforcement and intensify the problems of supply to youths, drunkenness, drink driving, liquor in the vicinity of dance halls etc. These problems would become much more complicated in country districts.'[38]

Eventually the government conceded to growing debate. As Walter Nash explained, in order to 'take the question out of the realm of party politics', a Select Committee on Licensing was convened from February 1960.[39] Something of the tone of proceedings is evident in a submission from Conrad Bollinger, Hugh Price and Jim Delahunty, who paid close attention to the liquor laws as drinkers and commentators:

> We are strongly of the opinion that what is urgently required at this stage is not a further crop of minor amendments (although some minor amendments are certainly needed as well) but a complete alteration of the basis of the law on this subject. The steady stream of amending enactments over the past three generations was no doubt partly necessitated by changing social conditions, but had the basic principles underlying the law been sound, amending legislation would hardly have been required at such a steady rate.[40]

The outcome was an extensive Licensing Amendment Act in 1961, with further revision and restatement in the Sale of Liquor Act 1962. Among the most important changes was the introduction of tavern licences, whereby licensed establishments were now able to sell alcohol without the requirement to provide accommodation. While the Alliance complained that it was 'regrettable that the Bill should set up straight out drinking shops with no other responsibility and when there is no evidence of public demand for it', the reality was that the need for accommodation, in place since the 1890s, had become a significant obstacle to establishing new hotels to relieve the problem of skewed distribution.[41] Investors were not enthusiastic about financing hotels with accommodation facilities when these were not needed. In contrast to the separation of alcohol and food in the tavern, restaurants were now licensed – although the number of licences was initially limited to ten for the whole country and alcohol was only available to diners who were having two courses or more. The number of licences was gradually increased over the next few years.[42] Further, despite lingering objections from Fred Young and the Hotel Workers' Federation, barmaids were permitted for the first time in half a century. Remarkably, perhaps four barmaids registered in 1911 were still working at the time of the law change in 1961.[43]

Finally, the powers of the Licensing Control Commission were significantly enhanced. It could appoint hotel inspectors, set minimum hotel standards and exert greater influence over the closure of hotels and the issuing of new licences. The limit on the total number of licences, effective since 1910, was also abolished. All hotel licences were now provisional, meaning that owners and managers had to prove that they conformed to the required standards or the licence may not be renewed. The National Council of the Licensed Trade was unimpressed. As it informed the select committee: 'It is indeed unfortunate that, without prior consultation and the vital opportunity of critical analysis, one of the largest industries in New Zealand should be compelled by legislation, practically overnight, to submit to the statutorily enforced "take over", which is implicit in the vesting in the Licensing Commission of the powers of "life and death" over the industry.'[44] Nevertheless, the Commission went about its task with renewed vigour.

Yet there were limits to liberalisation in 1960. Despite continuance polling consistently near 70 per cent, there was no move to abolish triennial licensing polls. Moreover, although members of the select committee visited Sydney in 1960 and returned a positive verdict on later opening hours there, they eventually recommended that 'conditions are not propitious for an immediate change from six o'clock to ten o'clock closing'. Cabinet also rejected calls for a referendum on the matter. As Paul Christoffel observes, the 1949 referendum cast a long shadow. The select committee could see no clear evidence that public opinion had shifted since that decisive result.[45]

Within six years, the position had changed entirely. Relative economic prosperity and increased purchasing power, long-haul air travel which gave more people overseas experience and brought new injections of popular culture into New Zealand, especially television, which began broadcasting from 1960, shaped broader horizons in a baby-boom generation that had not been so starkly shaped by war and economic depression. The slow growth of the international tourism industry during the 1960s also began to improve facilities and entertainment to meet the expectations of overseas visitors.[46]

The Presbyterian Church and others within the Alliance also began to shift from strict prohibition to advocacy of temperance. In the early 1950s it appeared that Presbyterians were as entrenched as ever. A 1952 debate in the pages of their *Outlook* magazine as to whether or not beer drinking was sinful generally concluded that it was and roundly condemned defenders of moderate consumption. However, by the end of the decade, a joint Methodist–Presbyterian statement on alcohol concluded that neither prohibition nor total abstinence was stipulated in the Bible, although strict controls were still needed to contain the alcohol culture that was blighting society. Gradually alcohol was recognised as a personal 'liberty of conscience' but with abstinence as the ideal. However, the position shifted further from 1960 in the wake of the publication of Conrad Bollinger's *Grog's Own Country*, which argued that temperance and prohibitionist support for six o'clock closing was wholeheartedly endorsed by the trade for its own economic reasons. By 1966 the church was supportive of later closing, as it saw no evidence that longer hours would encourage more drinking. Its approach was now to mitigate the problems of alcohol rather than to cling to a fading prohibition cause.[47]

In this context, and without an enduring prohibition bloc in Parliament, the usual political reluctance to provoke voters with innovative alcohol policy thawed somewhat. The National Party conference in July 1966 declared overwhelming support for changed hours without the need for a referendum. Yet even though Prime Minister Keith Holyoake claimed that 80 per cent of MPs supported change, doing so without a referendum was never seriously contemplated by National or Labour.[48] Nor would Holyoake allow a vote on the matter in conjunction with the 1966 election, insisting that to do so would 'distort and confuse' and could take people's attention from other more vital issues. In retort, the *Evening Post* argued that the triennial licensing poll was a distraction.

> With monotonous regularity the voters turn down Prohibition by a thumping margin. The total of votes in favour of Continuance has gone up as has the much smaller total cast for State purchase and control, but the Prohibition vote has dwindled noticeably. And yet when it is all boiled down the issue of Prohibition or no Prohibition is really what the voting is supposed to be about. All that the poll is doing now is to emphasise the obvious. The Prohibition cause at ballot-box level is as dead as the dodo.[49]

In May 1967 the National Government introduced legislation to hold a referendum in late September. The initial ballot paper proposal offered two options – the continuance of six o'clock closing, or the introduction of ten o'clock closing but on condition that the total weekly hours of opening would remain the same and that hotels would close for an evening meal break. The final form of the ballot removed these restrictions and provided a simple choice between six o'clock and ten o'clock.[50] However, as voters were also being presented with a referendum on the length of the term of Parliament, *Brewnews* speculated on the confusion that could arise. 'Such a delicious confusedness might result in closing Parliament when in session at 6 p.m. daily and opening the hotels ad lib for the next four years.'[51]

While the letters columns of newspapers were inundated, there was little sign of organised campaigning on either side as the referendum approached. The *Auckland Star* observed that this was not a vote about prohibition, but about changing habits. 'No one imagines that by merely changing some drinking hours from daytime to evening, and slightly increasing the daily total, a social habit for which New Zealand is notorious will be abolished overnight. But experience here and elsewhere strongly suggests that without that extension there can be no end to the pre-6 p.m. swill.'[52] The referendum on 23 September drew a 70 per cent turnout, significantly lower than the usual election turnout and suggesting that a portion of the electorate saw it as a foregone conclusion. Of those who voted, 64 per cent favoured ten o'clock closing –

A group at the Wakefield Hotel, Nelson, in October 1967 after the introduction of ten o'clock closing. Nelson Provincial Museum, NPM 9 GCW_L8003_fr9

certainly a clear majority, but also indicating that a portion of voters who did not support prohibition still supported early closing. As the *Evening Post* explained the outcome, 'The public's eagerness to get rid of present restrictions and inhibitions has all the appearance of a revolt. It has at last sunk home . . . that enormous harm has already been done through New Zealand's adherence to the furtive, anti-social drinking pattern which so appals the visitor from abroad.'[53]

In response, the government moved quickly to pass legislation allowing for evening opening from Monday, 9 October 1967, almost fifty years since early closing had first been introduced as a temporary war measure.[54] Yet not all in the trade were immediately happy with the change. In anticipation of later closing being introduced, hotel employers and unions had begun discussing pay and conditions for night work, but had not reached complete agreement by 9 October. The Hotel Workers' Federation therefore decided to initiate selective direct action targeting the main centres by withdrawing labour after 7 p.m. but continuing to work daytime hours as normal. Thus, the first night of later closing featured publicans running their establishments with the assistance of family members, a few non-unionists and a few who ignored the union. While Dunedin hotel workers voted against industrial action, Auckland and Wellington brewery workers threatened to cut the supply of beer to any hotel employing non-union labour. Finally, on 17 October a settlement was reached with the union securing some, but not all, of its demands.[55]

Despite union pressure, the introduction of later closing was not as chaotic as many had anticipated. The police reported very orderly conduct and made only two arrests across the main centres on the first night.[56] Fred Miller, a senior journalist for the *Southland News* and a member of the Invercargill Licensing Trust, responded in verse.

> The tumult and the shouting die,
> The captains and the kings depart;
> The 10 o'clock has settled down
> Although it had a hectic start.
> And now we are like other lands
> And free to drink within the law
> There is no swill at 6 o'clock
> Just like there used to be before.
> The Kiwis now can take good heart
> That they are not to be despised
> For they can say to all the world
> 'Hurrah! We now are civilised!'[57]

Indeed the following year, R. H. Wheeler suggested that the main trait of the new era was its ordinariness. 'Nothing seems to have happened. Pubs are not dens of carousing, and seem the same. The change is in the things that are missing: the tottery debouchment of customers on the footpaths outside the 6.15 doors; the impetuous post-closing traffic peak and the 6.30 demand on the eating houses – gone! The beery-cheery, post-six busloads have thinned.'[58] Although bar sales increased following the introduction of evening hours, such as New Zealand Breweries reporting a 9 per cent increase for its Auckland hotels in the year to October 1968, overall hotel profits went down due to higher labour costs and investment in hotel refurbishments.[59]

Writing in late 1971, Peter Cape detected some obvious changes to the hotel environment, not least in the shrinking length of the bar as drinkers no longer had such an urgent need to be near it. The previous bar space had now been replaced by 'islands' for standing drinkers as well as more tables and chairs. Carpet and panelling were also appearing in some refurbished hotels, along with provision of food and various other efforts to create 'atmosphere'. However, Cape also felt that some had adapted to evening hours with exaggerated standards of respectability. With reference to a friend who, although in a suit, was declined entry because his hair and beard were too long, he remarked:

> I'm happy that management is selective – and that the police, who also come into the picture, should also be on the watch for alcoholic malcontents – but I can't help thinking that if you've accepted that drinking is something that adults do then you need to regard drinkers as adults from the beginning. Tough lines on behaviour and even on dress are necessary, but one needs to keep in mind the fact that in the kind of situation the 1967 legislation permits one has to make one's judgements not before, but after, the event – and then only if there is an event to make a judgement on.[60]

The increase of women in public bars after half a century of almost total exclusion also posed challenges for publicans. In May 1967, prior to later closing, the commissioner of police, G. C. Urquhart, denied that any standing instructions had been issued about the serving of liquor to either women or Māori in public bars.[61] On the first night of later closing, some women said they were enjoying new-found drinking freedom while others admitted to being 'lost' in a previously male domain. As one observed, 'It is about time we were able to drink in the public bars, but there will have to be an improvement in them before women are prepared to make a habit of it.'[62] Yet as late as 1973, some still had little choice about embracing the habit as publicans in Christchurch at least confined women to the lounge bar. When a case eventually made it to court after a male patron was refused service on the basis that he was buying for a mixed group in a public bar, the magistrate observed, 'I had formed the opinion that the defendant has been endeavouring, for the sake of good taste and decorum, to provide amenities whereby women patrons may be entertained as becomes their station. . . . There is a good case for what the licensee is trying to do, in protecting women from coarser behaviour that may develop in public bars where drinking goes on for a long time.' Although the publican admitted charges of refusing to serve women, and the charge involving the male patron was proved, he was discharged without conviction.[63]

*

Later closing also drove a final round of brewery mergers which consolidated the DB–NZB duopoly by 1970, with no independent breweries remaining. Already in 1966 legislation imposing heavy additional tax burdens on overseas companies had prompted the international owners of the innovative Leopard Brewery to sell half of the business to NZB, which then closed its long-established Gisborne Brewery to allow Leopard more market share across the central North Island.[64] From 1967, however, it was DB that made the most significant moves, as it sought wider distribution beyond its Auckland base and more hotel outlets in anticipation of what it assumed would be increased consumption on hotel premises and a decline in its traditional packaged market due to later closing. In January 1969 DB launched takeover bids for both the Taranaki and Tui breweries, which were soon accepted. A month later Westland Breweries took over Morley's of Westport, the last independent brewery in the region, and the twenty-two hotels it supplied throughout Buller. In September 1969 DB then took over both Westland and Nelson breweries. The latter site was soon closed and eventually replaced with a new hotel. Thereafter Nelson brands were brewed in Greymouth.[65] But not all were pleased with the new arrangements. When Sir Henry Kelliher visited Westport in February 1971, he was presented with a hundred-signature petition asking for a better standard of beer from the Greymouth brewery. During the previous two months of high temperatures, the beer provided compared very unfavourably with that provided by Morley's, which had now been closed by DB. A DB 'official' was subsequently dispatched to inspect Buller hotel cellars and distribution systems.[66]

ABOVE & RIGHT: From the 1970s New Zealand Breweries/Lion pursued a more sophisticated emphasis on beer and food matching. Lion Technical Library

Beyond breweries, DB also significantly expanded its market, especially in the South Island, with the acquisition of various wine and spirit merchants, wholesale liquor outlets and hotels. By mid-1970, it controlled 84 of 161 hotels in Otago and Southland, and had taken over five wholesale firms in the south. By mid-1971 its total asset value had increased from $22.4 million to $51.4 million in two years.[67] Shortly afterwards a tourism division was established to coordinate the development of new hotels, initially in the South Island, to exploit the potential of a growing domestic and international visitor industry. By 1974 DB owned 53 hotels and leased 201. By 1980 the total was in excess of 280. A completely new brewery was built in Timaru and opened in 1976. Towards the end of the decade, DB was also investing in local wineries and importing wines for its various outlets to cater to shifting drinking trends.[68] Meanwhile, although NZB Chairman Sir Clifford Plimmer made the customary insistence in 1974 that they would prefer to sell beer to outlets rather than own them, increased competition from DB dictated that new acquisitions were necessary. Indeed, in 1970, when NZB owned 120 hotels and leased a further 80, it had established a hotel division under separate management from the brewing arm. In 1971 it had also secured majority interests in the Canterbury Seed Company and Canterbury Malting Company to secure a guaranteed supply of raw materials. In 1972 an export division was established to explore opportunities overseas.[69]

*

No doubt some of those who voted for ten o'clock closing imagined that the much-needed overhaul of New Zealand hotels would bring them something akin to the communal Rovers Return, which had become a staple of their *Coronation Street* viewing since the TV series started screening here in 1964.[70] But hopes for an English-style local had already been thwarted by the limits to licensing liberalisation. The Licensing Control Commission facilitated the transfer of some licences from the central city to new sites in the suburbs where licences were scarce. These taverns became more viable once evening hours were adopted as patrons, in part as a consequence of stricter drinking and driving enforcement, became more inclined to drink closer to home rather than closer to their workplace.[71] However, the Licensing Act 1948 had also allowed for 'site polls', whereby any twenty residents could object to a new licence on town-planning grounds such as the proximity to a school or church, and ask a magistrate for a poll. In 1953 grounds for such polls were broadened to a question of whether residents were in favour of licensed premises in the area under any circumstances. While a poll was not to be conclusive, as the then minister of justice said that would be a return to dry districts, the Licensing Control Commission had to have strong grounds to ignore it. Consequently, as the Justice Department submission to the 1974 Royal Commission on the Sale of Liquor in New Zealand summarised the 1960s, 'The path to the establishment of new licensed hotels was (as it still is) difficult and tortuous.'[72] The submission also condemned the polls and rights of objection and asked: 'Why do we insist on subjecting the existence of a community tavern to a poll when we do not think of doing so for TAB premises or rubbish tips or noisy and offensive trades? In these matters we rely on good town planning rules.'[73]

The consequence was that where licences could be secured for suburban taverns, the breweries compensated by making them very large – a phenomenon soon labelled 'booze barns' and quite at odds with the notion of the small local pub. In 1968 NZB developed a new tavern at Māngere Bridge in Auckland in which the four bars totalled 15,800 square feet or the equivalent of more than three basketball courts.[74] In March 1970 a summary of entertainment venues in Christchurch captured both the scale and 'atmosphere' of the recently refurbished Bush Inn tavern. 'It is indicative of the new line in hotel entertainment for the young. With a band and dance floor at one end of a long room, it has bars lining each side. The area is expensively carpeted and seats 580 but on late week nights it often holds 1300. A "with it" programme of entertainment features different musical groups each evening.'[75] Perhaps the most symbolic of the new ventures, given its proximity to Parliament, was the concrete brutalism of the Lion Tavern in Molesworth Street, Wellington, opened by Prime Minister Keith Holyoake in 1971, which also contained the head office for NZB.[76]

In response to numerous submissions to the Royal Commission regarding the desirability of small neighbourhood taverns, an NZB representative questioned their economic viability and was very sceptical as to whether small taverns could be run economically anywhere in the country. By this analysis, New Zealand was developing its own 'community tavern' model rather than trying to import a British version. It was claimed that some of the large taverns made more from food than beer and also promoted community links through such facilities as gyms and playing fields nearby.[77]

The size of the new taverns was no guarantee against frequent crowding, as new licences continued to lag behind a growing and shifting population. By the early 1970s there was a perception that many were no better than their predecessors. In an early 1974 television documentary, journalist Joe Cote graphically captured the apparent tone of proceedings.

> More and more the big taverns need bouncers and security men to keep the peace, and still on paydays and weekends the bottles and knuckles fly, the brawls erupt almost on schedule, and the parking lots become launching pads for motorists, few of whom could probably pass a breath test. Then there are the Sunday-morning left-overs: the vast parking lots sprinkled with the ground glass of hundreds of broken beer bottles; the warm sun drying out the pools of beer and urine. Inside the tavern they are sluicing out the spilled beer and vomit. For many people this is their local, their booze factory.[78]

While it was argued that in Britain the breweries often subsidised small corner pubs, in New Zealand the brewing and hotel divisions of the two major breweries were separate entities, with each needing to make a profit. The result, according to one experienced publican, was not healthy.

> It's a different type of drinking than that we were used to years ago. Taverns have come into their own with large scale entertainment to suck the public in. I think it's easier to get at booze now than it was before, because these vast places have been put up all over the suburbs and new-built areas. . . . The breweries want as much space as possible to get as many in as possible. Their prime object is to sell beer. I think they've built some taverns ridiculously large and that has caused a problem.[79]

In retort, NZB insisted that the location and size of taverns was generally determined by the licensing authorities and local ordinances, and not by them. However, location was also influenced by the breweries closing older establishments and transferring licences. A fortnight later, a *Listener* editorial suggested that most of the population did not have the means to enjoy clubs or licensed restaurants. 'So generations of debasement and degradation of a nation's drinking habits continue because conflicting public attitudes, powerful business interests, labyrinthian local and national laws, by-laws and regulations cannot be ordered to produce

TOP: Bob Thompson and Bob Rae comparing home brews on Campbell Island around 1960. P. G. Poppleton Collection, Alexander Turnbull Library, PA12-1424-040

BOTTOM: The DB Onehunga was typical of the new hotels being built by the two major breweries during the 1960s. DB Breweries pleasant and decent environments.' Legislative changes were aimed more at reducing abuses than eradicating the problem.[80] In 1979 both major breweries claimed that they were no longer inclined to build new taverns because they were 'fed up with running the gamut of objectors who have succeeded against them to the point where investment in taverns, particularly the neighbourhood concept, has been abandoned'.[81]

Running parallel to these tavern debates, the behaviour of many driving to and from such establishments also cast a disturbing shadow. Road deaths per 100,000 of the population hovered around 13 to 15 and sometimes lower from the late 1930s to the early 1960s. More powerful cars and increased alcohol consumption saw the figure balloon to 27.8 in 1973, when New Zealand road deaths peaked at 843 for a single year.[82] Although breath and blood tests for drivers had been introduced in 1969, with a legal limit of 100 milligrams of alcohol per 100 millilitres of blood for all drivers (twice the level allowable by 2014), traffic officers still required 'just cause to suspect' that a driver was under the influence, and there was no provision for random stopping until 1984. Delays in obtaining results also made prosecutions difficult to obtain.[83] But now there was a distinct shift from older notions of the 'drunken' driver to a recognition that one could be impaired without being drunk. A short film, *Drink Driving and Breath Tests*, shown on television in March 1969 stressed that it was 'aimed at the man who generally behaves and looks sober but whose seemingly modest consumption of alcohol is such as to affect his judgment and impair his driving skills'.[84] The Transport Department and police argued that the end of six o'clock closing had shifted the problem but in no way reduced it. The 6–7 p.m. hour still had a high rate of accidents, but there were now significantly more from 10 p.m. to midnight, especially on Friday and Saturday.[85] Not until the mid-1990s, following significant advances in scientific analysis of breath and blood tests, major improvements in vehicle safety, and a plethora of media campaigns, would the per-capita road toll return to the level of the early 1960s. By 2010 it was equivalent to that of the 1920s, and arguably much lower if one considers deaths per kilometres driven.[86]

Despite accusations of a 'ten o'clock swill', the breweries did embark on substantial, if not always sophisticated, endeavours to create entertainment venues and to tightly control those who entertained and were entertained within them. Unlike six o'clock closing, where patrons partook on the way home from work, evening hours required something to entice them out again once they had gone home. NZB's architect Colin Gunter observed in 1973 that they were aiming to attract customers, especially women, with comfortable seating, carpets, better acoustics, music and a 'total entertainment scene' in which bars were becoming themed 'stage sets'.[87] Under the regime of Richard Holden as national entertainment manager for NZB, an extensive tavern music circuit developed. With an annual budget in excess of $2 million ($28.5 million in 2022) and as many as 300 artists on the payroll, Holden contracted both local and overseas bands for residencies and tours.[88] He also had very firm ideas as to how bands should look and behave in the interests of attracting an audience to drink beer.

> We have to get away from the idea of four guys getting together and forming a group just because they liked each other, and instead get guys who are getting together to do a job. Musicians have to realise they are a product that can be sold. . . . There has been a tendency for musicians to play for themselves instead of the customers. There has been some good original music but a lot of original rubbish. They will have to realise that we're not in the musical genius business. We're in the entertainment business.[89]

Holden also created new bands to meet his standards of repertoire and image and adorned them with drink-themed names to reinforce the connection to NZB. One band, Distillery, with its combination of prohibition-era themes, old-time revival ballads, pop hits and comedy skits, was acclaimed by *Brewnews* as 'the first step in a trend towards a total entertainment concept incorporating decor, artists, lighting and food in selected smaller lounges – in other words, a full show'.[90] Distillery was followed by Beam before Holden ventured to Wales to recruit the members of Pilsener [*sic*].[91]

As Nick Bollinger recalls the life of a working musician on the circuit, 'This often meant dealing with grouchy publicans with permanent hangovers, who saw musicians as a necessary evil and told them so.' Signs and bouncers enforced strict dress codes and venues were frequently cleared at the end of the evening by means of a loud siren.[92] Some older barmen were not happy with the appearance of bands and claimed that they drove older drinkers away. Conrad Bollinger, who did not perhaps share the musical proclivities of his son, suggested that the environment had shifted too far in one direction. Although the public were demanding variety, such as some wanting coffee while others were having beer, others were

> demanding the right to have somewhere where there's quiet as well as somewhere where there's an orchestra making a noise like a Bucharest brothel in the next room. . . . there's nowhere where you or I could take granny. . . . I frequently want to go in and read something – an article in a magazine or a piece of poetry someone has just drawn to my attention – and I want to have a beer while I'm doing it! But you can't do that when there's something going BOO DOO GOU GOU at the other end of the room can you, and even less when it's right beside you. The decibels get pretty rockety over there.[93]

By the time Holden was promoted in the late 1970s to run the NZB export division, the dominance of the booze barns and tight brewery control over an entertainment circuit was being eroded as the range of licensed venues expanded dramatically. Theatre licences were granted from 1969, airport licences from 1970 and cabaret licences from 1971. The network of chartered clubs also expanded from 46 in 1948 to 365 by 1980. From 1976, legislative changes following from various recommendations of the 1974 Royal Commission also allowed for licensing of sports clubs and for restaurants to obtain BYO permits that allowed customers to bring their own alcohol for on-premises consumption. Licensed clubs soon became the most common outlet, with more than 2500 by 1987. There were over 1300 BYO permits in the same year, while the number of restaurants licensed to sell alcohol had increased from 198 to 760 in the decade since 1977. In sum, the number of licences per capita increased by 150 per cent from 1971 to 1987.[94]

*

With the emergence of a brewery duopoly by the 1950s that was completely in place by 1970, questions about its power in the market, as distinct from the power of the licensing laws to shape the market, came to prominence. After more than two decades, price controls were removed from alcohol in December 1961.[95] In June 1965, after the major breweries announced a price increase, the Hotel Association indicated that it had not yet decided how much of the increase would be passed on to consumers. This prompted the minister of industries and commerce, Jack Marshall, to advise the Association that any arrangement to increase the price of beer by the Association, or any advice it gave that had the effect of limiting competition, would be problematic. 'Competition between resellers of beer can be expected to provide reasonable protection for the consumer. If there is evidence of interference with this competition, the Trade Practices Act may be invoked.'[96] In Parliament the Labour opposition condemned the government for consenting to beer price increases by its silence, whereas it had loudly condemned Labour for its 1958 Budget. In response, Allan McCready, MP for Ōtaki, mocked Labour for focusing on beer prices when the world was facing problems in Southeast Asia and with the global economy. 'It brings the Labour Party right back in the eyes of the people to what they are — trying to secure a vote or two by attacking the Government on the price of beer.' He could cite the now leader of the opposition, Arnold Nordmeyer, saying in 1958 that beer was a luxury item no one had to drink.[97]

However, in October 1965 the examiner of trade practices determined that there was a case for the breweries and hotel owners to answer. A meeting of the Brewers' Association in May 1965 had apparently concluded that as the 'actual' price of beer, aside from tax and duty, had not increased since 1953 but costs had, there were grounds for 'suggesting' minimum price increases. The examiner judged this to be a breach of the Act. That the resolution of the Association was a 'suggestion' was immaterial, and the brewers were directed 'to discontinue the practice of fixing, nominating or recommending the prices or terms and conditions on which breweries should sell beer', to refrain from reviving this practice in the future and to pass a resolution that they would not.[98] In March 1966 this decision and the alleged agreement on prices by the Hotel Association were referred to a full hearing of the Trade Practices and Prices Commission, which found that the 'suggestions' of minimum prices had not in fact disadvantaged the public or limited competition. 'Licensees can still sell liquor at whatever price they like. They are not disciplined in any way if they do not adhere to the suggested prices. It is the opinion of the Commission moreover that some measure of guidance by the association is in the interest of the public.'[99]

Yet the sense that higher beer prices were unreasonable and unjustified did not easily retreat. In December 1969 the Hotel Association announced that there would be a price increase from Boxing Day, having delayed it until after Christmas as a gesture of goodwill to drinkers.[100] By late January 1970 the Ohai Miners' Union had imposed a boycott on the four hotels in the Ōhai–Nightcaps coalfield area in protest at a price increase throughout New Zealand. A fortnight later, hydro workers at Ōtemātātā unanimously declared a boycott of their local hotel in response to a jug of beer being increased to 43 cents, compared with only 40 cents at the Kurow Hotel just 18 miles away. Shortly afterwards, two hotels in Wellington were boycotted for five days. In Auckland, four hotels were boycotted by unionists from mid-April in response to what were regarded as unreasonable prices. By early May, two of the Auckland establishments were obliged to lay off staff due to declining patronage. Wellington unions then began further boycotts in response to a High Court injunction against the Auckland drivers' union for its refusal to deliver to the Auckland hotels.[101] Such machinations prompted the newly elected Labour MP for Manukau, Roger Douglas, to criticise the major breweries. 'This increase and the increase at the hotel end of the trade cannot be justified. There can be no doubt that New Zealand Breweries Ltd, and Dominion Breweries Ltd, have used their monopoly position to force upon the public a price rise equal to all price rises. In a single act they are likely to double their profits.'[102]

By mid-1970 beer prices were formally under investigation by the Price Tribunal. As in 1965, an investigation by the Trade Practices and Prices Division of the Department of Industries and Commerce determined that the late-1969 price increases were about twice what could reasonably be justified on grounds of covering increased costs. Moreover, the breweries had not provided all of the detailed financial information requested to assist the investigation.[103] Addressing the Tribunal in late June, both NZB and DB insisted that price increases had done no more than account for increased costs and were certainly no basis for the reimposition of price control on beer. The counsel for the Federation of Labour and Combined State Service Organisations argued that the breweries should be brought back under price control and be required to revert to their former prices immediately.[104] The matter remained unresolved; in early August the Tribunal abandoned its inquiry on the basis that it was futile to continue as the Hotel Workers' Federation had just secured a wage increase, thereby justifying a price increase. However, the minister of industries and commerce, Norman Shelton, evidently unimpressed with the workings of the Price Tribunal, immediately placed beer under price control again for the first time since 1961.[105] As *The Press* observed, this seemed a logical step. 'The hotel industry has given up all pretence – temporarily, at least – of being competitive in the sale of liquor; so it is some consolation to customers that a major part of liquor sales will be subject to another form of restraint.'[106]

The imposition of price controls did not soothe public disquiet. In mid-1975, the Combined State Services Association and the Campaign Against Rising Prices both criticised the breweries for requesting increases from the Price Tribunal when they were not willing to publicly disclose accounts to justify their claims of increased costs. Wage and salary earners were always required to justify their claims in public. Sir Henry Kelliher insisted that DB's profit margin had been 'seriously eroded' during the previous year due to unprecedented increases in direct costs.[107] The Tribunal eventually awarded increases amounting to $15 million extra revenue for the breweries based on recent consumption patterns. Ivan Reddish, chairman of the Combined State Services Association, was highly critical of the Tribunal for its lack of thorough investigation and the lack of involvement by consumers in shaping its decisions.[108] However, in some respects these debates were a little academic. With the exception of the decline in consumption immediately after the 1958 Budget, there is no discernible correlation between beer prices and consumption through the 1960s and early 1970s. Indeed, per-capita consumption continued to rise steadily to a peak of 133.9 litres in 1975.

If price did not determine consumption, brewery workers sometimes did. In July 1968, when NZB refused a 7.6 per cent pay claim by its Auckland drivers, they struck and were soon followed by others throughout the country to the point where some hotels ran out of draught beer. The strike petered out within a fortnight.[109] Significant reorganisation of the national brewery workers' union during 1970–71 heralded more sustained action. On 29 March 1971 the Auckland branch joined a twenty-four-hour strike in protest at the government's Stabilisation of Remuneration Act, which was designed to counter a 'wages explosion' by limiting any increase to no more than 7 per cent and no further increases for at least twelve months. By early June negotiations with the two Auckland breweries had broken down, as the union initially claimed an 18.5 per cent increase – later revised to 12.5 per cent – while the employers offered 9.5 per cent. A subsequent ban on overtime and adoption of work-to-rule meant a 25 per cent decline in production. Rolling strikes in Auckland during early July quickly spread throughout the country. A meeting of DB shareholders then decided to withdraw employee bonuses, effectively amounting to 5 per cent, leading the union to reinstate its original pay claim of 18.5 per cent. Workers were then effectively locked out of several breweries and employers threatened to issue

Nevile Lodge captures the dilemma facing beer drinkers with rising prices at the end of 1980. Alexander Turnbull Library, B-135-719

dismissal notices from 2 August amid suggestions that outside labour would be engaged to work the breweries.

Finally, following the intervention of Tom Skinner, president of the Federation of Labour, a settlement was reached on 11 August with an 11 per cent pay increase. The six-week strike had involved 577 workers and $15,000 had been raised in strike support funds.[110] Only in 1974, more than eighty years after the first union activity in the New Zealand brewing industry, was one national pay scale achieved across all breweries. Thereafter, although there was periodic industrial action, much of it was in support of other unions rather than a reflection of the brewing industry. The various union branches also engaged in wider political stands such as opposition to the Vietnam War, financial support for peace ships protesting French nuclear testing at Moruroa Atoll, and opposition to apartheid and sporting contacts with South Africa. The unions also briefly opposed the import of cheaper foreign beer, but eventually backed down. They also opposed the suggested removal of price controls amid concern that increased prices might lead to less demand and ultimately to job losses.[111]

*

If the price of beer and sometimes its supply was seldom a source of contentment, there was at least more scope from the early 1960s for discerning or adventurous consumers to detect an improvement in taste, as new products and packages gradually entered the market after two decades of little beyond the limits of what could be brewed at 1036. DB had responded to the introduction of import controls with the development in 1960 of DB Export, described as 'mellow smoothness', 'flavoursome excellence' and 'meeting the overwhelming approval of discriminating drinkers'.[112] NZB responded with Steinecker Lager as the seeming epitome of modern brewing in parallel with modern life.

> These ultra-modern, gleaming, copper-domed brew kettles were built in Germany specially for New Zealand Breweries limited. Nowhere is the modernity of the new Brewery more total and more spectacular than in the Steinecker Brewhall. All the equipment in it has been specially made overseas to provide for the widest variety of brewing art and the highest standard of taste excellence. The ultra-modern Steinecker Brew Kettle takes guess-work out of the brewer's calculations for this vitally important process so that he can be sure of producing beer of consistently high quality.[113]

But NZB soon encountered controversy. In February 1958, the company registered 'Steinecker' for use on its lager, ale and stout. In mid-1959, the Dutch brewer Heineken, which had registered a trademark in New Zealand in 1953, filed a notice of opposition on the ground that the Steinecker label was so similar to their own as to be likely to deceive or cause confusion with drinkers. The commissioner of trademarks then declined to register the trademark. When the case reached the Supreme Court in August 1962, counsel for NZB insisted that the use of 'Steinecker' was 'very much a New Zealand effort to get a word of Continental flavour' and that the word had been invented by employees of the company derived from a combination of the German word 'stein' (mug) and 'ecke' (corner). However, the commissioner flatly rejected this claim on the basis that Steinecker was both a German surname and part of the name of a firm that had previously supplied brewing equipment to NZB.[114] When Heineken's appeal was finally upheld in December 1962, Justice Leicester stated that while the name 'may not have been motivated by an intention to deceive or cause confusion with other lagers, it does seem to indicate an intention to clothe the New Zealand lager with the attributes of a lager brewed on the Continent, which attributes the respondent's lager does not necessarily possess. This in itself may cause confusion.' As to the labels, the Court noted a 'striking similarity' in design. While three experienced hotel managers, a bar manager, and a barman had all given evidence that there was no similarity or confusion, the judge concluded that there was 'likelihood of deception or confusion in so far as the normal retail customer requesting the beverage for immediate consumption is concerned' and that there was also sufficient similarity in the pronunciation of 'Heineken' and 'Steinecker'.[115]

LEFT: In 1960 New Zealand Breweries promoted its new Steinecker Lager with reference to continental tastes. However, legal action forced a name change to Steinlager at the end of 1962. Lion Technical Library

RIGHT: One of the earliest Steinlager advertisements from 1963, shortly after the name was changed from Steinecker. Lion Technical Library

NZB was allowed to challenge this decision in the Court of Appeal but was also thwarted. Justice North conceded that given the small amount of Heineken available in the New Zealand market, NZB's selection of a name was perhaps by accident and not by design.

> Even so, in my opinion, the appellant acted most unwisely in selecting a three syllable foreign word containing the letters 'einecke' when the whole of the German language was open to it. Visually the two words 'Heineken' and 'Steinecker' are not wholly dissimilar and phonetically in my opinion, they are very close indeed. Therefore, remembering, as I am enjoined to do, that allowance must be made both for imperfect recollection and for the effect of careless pronunciation and speech, I have come to the conclusion that the appellant has failed to discharge the onus which rested on it of satisfying the mind of the Court that the trade mark it desired to register was not likely to deceive or cause confusion.[116]

Justice Turner, no doubt informed by two bottles of beer that sat in the court throughout proceedings, emphasised the similarity in the labels but also issues inherent in ordering the two.

> Remembering the circumstances surrounding the sales of beer in New Zealand, I have reminded myself of the persistent, and sometimes loud, background noise current in New Zealand hotel bars, of the difficulties inherent in telephoned communications in the case of telephoned orders, of the tendency of New Zealanders to clip and slur final syllables, and to the identical second, and closely-similar first and third syllables of two words. In my view, in the case of the two words now under consideration, these factors must result in a likelihood if not of deception at least of confusion.[117]

Accordingly, NZB announced in late October 1963 that the name of its beer would be changed to Steinlager, under which it was to enjoy considerable national, and later international, success.[118] Its next venture into the 'continental' beer realm was decidedly less problematic, when in 1964 it entered an agreement with Skol International, itself a company established by Allied Breweries (UK), Labatt (Canada), Pripps-Bryggerierna (Sweden) and Unibra (Belgium), to produce Skol lager as a standardised global beer brand. Developed in Scotland and named with a clear Scandinavian overtone, it was marketed in New Zealand as 'The International Beer' and with the flavour 'of a light, dry continental lager, for discriminating drinking rather than lusty quaffing' – although with acknowledgement that it was brewed locally under licence.[119]

With hindsight, the next 'local' product to appear seems an extremely unlikely alliance. Guinness had been imported to New Zealand since the 1850s, and since 1934 there had been a reciprocal agreement whereby it was exchanged for Nelson hops. Initially there was no Guinness on the market after the introduction of import controls in 1958, but negotiations during 1959 secured a new agreement to export surplus hops in return for the same value of stout. This was not an ideal solution, and Guinness, as it had done in other countries, began to explore the possibility for contract brewing in New Zealand if more profit could be made than simply exporting. Negotiations with NZB from 1960, in part prompted by one of the Davis brothers based in London, did not amount to anything. Eventually Guinness's New Zealand agent pointed them towards the small and still independent Tui Brewery at Mangatainoka. Guinness executives visited New Zealand during 1964 to inspect the brewery and confirm the agreement. Guinness brewers were seconded to Tui, and several trial brews were completed on which the Customs Department agreed that no duty should be paid. Despite ongoing debates about the quality of the product, which undoubtedly owed as much to perception as taste, Dublin was evidently happy with the results.[120]

However, the Tui partnership had limitations. The brewery, although renovated, was not modern, and the company lacked marketing capacity. By late 1966 Guinness was only 4 per cent of the Tui sales volume and Tui's own stout was selling in larger quantities. Citing distribution issues and the need for stronger marketing, Guinness amicably terminated its relationship with Tui and reached agreement with NZB. The first Guinness was brewed in Auckland in September 1967, and an enhanced marketing campaign was evident in sponsorship of various sporting events in Auckland and events such as the New Zealand Rose of Tralee contest. However, investment and marketing efforts were never matched by sales. New Zealanders did not have a taste for a highly hopped, bitter stout, especially as NZB had its own, generally sweeter, stouts in the market at a lower price. The Guinness marketing operation was closed in the early 1970s and marketing responsibility shifted to NZB. In 1976 NZB gave notice of the termination of the brewing contract. From April 1977 Guinness supplied the small New Zealand market from Australia.[121] Other ventures to brew under licence, such as Tui producing Tennent's Green Label Strong Ale from 1967 and Waikato Brewery producing Bass Export from mid-1969, also seem not to have endured for long.[122]

Experiments with low-alcohol beer also drew mixed responses. The popularity of various weaker harvest ales and hop beers seems to have faded by the early twentieth century, with the notable exception of Innes & Co. of Hamilton, who were still marketing a hop beer under the name 'Kentish Ale' during the 1930s.[123] In July 1959, the MP for New Plymouth, Ernest Aderman, asked Prime Minister Walter Nash whether he was aware that dairies and other shops open seven days a week were selling large quantities of 'near beer' to children. 'Does the Prime Minister know that the strength of such liquor is only slightly below that of beer retailed through hotels, and that any child consuming two bottles of this liquor will have in his bloodstream virtually as much alcohol as is contained in a quart bottle of beer?' He claimed that sales of low-strength beer in Auckland and Wellington alone exceeded 100,000 bottles every month and were not subject to either duty or sales tax. The minister of health, Rex Mason, stated that no complaints had been received on the matter, but he would initiate an inquiry.[124]

Two months later the Dominion Federation of School Committee Associations entered the fray when its newsletter requested information as to whether 'near beer' was being sold in its various districts.[125] At a subsequent meeting of the Canterbury Education Board, a bottle described as 'Kentish ale' sat on the table in the midst of a spirited meeting. Mrs K. Greenwood, referring to the availability of the product in many shops and even one high-school canteen, declared: 'The police tell me they are worried about this near beer but can do nothing about it. It smells like beer, tastes like beer, and if one bottle of beer will affect a child, two of this stuff will do the same.' In retort, Mr W. Rosenberg exclaimed, 'Ah. Excellent stuff. I can heartily recommend it. But I do object to youngsters being charged 1s 3d for a small bottle.' Mr R. K. Baker said he would oppose any protest on the matter as New Zealand licensing laws were notoriously the worst in the world and any further restrictive move would make them worse. 'It is for parents to determine their children's handling of liquor. From the age of 12 my children have been allowed a small glass of beer on special occasions or a glass of wine when it is being taken by the family. In this way they learn to appreciate – and respect – liquor and learn what they can do with it and what they can't do with it. I have no fear of them coming to harm through liquor.' The Association agreed to seek a government analyst's report on the alcoholic content of near beer before proceeding further.[126]

ABOVE: The original Waikato Brewery was established in 1873. It was taken over by the Innes family during the 1890s and merged into New Zealand Breweries during the 1960s. *Brewnews*, Lion Technical Library

RIGHT: Lion Brown, which seems to have evolved during the 1940s from Speight's Imperial Draught Ale via the name Brown Bomber, was firmly established by the mid-1970s as the drink of the working man. *Brewnews*, Lion Technical Library

In early March 1960, the medical officer of health responded that one bottle would be enough to make a child aged two or three mildly intoxicated, while for an older child, aged about twelve, two bottles would be required. It would only cause a danger to the health of children if it was consumed in large quantities for a long time, and he doubted whether any children could afford to buy enough 'near beer' to become addicted to alcohol. Consequently, the Health Department would only take action if there was a danger to the health of children.[127] There the matter ended, without any indication of who had produced the beverage.

The next venture with 'near beer' came in Dunedin as a response to six o'clock closing. Powley's had apparently registered a label for non-alcoholic bottled beer, but there is no evidence that a beer was ever produced. In 1961 Speight's created a 'near beer' called Tom Thumb for sale at Forbury Park night race meetings which were unable to secure a licence. This was made by adding an infusion of hops, sugar, wort, caramel and a large amount of water to a small quantity of beer and strongly carbonating the mixture. Speight's historian Donald Gordon recalled, 'One had to be extremely thirsty to enjoy it.' Production ceased with the end of six o'clock closing as punters could now get a drink after the races.[128]

In mid-1973 Leopard entered the market with a 1.4 per cent 'hop beer' selling for 20 cents a can. Perhaps justifiably, they claimed that there was a demand for low-strength beer due to drink-driving campaigns, and that they planned to sell it in dairies and supermarkets. But the fact that it also appealed to teenagers was not denied. Immediately various schools and church groups protested the easy access to hop beer by children as this would encourage further experimentation and create a 'psychological effect' of conditioning them to drink beer.[129] Paddy Blanchfield, Labour MP for the West Coast, called on Parliament to condemn Leopard. 'Parliament should sympathise with so many parents and teachers who fear that the increasing practice of children drinking this beer with school lunches will start them down the road to the greatest social disease of our time – alcoholism.'[130] At the same time, the prime minister, Norman Kirk, received numerous letters criticising Leopard. One from the Avonhead Baptist Church perhaps not incorrectly referred to 'hot beer'. Another from a Christchurch couple concluded, 'PS We are both under 30 and not wowsers'.[131] Leopard responded that it was not trying to influence children and that hop beer probably contained less alcohol than some soft drinks. This inflamed soft-drink manufacturers, who denied that any alcohol was used in their products. In the interests of research, a Christchurch reporter drank eight cans of hop beer and admitted to feeling bloated and light-headed but not drunk – although he was 'on the way'. His consumption of hop beer did register on a breath-testing device.[132] As with its near-beer predecessors, the Leopard offering seems to have dropped quietly and quickly from the market and the pre-prohibition era notion that beer was acceptable for children showed no sign of revival.

The breweries encountered rather less controversy with their packaging innovations. In 1967 Waikato Breweries, now in partnership with NZB, introduced the 'Pleasure Pak', containing two dozen 12-ounce bottles of Waikato green label in a cardboard container with hand holes in the sides for easy carrying. In 1969 DB offered the 'cluster-pak' as a convenient way to carry six of its beers.[133] But the most significant development was canned beer. In 1936, the year after cans had first been introduced by the Krueger Brewing Company in Richmond, Virginia, hotels in Hamilton and Timaru imported cans for sale. NZB also imported samples for testing, but declared that the results were not promising.[134] Cans were first regularly used by Leopard from the late 1950s, and NZB imported a canning plant in 1963 and DB secured one the following year, although it seems that the cans were not manufactured in New Zealand.[135]

Contrary to a belief that later closing would increase on-premises consumption, which had certainly been the motivation for DB's final round of brewery and hotel acquisitions from 1969, the period witnessed a pronounced shift from bulk to packaged beer. In 1962, 32 per cent of beer available for consumption was packaged, almost all of it in bottles. By 1971–72, 46 per cent of available beer was packaged, and three years later that proportion had jumped to 58 per cent. Bottle stores were now reporting a large stock of unused half-gallon flagons as younger customers, including an increasing number of women, preferred bottled or canned beer to draught. By 1977, when NZB's total market share was 59 per cent, it produced nearly 70 per cent of all bulk (draught) beer and 50 per cent of packaged beer. On the other hand, DB, with 35 per cent of market share, produced just under 25 per cent of all bulk beer but 43 per cent of packaged – a clear legacy of its inability to secure more hotel access until the early 1970s. Leopard, half-owned by NZB, controlled the remaining 5.7 per cent of market share in equal portions bulk and packaged.[136] But the rise of bottled beer in particular posed another problem for the breweries as attention turned to littering and wastage. By the end of 1975 an anti-litter initiative by NZB included vans to collect empties from private homes in different parts of the North Island. Over the summer, 10 million bottles were collected for recycling, and 24 million during 1976 as a whole amid praise from the government and environmentalists.[137]

Changing drinking preferences also prompted new retail developments beyond the traditional hotel bottle stores that had satisfied the takeaway beer market. By 1970 supermarket-style bottle stores were emerging in which customers could walk around the shelves and choose their purchases. Typical of increasingly large new premises was the Carlton Hotel in Christchurch, which in late 1972 opened a bottle store and food bar either side of a covered three-lane drive-through.[138] But the most important change came with the expansion of licence types after 1976, which eventually removed the long-standing restriction on wholesale wine and spirit merchants selling only in amounts greater than 2 gallons. This cleared the way for the establishment of independent bottle stores and chains such as Super Liquor and Liquorland, which emerged by the early 1980s.[139]

In parallel with changing patterns of consumption, brewery advertising and marketing also evolved. The Medical Advertising Act 1942 had curbed some of the more extravagant health claims for beer by shifting the onus to brewers to prove them, as opposed to the likes of the Alliance having to disprove them. One response from DB was to collect overseas and local articles with scientific claims about beer, have these framed and put in bars.[140] Another campaign from the late 1940s onwards emphasised elderly people who attributed their long lives to drinking Waitemata. The most intriguing in this series featured Ihipera 'Bella' Patuwai.

> WAITEMATA. The Secret of Good Health and Long Life. At the great age of one hundred and nine years, still remarkably alert and active, Mrs. Bella Putuwai [*sic*], living in the Wakataki [*sic*] district, Wairarapa, looks upon her glass of Waitemata as her greatest comfort and enjoyment. 'I drink a bottle of Waitemata every day . . . sometimes even two', she added with a smile. 'It is food and drink to me and I have not been without it for the past 20 odd years.' Yes, there's something in Waitemata that adds immeasurably to the pleasure of living. Waitemata contributes to the genial good fellowship and comfortable well-being that make life worthwhile. Whether you live to be 109 or not, you'll find Waitemata will refresh and regale you.[141]

It is far from certain whether Mrs Patuwai, who purportedly lived to 116, ever knew she was part of this campaign. It is also suggested that she boiled her beer before drinking it, thus removing much of the alcohol. However, this advertisement is significant as seemingly the only brewery effort to specifically invoke Māori. Yet the period of dramatic post-war urbanisation had triggered fundamental changes to Māori society and brought previous rural dwellers into contact with Pākehā drinking cultures. In 'comic' portrayals such as W. Norman McCallum's *The Half-Gallon Jar*, the narrator 'Hori' and other Māori characters were stereotyped as happy-go-lucky beer drinkers.[142] In a formal sense the Sale of Liquor Act 1962 removed any discrimination by stating:

> No licensee or manager and no employee of any licensee shall refuse to admit any person to any licensed premises or any part thereof or order any person to leave any licensed premises . . . or refuse to supply accommodation, meals or liquor to any person on any licensed premises by reason only of the race, colour, nationality, beliefs or opinions of that person.

But amid various reports of increased and often problematic drinking in new urban settings, older responses remained. In 1965, a Christchurch publican was fined £5 for refusing to serve a Māori woman amid a lingering view that the presence of Māori in hotels was disruptive.[143] Such views gained renewed momentum during the booze-barn debates of the 1970s, and amid reports of frequent racial tensions between different Pasifika groups, which undoubtedly fed into a growing hostility to Pasifika migrants and 'overstayers', as unemployment rose and the National Party harnessed elements of moral panic leading up to its 1975 election victory.[144] As Justice Speight assessed the situation in October 1975, 'It is easy to see where the blame lies. Not primarily with these young immigrants, but with the evils of our so-called industrialised society where high wages, combined with the excessive consumption of alcohol, in a city environment, is a high risk even to those accustomed to it – and is disastrous for those who are not.'[145] In November, the New Zealand Liquor Industry Council fully funded a conference in Rotorua on 'Maori and Polynesian drinking'. This was strongly opposed to alcohol on marae, and discussed various social problems associated with Māori drinking. Against this background, the breweries were reluctant to highlight or specifically target Māori in their advertising.

Amid the slow liquor liberalisation and relative cosmopolitanism of the 1960s there was an appeal to Pākehā via European or 'Continental' sophistication in products and advertising themes. NZB promoted its Bavarian Bitter Old Style Beer, evidently the most popular bottled beer in Canterbury, 'as providing hearty fellowship in a New Zealand beer', while Leopard waxed lyrical on its 'Continental' option. 'ONE OF THE THREE BEST BEERS ON THE MARKET . . . Light, lively Leopard Continental, the lager with the joie de vivre of the continent. One sip and you'll agree it's one of the three best beers on the market. The other two? Leopard Canned Lager Beer and Leopard Double L of course.'[146] On a related theme, DB played on the endorsement of its beer by international experts. In 1964 it trumpeted, 'DB Export wins BEST LAGER BEER AWARD at Brewers Exhibition, Olympia, London!' In 1968 it announced, 'That DB Export is the world's best beer has been confirmed by twenty-five brewing judges at the International Brewing, Bottling and Allied Trades Exhibition at Earls Court, London, 1968. DB Export secured the supreme award against all-comers. Proof that New Zealanders are good judges of good beer when they say "DB EVERYTIME."'[147]

At the other end of the sophistication spectrum, Lion Brown, which seems to have morphed from Speight's Imperial Draught Ale via the name Brown Bomber, inspired by

TOP LEFT: From the 1950s both major breweries drew international comparisons to emphasise the quality of their products. *New Zealand Licensee*, February 1954

TOP RIGHT: An early 1963 advertisement for Bavarian Bitter. Lion Technical Library

BOTTOM LEFT: A 1964 advertisement for Leopard Continental Lager. Lion Technical Library

BOTTOM RIGHT: A 1971 New Zealand Breweries advertisement from its extensive 'mighty' campaign for Lion Brown. Lion Technical Library

servicemen during the 1940s, was aimed firmly at the working man via its 'mighty' campaigns from 1970. 'Spring is sprung! Boats to paint, concrete to pour, ground to dig – better mastermind a Lion Brown working bee!' was followed by an appeal to gardeners. 'Mighty relaxing! Flowers or veges? Annuals or perennials? Problems! Problems! Have a quiet drink and think. No problems there . . . Only Lion Brown tastes that good!'[148] Indeed, such was the strength of the Lion brands, Brown and Red, and the Lion symbol associated with them, that in 1977 NZB made the decision to rebrand the northern division of the company as Lion Breweries.[149]

The strongest association from the late 1960s via both advertising and sponsorship was between sport and beer. In its efforts to build a New Zealand market, Guinness sponsored a variety of sports in Auckland during the late 1960s, donating cash prizes and trophies, including a premier grade rugby-league trophy that subsequently disappeared without trace. Among its innovative sponsorships were motor-paced cycle racing and single-wicket cricket tournaments. Its most significant sponsorship was of yachting, including the 18-footer *Guinness Lady*, which enjoyed considerable success in New Zealand from 1968.[150] From the early 1970s, DB sponsored a range of *Annuals* for cricket, golf, rugby, soccer, racing, trotting and yachting as well as *DB Sporting Records of New Zealand*, published in 1976. Likewise, the Lion symbol or specific NZB brands appeared in association with everything from rugby union and league match programmes to the New Zealand Olympic team and surfing. The breweries also provided financial backing, frequently as loans, to chartered clubs and the many sports clubs that secured licences from the mid-1970s.[151] If the definition of competitive sport is broadened, NZB sponsored the Victoria University drinking horn in 1975 with 48 gallons of beer for contestants and spectators as Law A won the teams event and made an attempt on the New Zealand record.[152] But three years later, management decided that students would not be allowed to attempt to drink Dunedin's Lion Tavern dry as a capping-week stunt. A public relations staff member stated, 'No doubt there would be a profit involved, but we are not interested in that type of profit.'[153]

The breweries also continued their long tradition of community philanthropy. NZB supported the Crippled Children's Society from the early 1960s, and by the 1970s was making donations to groups as diverse as IHC and the Polynesian Education Foundation. In 1968 NZB established a charitable fund from which investments of $13,500 ($270,000 in 2022) per annum were made to various causes. In 1974 the Salvation Army received $25,000 ($320,000).[154] Both major breweries also lent their support to cultural activities. In 1959, NZB provided £5000 ($260,000) sponsorship per annum for five years to the financially struggling New Zealand Opera Company. As Fred Turnovsky, its board chairman, explained, 'I was selling respectability – a valuable asset for those in need of it, such as breweries – which would earn the sponsors the gratitude of a public whose enjoyment of opera was in no conflict with its taste for a glass of ale.'[155] However, not all approved of the sponsorship. One 'Indignant Shareholder' wrote to *The Press* demanding to know whether the 'gift' had been authorised by a general meeting of shareholders. As many were apparently widows or people depending on dividends, it was argued that the directors should raise the rate of dividends rather than 'play the role of philanthropist'. The chairman of directors, M. O. Barnett, replied that the expenditure was a very small portion of the breweries' annual advertising budget. They frequently helped sports clubs and others through sponsorship and he believed when the opera company next performed in Christchurch, if the writer of the letter attended he would go home 'very satisfied and proud of the company in which he claims to be a shareholder'.[156] NZB then used its *Brewnews* magazine to regularly promote the programme of the Opera Company.[157]

For its part, DB sponsored the Kelliher art competition from 1956 until 1977. This was the first national art competition, with an initial single prize of £500, and prizes for second and third place added the following year. The focus was on more traditional styles of New Zealand landscape painting, with 'modern' art discouraged. As Warren Feeney argues, emphasis on the rural idyll rather than the increasingly urban reality of most people positioned DB as an ally of 'ordinary' New Zealand. In his speech at the 1963 awards, Henry Kelliher lamented that 'it is becoming increasingly difficult to secure good paintings that really do justice to our glorious scenery. . . . Much so-called modern or abstract painting has no natural nor national characteristics whatsoever and makes no contribution to the heritage of our homeland. Every artist who is endeavouring to depict the beauty of nature is making a definite contribution to the cultural life of the community.'[158] While modernists and others complained, the competition certainly brought a wider audience to art. Artists were also supported as DB commissioned paintings for its hotels emphasising New Zealand landscape, fishing, racing, physical labour and other 'national characteristics'.[159]

Yet no matter how the breweries diversified their offerings and built their social capital, there was always a community, especially among the home brewers, who felt that New Zealand was being sold short with bland, mass-produced beer. A manual from 1960 insisted that home brewing was about more than making stronger beer at a lower price. 'It admits to a brotherhood where camaraderie exceeds that of fanatical gardeners and chess players, and for some not very obscure reason weakening friendships with non-brewers show a quite spectacular revival.'[160] In 1972, *The Brewers' Bible,* originally published as *Get Lushed on Your Own Grog*, began with a dedication: 'This book is dedicated to the beer barons who, having watered and artificially coloured the beer, are now initiating the same procedure in those vineyards they control. May they all die from cirrhosis of the liver.' The trade had tried to curb a long tradition of home brewing in New Zealand by restricting the hop supply. In turn it had sacrificed tradition to 'stifling uniformity' and only an elite were able to afford premium imports. 'Like the demise of fresh bread at a country bake house, beer for the average New Zealander has become a standardised bastard product of a system with no values save the material.'[161] During the next decade the two major breweries would take their corporate endeavours and rivalry to a new level, but the resistance of critics and home brewers would also take practical and dynamic form. More adventurous consumers would not be disappointed.

A Lion Red advertisement from the late 1970s emphasising the strong connection between the brewing industry and sport. Lion Technical Library

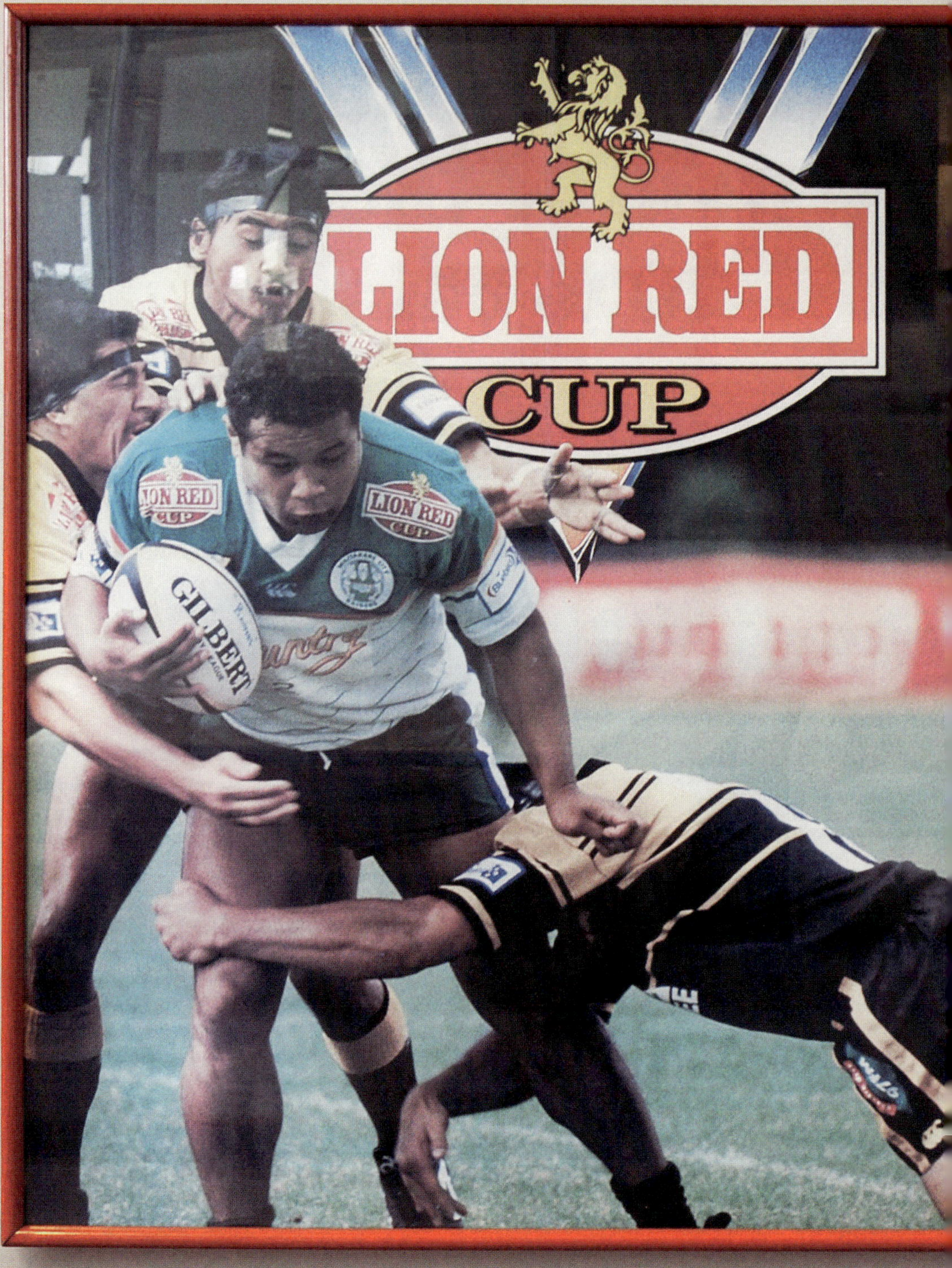
LION RED
CUP
GILBERT

During the 1980s and 1990s, rugby league became a major source of sponsorship rivalry between the two major breweries. Lion Technical Library

TEN

Challenging the Big Two: New Beer and New Rules, 1980–1990s

In June 1979, Frank O'Flynn, Labour MP for Island Bay, criticised the National Government as supporters of big business at the expense of small business: 'returning to my student days, we had then about 20 or 25 breweries in the country making very good beer. Now we have two large combines turning out chemical stuff in bottles.'[1] Five years later, a column in the *Listener*, while not as explicitly critical of the standard of New Zealand beer, condemned the way it was served in most pubs. 'The first mistake of the brewers and bottlers is to serve their beer ice cold. Cold beer is all very well after a hard day's work in the hot sun, but the majority of New Zealanders are city dwellers who work indoors, and anyway this is the age of unemployment.' It went on to argue that 'cold beer is tasteless beer' as one's taste buds could not sense flavour 'in the arctic conditions introduced to the mouth by a draught of Kiwi Brown'.[2] This was perhaps no bad thing given the reputation of the taste. An unnamed beer expert visiting in 1988 observed that 'The average New Zealand beer would be best kept for casseroles and then only with cheap meat.'[3]

As Terry McCashin's biographers put it, 'The breweries produced the kind of beer that they did because people drank it. People drank the kind of beer the breweries produced because there was no alternative.'[4] Change was coming, and the 1980s became synonymous with the beginnings of a crusade against bland and maltreated beer, as other possibilities emerged from new breweries variously described as 'boutique', 'independent', 'micro', 'mini', and later 'craft'. Yet this is a deceptive history. The 1980s were as much the decade in which DB and Lion fully embraced corporate expansion, including overseas, and engaged in an intense rivalry to capture a shrinking beer market. Out of this, and amid significant liberalisation of the liquor industry from 1989, came the repackaging of long-established and distinctly regional products such as Speight's and Tui as enduring national brands. While some mocked these beers and the strategies used to promote them, many more indulged and long remained suspicious of the new 'premium' offerings that were yet to capture a market niche.

*

The last quarter of the twentieth century witnessed a sharp decline in the volume of beer consumed. Per-capita intake had increased by about 75 per cent from 1945 to a peak of 133.9 litres in 1975. By 1994 it had fallen to 102.1 litres. While New Zealand was still ranked in the top five beer-drinking nations in 1977, by 1994 it had fallen to tenth.[5] There were a number of factors at play. Firstly, notwithstanding the problematic beer barns of the 1970s, experiences derived from overseas travel, images conveyed by television, and efforts to position New Zealand for the international tourist market all led to a decrease in the worst excesses of swilling. While people drank more frequently, they drank less. Secondly, although some enjoyed relative affluence and aspired to greater sophistication, for others the economic boom and high employment of the 1960s gave way after 1972 to a more volatile economy. Oil shocks in 1973 and 1978–79, declining export prices, rising inflation and unemployment, followed by rapid free-market restructuring under the Fourth Labour Government from 1984, inevitably placed some constraint on disposable income.[6]

Yet those who did spend on alcohol were less likely to buy beer. In 1960 beer comprised 74 per cent of pure alcohol and 95 per cent of the liquid volume available for consumption, spirits comprised 19 per cent of the pure alcohol and wine only 6 per cent. By 1990 beer had fallen to 58 per cent of the pure alcohol and 86 per cent of the liquid volume, spirits had increased slightly to 20 per cent of pure alcohol and wine dramatically to 22 per cent.[7] From perhaps 60 vineyards producing 40 million litres at the beginning of the 1980s, there were 238 vineyards in 1996, and in 2006 530 producing around 90 million litres. The fortified wines still common in the early

Although the road toll had declined sharply from the 1970s, health organisations maintained strong campaigns against drink-driving, emphasising host responsibility and especially the provision of food with beer. This poster was produced by Healthlink South in Christchurch in 1994. Alexander Turnbull Library, Eph-D-SAFETY-Road-1994-01

DB, which had always produced a larger proportion of packaged beer than Lion, quickly embraced the vogue for larger cans during the early 1980s. DB Breweries

1960s gave way to the dominance of sauvignon blanc and a reduction in the average strength of wine from 15 to 12–13 per cent alcohol by volume by the early 1980s.[8] But brewers were inclined to attribute the initial growth of wine at least to an unfair tax regime. While the Brewers' Association referred to a 'crippling burden of taxation' in 1982, it claimed that increases in taxation on wine had been lower than the rate of inflation, with the result that, in terms of pure alcohol, wine was becoming cheaper than beer. It was simply not sensible to tax the weakest alcoholic beverage while allowing stronger wine to expand unhindered.[9]

There was also a marked shift in where and how beer was consumed. As drink-driving laws tightened, and perhaps also as television and other home entertainment options expanded, the proportion of drinking at home increased. Draught brown beer on tap declined sharply from the 1970s. Canned beer, popular from the early 1970s, outstripped bottled by 1990. For a time, Lion imported cans as its existing canning plant could not meet the competition from imports – of which more shortly. It then entered a partnership during the late 1980s to establish a can manufacturing plant in Auckland and consequently reduced can cost by one-third. Can sales doubled in the two years to 1990. DB established a new canning plant in 1991. However, bottles then enjoyed a resurgence as new, less light-sensitive hop varieties allowed more beer to be sold in clear or green bottles, which came to appeal to the premium, image-conscious beer drinker more than traditional brown.[10]

Ironically, as demand for beer contracted, both major breweries embarked on substantial corporate transformations. In 1965 Douglas Myers, a grandson of Sir Arthur Myers, who was in turn the nephew and protégé of Louis Ehrenfried, returned to New Zealand to work for Campbell & Ehrenfried, which remained a leading firm in hotel ownership and liquor importing. Cambridge and Cornell University-educated, Myers became managing director in 1970 and launched a successful takeover offer for the mostly family-owned firm in 1972. Always critical of what he perceived as lethargy and lack of ambition in New Zealand business, he set out to restructure and modernise hotel ownership and challenge the strong influence of the Hotel Workers' Union. By 1971 Myers concluded that Campbell & Ehrenfried was overcapitalised and decided to sell many of its cash-draining hotels. This effectively forced NZB to buy them, otherwise they would be taken over by DB, which was then in a phase of aggressive hotel acquisition. Myers then used the sale funds to create New Zealand Wines and Spirits as a joint venture between Campbell & Ehrenfried and NZB. But relations within these interlocking companies became increasingly fractious, especially between Myers and the NZB (now Lion) chairman, Ralph Thompson, whom Myers regarded as bullying and a business conservative. In 1980 Myers exercised a clause to be bought out of New Zealand Wines and Spirits, a move which triggered a protracted court mediation to determine the value of the company. Without waiting for a resolution, Myers borrowed against the anticipated funds from the buyout and in mid-November 1981 acquired almost 20 per cent of Lion shares. Working with another large shareholder, Fraser & Neave of Singapore, which had been involved in New Zealand brewing since its part-ownership of Leopard from the mid-1950s, Myers removed Thompson and made himself managing director. The Lion head office was then moved from Wellington to Auckland. While Wellington had been beneficial in terms of the need to lobby politicians against temperance, Myers argued that head office now needed to be in Auckland at the heart of business action.[11]

The climate for Myers to thrive was ideal. The Fourth Labour Government elected in July 1984 responded to New Zealand's declining economic fortunes since the mid-1970s, especially the impact of oil shocks in 1973 and 1979, with rapid and far-reaching free-market reforms which became known as 'Rogernomics' in reference to Minister of Finance Roger Douglas. The tightly regulated economy was swept aside as government departments were corporatised and in some cases privatised. The financial market was deregulated, foreign exchange controls were removed, and corporate capitalists were in buoyant mood, at least until the October 1987 stock-market crash. Myers wanted to move further and faster than the government, and was a founder of the pro-free-market Business Roundtable in 1985. As he told an interviewer in 1986, 'New Zealanders have suffered from mollycoddling by politicians with their cradle-to-grave philosophies. Even the thickest New Zealander can't go believing that New Zealand is contributing anywhere near its potential.'[12] It was surely symbolic of his uncompromising approach to business that he tweaked the Lion logo to reinstate the testicles that had been castrated by a predecessor. 'They had actually removed the balls – it was inconceivable.'[13]

Throughout the 1980s, Myers moved steadily to streamline Lion operations and especially to curb union influence. In his view, the national brewery workers' award effectively stymied competition between Lion and DB and perpetuated very inefficient breweries. The Auckland workforce was reduced from 396 to 123 during the 1980s. Lion also closed its breweries in Wellington, Palmerston North, Hastings and Waikato between 1989 and 1997, concentrating brewing in Auckland, Christchurch and Dunedin. In 1988, Lion Nathan was formed following a merger with L. D. Nathan & Co., a company founded in the 1840s that had become New Zealand's largest retailer by the early twentieth century and expanded into supermarket ownership. Indeed, the merger was in part driven by a belief that alcohol would eventually be sold in supermarkets.[14] In 1990 Lion Nathan acquired a half-share in Natbrew Holdings, which owned Alan Bond's extensive brewing assets in Australia, including Castlemaine Perkins, Toohey's and Swan Brewing. The other half was acquired during 1992–93. In 1994 Hahn Brewing in New South Wales and the South Australian Brewing Company were added. In 1998, in anticipation of his retirement to London, Myers sold most of his Lion Nathan shareholding to Kirin Brewery Company of Japan. As Kirin further expanded its interest in the company during the early 2000s, it created the fourth-largest brewing alliance in the world. The company also acquired controlling interests in various Australian and New Zealand wine companies.[15]

Change was also afoot at DB. In 1981 Brierley Investments Ltd (BIL) acquired an interest in the company. The following year Sir Henry Kelliher retired after more than half a century as managing director, and in 1983 Morton Coutts, who had driven DB's substantial technical innovation since its foundation in 1930, also retired. During 1986–87 the BIL-controlled Magnum Corporation, which had related interests in liquor and other retail, took control of DB. From 1989 Magnum expanded its operations to Australia before establishing a joint venture with Asia Pacific Breweries of Singapore under the name DB Group. The brewing division became DB Breweries. With Brian Blake as managing director from 1993, DB embarked on a rationalisation similar to that conducted by Myers at Lion. While Kelliher had been essential to the growth of the business, there was a sense that his longevity had ultimately stifled innovation. In 2004, shortly after the death of Morton Coutts at the age of one hundred, DB Breweries left the New Zealand Stock Exchange as Asia Pacific Breweries completed its buyout of shareholders.[16] Both DB and Lion, which could trace their origins to small local breweries, were now multinational brewing giants.

Meanwhile, the move from draught to packaged beer, including a growing tendency to drink bottled beer on the premises, prompted both breweries to sell off less profitable hotels from the 1980s. Lion reduced from somewhere in the vicinity of 250 to less than 50 hotels by 1994, and DB sharply reversed its 1970s acquisitions. This enabled both breweries to claim that they had given up their duopoly, but the reality was that they had swapped ownership for sole-supply agreements and favourable pricing structures with new hotel owners – arrangements no different from earlier tied-house tactics. Lion and DB still controlled 70 per cent of retail and takeaway outlets by 1991.[17]

At one level, as with all such corporate machinations, most consumers will have been oblivious to the transitions at head office. In other respects, they witnessed significant change. With a declining beer market, Lion and DB became locked in a frenetic and periodically acrimonious battle for market share. Although Lion had always held a substantially larger share, it was certainly threatened by the success of DB in the South Island, especially during the 1970s, as it completed brewery mergers and acquired hotels in the wake of ten o'clock closing. DB marketing was more effective, and DB Draught, as the most consumed beer in New Zealand, became something of a 'national brand'. DB marketing capability was further enhanced under Brierley's influence. Although Lion held an overall 55 per cent market share during the 1980s, and 62 per cent in Auckland, DB held 54 per cent of the South Island and rural areas. It produced more brown beer, whereas Lion provided more light-coloured products.[18] Subsequently, as one brewery launched a new beer to capture a perceived market niche, the other moved quickly to counter. Likewise, as one sought to boost its profile through sports sponsorship, the other tried to outmanoeuvre or exceed it. Increased volume for one generally came at the expense of the other. Yet as Lion's managing director, Kevin Stratful, observed of the pressure to succeed in this setting, 'Let's be realistic – we are only three million people, of which only half of that number you can get near an alcoholic beverage and only half that number again are interested in changing their beer brands.' It was now essential to define 'segments' in the market where customers could be persuaded to try different products, but products that were not too expensive to produce.[19] The result for the adventurous consumer was a greater range of frequently cheaper beer – if not of beer styles.

In a 1987 analysis entitled 'The Beer War', Tom Hyde observed that since Douglas Myers' takeover of Lion and the retirement of Henry Kelliher from DB, 'hostilities between the two camps have escalated beyond a mere domestic dispute over beer and hotels to an all out war between two large multinational corporations'.[20] This was especially the case from 1982 to 1986, when DB was under the management of Jim Fletcher, from the third generation of New Zealand's largest construction firm, Fletcher Challenge, who had much in common with the corporate outlook and aspirations of Myers. In late 1985, DB launched newspaper ads featuring rugby players in a changing room praising DB and pouring scorn on another beer in a white can. Given that Steinlager was the only local beer in a white can, Lion accused DB of illegal attack advertising and lodged a complaint with the Association of Accredited Advertising Agencies. Lion won the case, but DB refused to withdraw the ad, as the agency only had the power to make guidelines and could not compel withdrawal. The breweries then accused each other of plagiarism when both released ads using old-style news headlines. Lion then took a High Court injunction forcing DB to change the name of its DB Double Red to Double Bitter, as the name and label were judged to be too close to Lion Red.[21]

By mid-1988 there was a 'vicious' canned-beer price-cutting war between DB and Lion as they scrambled for market share against a sharp increase in the amount of imported Australian cans. Lion claimed that its price reduction was due to cheaper can production from its new canning plant and more efficient work practices. DB replied that Lion's prices were a desperate attempt to compete.[22] By early 1992, Lion was reportedly offering independent liquor retailers between $2000 and $7000 of free beer not to stock the new DB Bitter natural beer. While Lion did not confirm the report, it argued that 'incentive programmes' were a normal part of the industry. However, retailers suggested that there was a difference between incentives and 'blocking tactics'. Lion also dismissed the new DB offering as a 'knock off' of its own Red Band natural beer.[23]

By the end of the century, competition was so fierce that it was taking a toll on jobs and wholesalers. DB had made at least twenty-five staff redundant at its Tui and Mainland breweries, while the company holding the franchise for its Five Star Liquor chain had been forced into voluntary liquidation. For its part, Lion mused that its production levels were much closer to its market requirements than DB's and they were adjusted continually. But it conceded that the closure of the Hastings brewery had been part of a retrenchment strategy.[24] By 2007, the two were still arguing in the High Court as DB sought to prevent the launch of Sundance Summer Ale in opposition to its Monteith's Summer Ale. DB objected to the use of the term 'summer ale', but Justice Harrison noted that the packaging of the two products was quite different and unlikely to cause confusion. A further claim that using the name 'summer ale' was trading off the goodwill of Monteith's was also rejected on the ground that 'summer ale' was a generic beer style originally made popular in the United Kingdom.[25]

The sporting arena was similarly contested. In 1986, with Steinlager ensconced as a major sponsor of New Zealand's America's Cup sailing challenge in Fremantle, DB began to promote DB Export in Perth as 'The New Kiwi Challenger'. Under threat of legal action, DB eventually withdrew all material linked to the Cup campaign.[26] However, DB had its opportunity in 1990 when Lion rejected a chance to lead sponsorship of the Commonwealth Games in Auckland. By the end of 1990 the focus shifted to horse racing, as Lion established a marketing team in Christchurch in an attempt to reclaim the South Island from DB. Both breweries launched summer campaigns around Christchurch Cup Week with the Lion Brown New Zealand Cup and the DB Draught New Zealand Trotting Cup.[27]

At the same time, although Lion had a long-standing sponsorship association between rugby league and Lion Red, those preparing the ultimately successful bid for the Auckland Warriors to enter the Australian Winfield Cup competition were unable to pin down a naming sponsorship commitment from Lion. They then approached DB. As journalist Chris Mirams recalled, 'Almost instantly, the breweries started leap-frogging the other's bid. Madness prevailed. As soon as one company put an amount on the table, the other countered.'[28] Eventually DB agreed to a sponsorship of $1.5 million per year, which prompted Kevin Roberts, chief operating officer for Lion, to observe to the press, 'The money got crazy. I've never seen anything like it in Australia or New Zealand. It's completely over the top and, quite frankly, unaffordable. I don't know how the sponsor can get good value for money out of it.'[29] As events transpired, DB held the major sponsorship for four seasons and maintained a secondary sponsorship for one more before Lion Red was installed as the 'official beer' of the rebranded New Zealand Warriors. DB then took over from Lion as a major sponsor of Wellington Rugby and the Hurricanes Super Rugby team. Lion responded that it was not unhappy about the deal. 'We had an opportunity to sponsor if we wanted to pay that sort of money, but we did not think it was worth it.'[30] Matters were further complicated in 2004 when issues emerged over 'pourage' rights at venues for international fixtures where DB sponsored the provincial team but Lion sponsored the All Blacks. It did not help that a sponsorship

TOP LEFT: During the 1980s breweries became more actively involved in campaigns against drink-driving. Lion Technical Library

TOP RIGHT: The association between rugby and beer has always been a strong focus for the major breweries. Lion Technical Library

BOTTOM LEFT: This advertisement from the 1970s touts Lion as sponsors of the Olympic team. Lion Technical Library

BOTTOM RIGHT: Lion's sporting sponsorship during the 1970s and 1980s extended well beyond rugby and racing. Lion Technical Library letter from the New Zealand Rugby Union intended for Lion was accidentally sent to DB. Very delicate negotiations were required to resolve the matter.[31]

The rivalry also unfolded in the potential market for low-alcohol beer. As previous chapters have noted, once the traditions of harvest ale and hop beer faded, the twentieth century had not been kind to various efforts to produce lower-strength beer and consumers had generally treated it with contempt. But the 1974 Royal Commission had paved the way for a more sympathetic environment with the creation in 1976 of the Alcoholic Liquor Advisory Council (ALAC), 'having as its primary objective the promotion of moderation in the use of alcoholic liquor and the reduction of the personal, social, and economic evils resulting from the misuse of alcoholic liquor'. The nine-member Council comprised representatives from Treasury, the departments of Health and Social Welfare, and six members appointed by the governor-general on the recommendation of the minister of health.[32] Other organisations such as the Auckland-based Alcohol Healthwatch followed during the 1980s, along with the development of more systematic and methodologically robust surveying of drinking habits.[33] All created a climate of greater acceptance for low-alcohol beer, and Lion L.A. was released in December 1979. It has also been suggested that the quid pro quo for relaxing price controls on beer during the early 1980s was a commitment by the breweries to support campaigns for moderation being developed by ALAC and others.[34]

One advantage of low-alcohol beer, usually regarded as being under 1.7 per cent alcohol by volume, was that it could be stocked in supermarkets. However, sales growth was slow due to an inability to advertise or explain the nature of the product on television and radio, where any alcohol advertising was still not allowed. When Lion launched McGavin's Light around 1988, its campaign slogan 'The only light beer that doesn't taste like weasel's . . .' was deemed to be in bad taste and banned by newspapers. At the time there were three standard processes for producing low-alcohol beer: brewing to low strengths; boiling off extra alcohol; or reducing it under vacuum. DB, which claimed that people could not tell the difference between full-strength beer and its Kühtze Light, also suggested it had a 'secret' brewing process that obviously worked because others were copying it – which rather casts doubt on its secrecy. Although the breweries believed that low-alcohol beer tended to be drunk by 'trend setters' who were more likely to be lager drinkers than conventional brown beer drinkers, the generally lower strength of New Zealand beer compared with most countries meant that consumers did not initially feel the need to follow the trend. By 1993 it was estimated that only 4 per cent were drinking light beer, well behind other countries, including Australia at 20 per cent.[35]

A significant increase in light beer consumption was predicted following the introduction of new drink-driving laws from 1 April 1993, which included random breath-testing and a particular targeting of the under-twenties. A variety of mostly lager styles was available, including some imports and other international products such as Tennent's Light, brewed under licence by Lion. Independent Liquor, of which more shortly, also offered bottled shandy. Some distinction was now drawn between 'light beer', generally around 1.5 per cent alcohol by volume, and 'reduced alcohol' beer at perhaps 2–2.5 per cent alcohol by volume.[36] But Lion sharply disputed a claim from DB that its Mako Special Bitter had captured more than half of the low-alcohol market. Managing Director Doug McKay said this was not the first time DB had misled the public over market share, and observed, 'I get a little annoyed with these "facts" where they get beyond the bounds of moral decency.'[37] Ultimately, neither brewery made real progress in this round of hostilities. By the end of 1994, consumption of beer under 2.5 per cent alcohol by volume had fallen by 30 per cent from the previous year and comprised only 2.3 per cent of a market that was dominated by beer in the 2.5–4.3 per cent range. At the other end of the spectrum, although only 0.4 per cent of the market, beer over 5 per cent was showing pronounced growth.[38]

The rivalry between Lion and DB was also exacerbated by the fact that the shrinking beer market now included more imported beer. At the end of the 1970s the General Agreement on Tariffs and Trade (GATT) round allowed greater international access to the New Zealand market. While this initially created little interest from Europe due to high freight costs, the Closer Economic Relations agreement signed with Australia in 1982 enabled a large amount of canned beer to cross the Tasman – cheaper due to lower can production costs in Australia and brewed slightly stronger for what was described as a New Zealand 'premium' beer market. Indeed, such was the impact from Australia that Leopard, long a leader in the canned market, submitted to financial pressure and became a wholly owned subsidiary of Lion.[39]

Although extra duty imposed on imports from August 1981 increased prices, this was soon offset by a price war between wholesalers.[40] However, in a competitive environment, importers tended not to take risks, and most imports were lagers and not especially exotic. The partial exception, at least by contemporary New Zealand standards, was in some of the beer range available in themed pubs. As Lion and DB sold off establishments from the 1980s, new owners looked for points of difference, with Irish- and English-themed pubs the most common, as they sought to replicate what people had encountered on their overseas travels.[41] This led to the return of Guinness on a larger scale. New Zealand Wines and Spirits had imported both bottled and canned Guinness since the late 1970s. From the mid-1980s a quiet start was made to reintroduce draught Guinness, with kegs imported from Ireland. While they could have been imported more cheaply from Australia, it was judged that consumers would prefer Irish to Australian Guinness. There was also a more sophisticated emphasis on the way Guinness was served than had been the case during the 1960s. Local tastes dictated that it needed to be slightly colder than in Ireland, but warmer than most New Zealand beer. It was also important to have the right mix of CO_2 and nitrogen from the kegs to produce the famous creamy Guinness head. The success of this venture gradually evolved to renewed brewing of Guinness in New Zealand from 1990. After lengthy technical investigation, Lion settled on production from its Christchurch brewery. By 1994, 380 outlets were selling locally produced draught Guinness.[42]

In one area, overseas exports, Lion could unequivocally claim dominance over DB. Although the manager of an unnamed 'large Auckland brewery' had stated in August 1957 that they would not take up an export opportunity to the United States because they were still rationing supplies and having difficulty meeting New Zealand demand, it was evident that the Department of Industries and Commerce and various New Zealand trade commissioners overseas were exploring the possibilities for markets in India, Malaya, Singapore, South Africa and the United States. No mention was made of the closest market in Australia. In late 1962 Bernard O'Donnell completed a seven-month global business tour for NZB and returned convinced that New Zealand brewing was the best in the world with ample opportunities for export.[43] By the following year, small quantities of Steinecker were being exported to the Pacific and a consignment of Bavarian Bitter was sent to Scott Base in Antarctica, where it proved so popular that 'it was quaffed by the boys in the record time of one hour'.[44] From 1966 NZB's Zealandia Light Beer and Zealandia Half and Half, a blend of lager and stout, were apparently popular in selected supermarkets on the West Coast of the United States. At its 1966 annual general meeting, NZB emphasised that it had accepted a challenge from the government to diversify exports in conjunction with the New Zealand Manufacturers' Federation. Aside from the United States, markets were being explored in the Pacific and Asia and small quantities had been dispatched to London. The export volume had increased threefold in three years to around 1.7 million litres by late 1967 – 80 per cent of which came from NZB.[45] In 1971 NZB and Leopard exhibited beer at the Pacific Fine Food and Beverage Show in Los Angeles in an effort to develop Canadian outlets as well. It was also suggested that new seedless hops being developed at the Riwaka research station would remove some of the bitter after-taste from New Zealand beer that had posed a barrier to exports.[46] In 1979 Steinlager was voted best beer in the world for the third year in a row at a blind tasting of twenty-nine entries in Washington, DC. Leopard Lager placed

During the 1980s Steinlager accounted for the vast majority of New Zealand beer exports and was a stronger international than local brand. Lion Technical Library (left); *Brewnews*, Lion Technical Library (right); Lion Technical Library (bottom)

fifth, just ahead of Heineken, while it was reported with evident glee that the Australian entry, Fosters, was not in the top ten.[47]

However, there was no consistent strategy to any of these initiatives until the mid-1980s. Thereafter a concerted effort was made through Steinlager's 'They're drinking our beer. Here.' campaign, with the beer reportedly in twenty-nine countries by 1986, although DB argued that this claim was exaggerated as the total export to one of them, Sri Lanka, was only 480 bottles. Steinlager was arguably more popular overseas than in New Zealand until the mid-1980s. It gained more local traction from 1986 after winning 'best beer' at international beer awards in Burton-on-Trent. Yet its subsequent sponsorship drive via the All Blacks, the America's Cup and Peter Blake, who skippered *Steinlager II* to victory in the 1989–90 Whitbread Round the World Race, retained strong international exposure. While DB made some effort to push Kiwi Lager in Australia to capitalise on New Zealand's visibility through the America's Cup campaign, by 1992 97 per cent of the export volume was still Steinlager.[48] By late 1995, with Japan emerging as the leading destination, beer exports amounted to 20.4 million litres worth $25.4 million, a 36 per cent increase on the previous year. By comparison, wine exports were worth $42.4 million. In another international innovation during 1995, Lion acquired a 60 per cent share of the Taihushui Brewery in Wuxi, a city of 5 million in China's Yangtse River Delta.[49]

While the shrinking market and changing patterns of consumption were undoubtedly a challenge for the breweries, in other respects, the 1980s and especially the 1990s brought dramatic liberalisation of the licensing laws. From December 1976, hotels and taverns were able to obtain permits to close at 11 p.m. on Friday night, Saturday night and Christmas Eve,

Sponsorship of successful international yachting campaigns was crucial to the promotion of Steinlager as an international brand. Lion Technical Library

and 12.30 a.m. on New Year's Eve. There was a 150 per cent increase in licences during the next decade, although some doubted the actual increase, as many newly licensed sports clubs had undoubtedly been selling alcohol illegally since the 1960s. There was certainly an increase in the likes of restaurant licences. In 1981 the Licensing Control Commission belatedly tackled the 'booze barn' phenomenon by imposing a size limit on new taverns. Some of the existing establishments were also partitioned into smaller bars and dining areas to attract different customers

In the context of a rapidly emerging free-market economy from 1984, the Fourth Labour Government established a Working Party on Liquor, chaired by prominent public servant and former ombudsman Sir George Laking, to report to Minister of Justice Geoffrey Palmer. The Working Party comprehensively rejected the 'availability theory', a staple of the temperance and prohibition movement from the eighteenth century, which held that more outlets and longer opening hours led to more drinking. To the contrary, it pointed out that the significant increase in outlets since 1977 was in parallel with an overall decline in alcohol consumption. Its draft Sale of Liquor Bill provided for a much simplified licensing system with only four options as opposed to the twenty-nine or so then in existence. Licensing decisions were also to be transferred from licensing committees to local government and a new Liquor Licensing Authority. Many of the barriers to new outlets, such as remaining restrictions on dry districts, rights of local objection, and the 'need' test that protected existing outlets from competition, were also recommended for abolition. With the exception of some provisions for dry districts, which lingered until Eden and Roskill (Auckland) and Tawa (Wellington) voted wet in 1999, everything else was accepted and adopted in the Sale of Liquor Act 1989. This prompted the most dramatic expansion in alcohol availability ever seen in New Zealand. The per-capita number of licensed liquor outlets increased by 64 per cent between 1990 and 1995 and had doubled by 2004. There was also an end to legislated standard hours of sale, with the hours that each outlet could open now being determined by the conditions for its particular licence as set by the local licensing authority.[50]

The 1989 Act also dispensed with the triennial licensing poll. The 1974 Royal Commission had questioned the relevance of the poll, not least because it doubted whether voters really understood what would happen in the very unlikely event that prohibition or state purchase and control actually won. There were even reports that 'National Continuance', as the option for the status quo appeared on the ballot paper, was somehow linked to the notion of perpetual National Party rule. While some wished to retain the poll as some sort of barometer to measure resistance to liberalisation of the licensing laws, it was clear by the 1980s that most people, including the 1986 Working Party, regarded it as an anachronism. Indeed, although support for prohibition had rallied to 21 per cent in 1987 from a low of 13 per cent in 1969, it is likely that some of this vote was more a reflection of concern with such things as drink-driving or proposals for liberalisation than commitment to prohibition. Certain young voters were also attracted by the prospect of lucrative opportunities for provision of moonshine under prohibition. When the Sale of Liquor Bill was introduced to Parliament in July 1988, Minister of Justice Geoffrey Palmer stated that the poll had not served any useful purpose for a long time, there was no prospect of prohibition being approved, and more important matters were not the subjects of similar polls. As the Bill progressed, there was almost no discussion of the poll and it disappeared without a whimper.[51]

Not all went in favour of the beer drinker. Wine was allowed to be sold in supermarkets, except on Sunday, but beer was not. Certainly, this was not a moral stance from the Working

Party, which bluntly rejected certain types of argument against supermarket alcohol sales. 'We give no weight to the protestation of some, usually men, that liquor sales should be banned in supermarkets to protect housewives who would otherwise frivolously spend their housekeeping allowance on liquor at the expense of groceries. This sexist argument has a glib superficiality that will not stand up to scrutiny.'[52] Rather, the view seemed to be that wine was an acceptable supermarket item because it was a usual accompaniment to food, while beer and spirits were not. It was also clear, although the Working Party had no time for measures that stifled competition, that the breweries were not in favour. In 1994, a supermarket representative observed, 'We can't see any downside in selling beer in supermarkets. Obviously we believe we can sell beer cheaper than the normal retail outlets as has been noticed in selling wine.'[53] For the breweries, this threatened the viability of their existing retail outlets. Price cutting by supermarkets could also expand the market share for imported beer.[54]

In 1995, National MP Max Bradford introduced a private member's bill to allow beer in supermarkets, and argued that successive surveys had shown that around 65 per cent of the public supported the move. This was opposed by the breweries because of anomalies such as allowing beer and wine sales in supermarkets on Sundays but not in other retail outlets.[55] For DB, Brian Blake took exception to a *National Business Review* 'brewery humbug' editorial, which accused them of trying to kill the bill. He insisted that all they desired was consistency and a comprehensive rather than piecemeal review of the Sale of Liquor Act. 'To claim that DB is afraid of competition is quite wrong; we welcome further deregulation of the industry but we do want to see a level playing field. We are certain that beer will be sold in supermarkets in the near future but believe that it should be done without commercially disadvantaging other outlets.'[56] Nevertheless, DB had been moving to reduce its direct exposure to any change in retailing by developing franchising and rationalising its outlets.[57] Bradford's bill eventually lapsed.

The tide of further reform was inevitable. Although several private members' bills had tried to ban all alcohol advertising during the early 1980s, sponsorship advertising with alcohol was permitted – not least being the association between Steinlager and the All Blacks that began in 1986 on the eve of New Zealand's hosting of the inaugural Rugby World Cup in 1987. Alcohol brand advertising was allowed on television from early 1992, a move with important consequences to be discussed shortly. To balance this, free broadcasting time was also allocated to health promotion messages.[58] In 1997, the Liquor Review Advisory Committee concluded that the Sale of Liquor Act had been beneficial, and recommended further reforms. In an amendment to the Act in 1999, supermarkets and grocery stores were allowed to sell beer, alcohol sales were permitted on Sundays and the drinking age was lowered from twenty to eighteen.[59]

*

If legislative change took time to produce all of the liberalisation that beer drinkers hoped for, by other means the 1980s and 1990s brought considerably more variety than they had been used to. But it is important not to mythologise what happened. The notion that the 'big two' were suddenly challenged by Terry McCashin's Mac's Brewery in 1981 and New Zealand was thereafter set on the inevitable path to a craft beer paradise is far too simple. For one thing, many of the early 'competitors' would be best described as 'independent' or 'micro' rather than 'craft', and their market share remained tiny as Lion and DB sought to obstruct or outflank them and consumers were slow to adapt.

In 1978, former All Black and entrepreneurial publican Terry McCashin went to the United Kingdom to explore rabbit farming in the context of a need for diversification that was confronting the New Zealand farming sector during the 1970s. Instead, he was more captivated by the variety of beer on offer in Derbyshire and toured a local small brewery attached to a pub.[60] That there was a small brewery to tour perhaps reflected important changes in Britain. As with New Zealand, and indeed much of the beer-drinking world, British brewing had witnessed mergers, closures and industrial concentration to the point where it was dominated by the 'Big Six' – Allied, Bass, Courage, Scottish & Newcastle, Watney and Whitbread. They produced 82 per cent of beer in the United Kingdom by 1972, controlled 56 per cent of all pubs, and produced pressurised keg beer that was a testament to efficiency and homogeneity rather than tradition and variety. Guinness produced another 9 per cent, and the remaining 9 per cent was in the hands of around 85 small breweries – a number that had declined from 380 in 1952. But in March 1971 what became known as the Campaign for Real Ale (CAMRA) was formed as a consumer organisation to encourage the preservation of traditional beer, cider and perry styles and to promote and protect the cultural heritage of pubs and clubs. Its first national real ale festival in 1975 attracted 40,000 visitors. The Big Six soon responded by reintroducing some cask ales, but the more significant development was the revival of independent breweries, with eighty-six new establishments by the end of 1981.[61] While the traditions of real ale, or 'beer in the wood', had long departed from New Zealand brewing, the sense that there was an alternative to mass production certainly inspired McCashin and those who followed.

Prime Minister Robert Muldoon, a strong advocate of small business but perhaps less renowned as a beer drinker, at the opening of Mac's Brewery, Stoke, in September 1981. Nelson Provincial Museum, NPM 12 NM_6772_fr_9

The genesis of the McCashin brewing venture was not straightforward. It began with the revival of the Rochdale cidery in Stoke near Nelson. Brewing equipment was assembled from the likes of dairy factory closures and scouring the country for anything surplus to the requirements of DB and Lion. After some difficulty, a brewer was found when Englishman Jim Pollitt, who had been head brewer at Tui during its Guinness era, was enticed back to New Zealand from Carlsberg and allowed to stay after direct intervention from McCashin's local MP, Doug Kidd, with the minister of immigration, Aussie Malcolm.[62] In response to the Lion and DB monopoly on the supply of hops, malt and bottles, McCashin employed his own maltster and also threatened to go to the media if obstruction continued over his bottle supply. But perhaps his shrewdest move was to invite the prime minister, Robert Muldoon, to the official opening of the brewery on 26 September 1981.[63] Although Muldoon was more a wine drinker, McCashin recalled, 'He came down and he was not a man to be messed with and I'm sure Lion and DB gave me a bit of leeway.'[64] Indeed, Mac's Brewery was precisely the kind of local entrepreneurship that the National Government claimed to encourage.[65] With the exception of the international investment in the existing Leopard Brewery in 1956, Mac's appears to have been the first company to secure a new brewing licence in New Zealand since the establishment of DB in 1930.

The initial Mac's offering was a natural, lightly hopped malty ale that was somewhat stronger than other beer on the New Zealand market. A lager and a dark ale were soon added, along with Southtop as a staple 'New Zealand' beer which did not challenge local taste buds. With hindsight, being pressured by the big breweries at the outset, which had forced Mac's to use their own malt and an alternative hop supply, had actually enabled Mac's to produce more distinct products. Beginning with individual customer sales direct from the brewery, Mac's supplied a small number of pubs in the upper South Island before securing a toehold in Wellington, Auckland and later Christchurch. However, progress was limited in that Lion and DB controlled the vast majority of distribution and outlets. But efforts by distributors to price Mac's above Steinlager to curb sales did not work when it became apparent that drinkers were prepared to pay more for what they perceived as a quality and somewhat different beer. In particular, a market developed among young, urban professionals with experience overseas who were willing to try something new and enjoy the status in being seen drinking a 'premium' product. In 1985 Mac's was also assisted by a nine-week brewery strike, which dried up Lion and DB supplies in some establishments and obliged them to acquire Mac's to tide them over. Some happily continued with it after the strike. By 1991, Mac's had secured about 1 per cent of the market and had established two brew bars in Christchurch.[66]

Mac's hardly prompted a rush to imitation. It seems that it was five years before another brewer, Auckland's Shakespeare Tavern, started producing its own beer. Thereafter new breweries appeared at the rate of three or four a year. Among those that have endured well into the twenty-first century in one form or another are the Sunshine Brewery in Gisborne and the White Cliffs Brewing Co. in Taranaki, both established in 1989, Pink Elephant from Blenheim, established in 1990, and Emerson's from Dunedin (1992). Eleven new breweries were established in 1995 alone, four in 1996, three in 1997 and four in 1998, by which time there were fifty-four independent breweries operating in New Zealand. Of these, twenty-three operated as brewpubs and five others were jointly owned by groups of publicans and produced beer for their own establishments. By comparison, Australia, with a population almost five times that of New Zealand in 1998, had only twenty-three such breweries.

Sydney, with a population similar to New Zealand, had only one.[67] For reasons that are not immediately apparent, Australia has continued to lag behind New Zealand in the scale and sophistication of its brewery revival.

Some breweries followed the Mac's path of repurposing dairy equipment. The reduction in tanker delivery to hotels, to be discussed shortly, also created a surplus of tanks for conditioning and storage. For others, the rolling back of import controls from the mid-1980s made it feasible to import brewing systems. Yet such progress was in no danger of

DB Export, created just before the 1958 'Black Budget', has endured as a major DB brand. DB Breweries

flooding the market, because most new establishments had a brewing capability of between 1200 and 2500 litres and tended to be operating at capacity. Hence, in the manner of the nineteenth century, these breweries initially survived on small local markets and there was little distribution further afield. During the 1990s, the most prosperous small breweries were judged to be those associated with a particular pub or restaurant. Those without such an outlet, and without easy access to bottle stores controlled by the big breweries, were often reliant on over-the-counter sales in plastic riggers or being able to place their kegs in local clubs, bars and restaurants that were relatively free from the control of Lion or DB. Within the small New Zealand market, and the minute portion of it that independent breweries secured, sophisticated bottling equipment was often prohibitively expensive and unlikely to produce a return on investment.[68]

There were clearly different motives for establishing new breweries during the late twentieth century. Some, especially brewpubs such as the Loaded Hog chain or Nelson Bays Brewery, established in 1993, sought to compete on price by producing draught beer at prices cheaper than or comparable with the major players. To this end, there was some homogeneity of styles across the smaller breweries during the 1980s and 1990s, as most produced an amber draught, a lager and a dark, which was frequently a dark lager, in the standard 4 per cent alcohol-by-volume range and not substantially removed from the styles of mainstream offerings. As Michael Donaldson put it, 'Many of the new breed of brewers did little more than create handmade versions of the big-boy beers. They followed an established pattern of creating three products – colloquially known as yellow, amber and black.'[69] While some also branched out to strong lager or stout and all emphasised that their beers were natural and chemical-free, the knowledge and ingredients to make anything too exotic were not readily available. Nor did most wish to push the boundaries of flavour too quickly in a 'novice' market for new beer.[70] However, some certainly did offer beer at a higher price and sought to compete on taste and variety by producing beer styles not otherwise available in New Zealand. Among them was Pink Elephant, where Roger Pink specialised in traditional real ales familiar from his English upbringing.[71] Another was Richard Emerson, who first encountered the possibilities of beer on a family visit to Edinburgh in 1983, refined his interest as a talented home brewer, and worked at Matilda Bay Brewery in Western Australia and then in Britain before returning to establish his Dunedin brewery in 1992.[72] What all of these brewers surely had in common, given that there was not an organised consumer momentum of the likes of CAMRA demanding different beer, was that in an increasingly deregulated society in which more New Zealanders were travelling overseas or receiving influences from it via television and imported products, there was greater scope and reward for offering variety.

Irrespective of motive, this was not an easy environment to enter and certainly not a licence to print money. As Ed Franken, director of Ōnehunga's Newbegin Brewery, explained in 1989:

> It's a very fragile life. We have to work damn hard to make a go of it. You need a big pile of money to play the game – it's not like groceries where you don't need much capital equipment. In this business you need a lot of plant and a lot of working capital. I'd be very surprised if many others entered the game in the future. If they were contemplating it, I'd suggest first they come and buy us out and learn by our mistakes – because, like everybody else, we've made a lot of them.[73]

One mistake was to assume that big breweries could be replicated in miniature, when in fact they had years of experience, resourcing and brand loyalty behind them. Small breweries had to attract customers by word of mouth and had no advertising budget. Indeed, Newbegin had already been restructured precisely because some of its original investors had not made the returns they imagined.[74] By 1994 Newbegin had diversified to the production of home brew kits consisting of a canned wort and hop concentrate to which the aspiring brewer needed only to add water and yeast. They were also providing commercial micro-brewery systems with a capacity of up to 2000 litres. As many as 850,000 brew kits had been imported in 1991 – enough to brew 17 million litres or 5 per cent of total beer consumption in New Zealand. As specialist home brewing suppliers emerged, it was estimated that only a very small percentage of home brewers were still doing their own malting. Indeed, the ease with which home brewing could now be done, compared with an earlier period when many brewers had to produce their own malt, certainly encouraged more to experiment and some of them to then move on to establish new breweries.[75]

Although it is difficult to assemble a complete list of this new wave of New Zealand breweries, it seems that a majority did not last beyond a decade and some that did were subject to changes of ownership and name. Aside from the difficulty of trying to break into a very competitive and shrinking beer market, some claimed that the big breweries were actively stifling them. In 1989, DB Chief Executive Graeme Hawkins stated: 'We've taken the position that competition from the independents is good – anything that makes beer more interesting is good for the industry as a whole.' He suggested that the small breweries had 'product niches' that were difficult for commercial brewers to fill.[76] Yet bottle stores apparently insisted that they could not stock new beer brands without head-office permission. They also repeated the practice experienced by Mac's with high pricing strategies designed to discourage purchasing.[77] Although this was a very small market share, it seems that any share was worth fighting for in a tight market.

In a move that can be regarded as somewhat ironic after half a century of efforts to rationalise and standardise brands, including NZB's calamitous Lucky endeavour, the two major breweries also fought beer with beer. From late 1987, DB produced a 'brewer's selection' series including Stratton's Strong Ale and later Duncan's Noble Hop Lager. When questioned as to why the pioneers of continuous fermentation uniformity were now allowing their brewers to experiment, Graeme Hawkins explained, 'It's a statement that we can do that sort of thing – and do it very well. It's good for company morale, good for the brewers and all part of the culture of the industry.'[78] In another strategy, Flame beer was released in 1995 by Black Dog, accompanied by a website asking 'tired of the boring beers from the big boys?'[79] It was quickly pointed out that Black Dog listed the same office address as DB, and the new arrival was treated with some scepticism.

> DB Breweries has created an alter ego as it tries to take possession of the boutique-brewery market. But not everyone is taken in by the brewery-within-a-brewery ploy. And with a launch function using scary-looking PVC-wearing women, you might think subsidiary Black Dog Brewery (BD – geddit?) is DB's evil twin. As you have come to expect, the beer is meant to have been brewed by a new and ingenious process. This time, it's called 'extreme brewed, crash cooled' because apparently the beer is heated up then chilled. Although one is loath to use the expression these days, the packaging looks as if it's an appeal to Generation X.[80]

As Alan Knight from Hamilton's small Kahikatea Brewery observed, 'They're sufficiently threatened by us now, that they've begun to put out products that look as if they're made by independent breweries, but aren't.'[81] For its part, Lion made some effort to invoke its history with a renewed emphasis on earlier local brands such as McGavin's and Ward's. But its more fundamental contribution, arising from Doug Myers' engagement of American brewing consultant Chuck Hahn during the mid-1980s, was to abandon continuous fermentation and return to batch brewing as part of a clear campaign to emphasise that its beer was of better quality than that offered by DB. The days of beer delivery by tanker soon came to an end.[82] To reinforce the point that the independent brewery movement was still in its infancy, Hahn believed that it would be up to the major breweries to produce and market niche products and to educate the public about the wider possibilities of beer in hope of reversing the drift to wine. As he claimed in 1994, 'Nowhere else in the developed world do premium beers occupy such a small percentage of the market – less than 5 per cent. In Europe and the US, premium beers have a 10 per cent share. In the US, 200 small brewers have sprung up in the last five years. . . . Here, the large brewers are looking to fill the gap.'[83]

If Hahn saw the big breweries as the key to a diverse beer market, others saw an opportunity to leverage against their power. At the annual conference of the Hotel Association in September 1994, publicans told their national executive to find alternative suppliers such as small breweries or imported beer to punish DB and Lion for raising prices in the North Island. Ongoing boycotts of the two breweries were also threatened. The Chartered Clubs' Association had passed a similar resolution. Both felt that a price increase was misjudged on the eve of the Hotel Association conference and because it preceded the annual increase in excise duty on beer which followed a review of the consumer price index. The South Island had not been subjected to price increases precisely because of the stronger position of small breweries as alternative suppliers. These included Otago Draught Ale Ltd, which had been established in 1991 to provide cheaper beer to independent pubs in Dunedin. However, it doubted whether its 10,000-litre capacity would enable it to become a significant contributor to the market.[84] While the boycott threat amounted to nothing, it did reinforce to DB and Lion that they could no longer take dominance for granted.

Just as the big two were trying to outflank the small breweries and each other, a threat of greater magnitude emerged. In the early 1980s Michael Erceg abandoned an academic career as a mathematician to join his family's Pacific Vineyards wine business in Auckland. Following disagreement over the direction of the company, Erceg established Calypso Beverage Company in 1987, which soon evolved into Independent Liquor. Although the company was best known for its ready-to-drink pre-mixed spirit-based beverages (RTDs), it began brewing at Papakura from 1990 with barley sourced from Australia due to Lion and DB's malt monopoly. By late 1991, it had secured 3 per cent of the canned beer market through cut-price offerings and acting as a distributor for Australian brands that had encountered distribution difficulties with DB and Lion. Later, Independent secured licences to brew international lagers such as Carlsberg, Tuborg and Kingfisher while expanding its own budget range. By the time Erceg was killed in a helicopter crash in November 2005, Independent Liquor was the third-largest liquor distributor in New Zealand and had an extensive export network. After a period of private equity ownership, the company was sold to the Japanese Asahi brewing group in 2011.[85]

Such competition made advertising and the establishment of brand loyalty more important than ever. As Lion's marketing director, Tony Balfour, noted in 1996, 'The old myth of cradle to grave brand loyalty no longer applies. A typical beer drinker today buys four different brands and feels loyal to all of them.'[86] Certainly, there was a long tradition of newspaper and some poster advertising by breweries, and both DB and NZB had run various sophisticated and memorable themed campaigns throughout the twentieth century. At other times, the approach tended to the bizarre. The black tie, celebrity launch of DB's Kiwi Lager in 1986 ended with all present being asked to stand and sing to the tune of 'God Defend New Zealand':

> Grog of nations,
> One's taste sweet,
> Kiwi Lager can't be beat.
> So clap your hands,
> And stamp your feet,
> Grog defines our beer land.[87]

With beer advertising allowed on television from 1992, the stakes were raised considerably. Lion initially faced something of a dilemma. Much as it wished to have a national brand to emulate the profile that DB Draught had enjoyed during the 1970s and earlier 1980s, Lion Brown had limited profile beyond Wellington, and Lion Red, notwithstanding its strong association with rugby league, was very much an upper North Island beer. Although the 'red blooded' advertising campaign provided some national impetus for Lion Red, the brewery moved to a more regional focus. Ward's was rebranded as Canterbury Draught in 1990. Further south, despite the fact that it would have made economic sense to close the Dunedin brewery as had been suggested during the 1980s, Speight's was reinvigorated and began to achieve momentum beyond Dunedin. Speight's sponsored the Dunedin senior rugby competition from 1977, and succeeded Steinlager as the sponsor of the Coast to Coast multisport event from the mid-1980s. A 1982 internal competition had also produced the company slogan, 'Pride of the South', although this actually placed second in the original selection. Perhaps most important was sponsorship of the Otago rugby team as it enjoyed an on-field renaissance from the late 1980s, which culminated in winning the National Provincial Championship in 1991. In turn, an association with Dunedin student culture helped to move demand for Speight's around the country as some of its new devotees completed their studies and moved elsewhere.[88]

But the key to national success was the 'southern man' advertising campaign, launched in 1992 as Speight's became the first beer brand to be advertised on television. Set in the high country of the South Island, the campaign emphasised the loyalty of laconic 'southern men' to their locality and its beer. This drew on a long tradition that associated the core of New Zealand's national character with rural values and authenticity, in contrast to the supposedly effeminate and increasingly corporate urban setting. In themes of resistance to the long-standing northward drift of the New Zealand population, clear differentiation was established between the 'real men' of the south and the 'yuppies' of the north, and the political and economic influence of Auckland in particular.[89] It was perhaps no coincidence that the campaign gained momentum at a time of north–south tensions in New Zealand rugby following an inept All Blacks performance at the 1991 Rugby World Cup, which many traditionalists attributed to 'corporate' interference by Auckland's John Hart, a Fletcher Challenge executive, with the coaching of rugged Canterbury farmer Alex 'Grizz' Wyllie. In 1992, Otago's Laurie Mains became the All Blacks coach. Speight's success prompted further Lion interest in provincial rugby sponsorship during the 1990s, and it became something of a national brand for Lion, even as Steinlager continued as sponsor of the All Blacks.[90]

TOP LEFT: Beginning with its northern operation in 1977, New Zealand Breweries began a full rebranding to Lion. Lion Technical Library

TOP RIGHT: A special bottling supplied to the Waikato Racing Club in 1980. Lion Technical Library

BOTTOM: Brewery worker Mohammad Khan supervises the filling of beer cans at the DB Waitemata Brewery, Ōtāhuhu, 1990. DB Breweries

The transformation of Tui from a local to a national brand was one of the most significant and enduring campaigns of the early twenty-first century. DB Breweries

On the other side of the market, DB had already pursued similar rural themes through the DB Draught Clydesdales from the mid-1980s. Images of the team of working horses pulling a wagon of beer kegs, which became a prominent DB logo, embodied, in the words of Melissa Kennedy, 'the qualities of honesty, dignity and strength inherent in the heritage of the brand and the proud spirit of the Clydesdale'.[91] In other respects it was apparent that DB had an image problem. As outlined earlier, it had lost a number of skirmishes with Lion during the 1980s and the momentum that had produced South Island dominance during the 1970s was scarcely being maintained, especially with the rise of Speight's. The solution was a brand overhaul, emphasising products without DB in the name. In 1991 it offered a new range of Double Brown cans without a DB logo in a deliberate effort to target non-DB drinkers in an Auckland market dominated by Lion.[92]

However, the most enduring success came via Tui East India Pale Ale, a brand that was still little known outside Wairarapa and Hawke's Bay by the early 1990s. From 1994, Tui's 'Yeah right' billboards, which offered sarcastic commentary on current issues – such as 'I don't mind paying for Auckland's roads. Yeah right' – slowly began to appear in the central region of the North Island. There were none in Auckland until 2001 or in the South Island until 2004. Despite periodic complaints to the Advertising Standards Authority about the content of some billboards, the campaign has endured for more than a quarter of a century, spawned three books and transformed the Tui Brewery at Mangatainoka into a popular beer tourism destination.[93] But DB evidently did not regard imitation as a sincere form of flattery and threatened legal action in 2010 when the Bethlehem Community Church in Tauranga put up a billboard reading 'Atheists have nothing to worry about! Yeah Right'.[94] Given that the church had been using such billboards for several years without complaint, and that DB had drawn numerous complaints to the Advertising Standards Authority for 'Let's take a moment this Christmas to think about Christ. Yeah right', it quickly backed down and agreed to assist the church with new advertising.[95] Other successful Tui campaigns included 'beer plumber', an online prank video in which a house was plumbed with kegs of Tui, and 'catch a million', in which any spectator wearing a Tui T-shirt at a one-day or T20 international cricket match during the 2013–14 season would win $100,000 if they cleanly caught a six with one hand. Two spectators succeeded.[96] Perhaps in an attempt to deflect from its gender stereotype, the annual Miss Tui contest pitched to resourceful women: 'Wanted: A woman who can swing an axe, hook a fish, apply lipstick without a mirror and loves Tui more than her hair straightener, she is out there somewhere!'[97]

DB's other major campaign of the 1990s, the launch of Monteith's as a 'craft' beer range, became a victim of its own success before rebounding to become a staple of the New Zealand beer market. Beginning in 1990, but more comprehensively from 1995, the DB brewery in Greymouth followed the directive of Brian Blake that the Monteith's brand should 'operate like a craft brewery'.[98] The beer became very successful and was popularly associated with the West Coast of the South Island. It invoked its 'history' with such slogans as 'brewed for right now since 1868'.[99] Of course this was deceptive in that Monteith's Phoenix Brewery, established in Reefton in 1868, was merged into Westland Breweries in 1927. The beer, advertised as the 'Boast of the Coast', did not appear after July 1928 and the Monteith's company was formally wound up in October 1936. The Reefton brewery was closed in 1960 and Westland Breweries was taken over by DB in 1969.[100] In sum, the 'continuity' evident in the branding of Monteith's was tenuous at best and certainly not Greymouth-based.

By the turn of the century, the Greymouth brewery was operating at capacity and required a significant and costly upgrade to meet growing demand. Therefore, an economic decision was announced in March 2001 to close the brewery and shift all production to Auckland, which was also more logistically convenient to launch an export drive.[101] Yet the success by DB in aligning Monteith's to the West Coast meant that its decision triggered an immediate backlash and calls to boycott all DB products. As one Hokitika beer drinker put it, 'Some bean-counter in Auckland seems to have decided they can make more profit up there, but it could be a short-term thing for them, because I think it'll destroy the brand. The brand is West Coast.'[102] The *Waikato Times* added, 'To Coasters, Monteith's isn't a brand. It's a beer, and it's theirs. They were making it long before DB bought the brewery in 1969. It's an international award-winning success story in a region that knows the meaning of the word downturn.'[103] Veteran blues singer Midge Marsden and his band Legal Tender wrote 'It's Not Closing Time Just Yet' as a protest song.[104] Before long a campaign was orchestrated by a Wellington public relations consultant, Gerry Morris, who quickly summed up the gravity of the situation. 'It's like taking Christianity out of Bethlehem and moving it to Disneyland. You would never see Guinness taking their brewery out of Dublin.'[105] While many reacted to a decision that put profit ahead of people and community, and local Labour MP Damien O'Connor suggested that DB stood for 'Dumb Bastards', it was soon apparent that the reaction was a focal point for a wider debate about global business practices in which the ordinary drinker was subservient to corporate greed. A link was also made to DB's 'foreign' ownership by Asia Pacific Breweries.[106]

DB quickly backed down, announced that it had underestimated the public connection with the West Coast branding of Monteith's and agreed that a 'significant volume' would continue to be brewed in Greymouth.[107] Damien O'Connor now declared that DB stood for 'Decent Blokes' and that everyone had won: 'Greymouth retains the brewery, the West Coast retains its brand and the image with it, and DB gets the endorsement of our region. It was always going to require some compromise and I don't think there could have been a better outcome.'[108] As *The Press* reflected:

> The closure of the brewery would only have cost 14 jobs, but it exemplified a trend provincial New Zealand knows too well. Local industries are deemed uneconomic by their big business owners and either shut down or shifted north to Auckland. Underlying the reaction to DB's threat was the visceral distaste for the shallow glamour of Auckland, normally most visible on the rugby ground. Whether DB had not anticipated the intensity of the public flak, or whether it simply felt that it could safely ignore it is unclear. A cynic might say that the entire episode was a fiendishly cunning vehicle for the giant brewer to show its responsiveness to public opinion, and increase sales of Monteiths. It is certainly true that Monteiths has received publicity that money could not buy. Yet as a deliberate ploy, such a Lazarus gambit was surely beyond the creativity of most Queen Street publicity brokers.[109]

However, the outcome was not as harmonious as some imagined. There were eventually five redundancies from a brewery staff of fifteen, all of whom had been public faces of the campaign opposing the closure.[110] There was also further controversy and renewed calls for a DB boycott a few weeks later amid claims that the brewery was selling Auckland-brewed Monteith's on the West Coast. A local connoisseur said the 'imposter beer' was different to 'true Monteith's' in every way – taste, head and smell. While Brian Blake argued that brewing at both plants was necessary to keep up with demand, Coasters were given advice on how to identify Greymouth-brewed Monteith's from an Auckland batch.[111] DB's final loss in the public relations campaign came the following year when Gerry Morris won an international award for his campaign against them. He promptly donated his US$2500 prize to the Greymouth Parents Centre and invited DB to match his contribution. The brewery declined on the basis that its sponsorship budget for local communities was fully committed.[112]

Following from this partial consumer victory, attention turned to the place-of-origin claims of other beer, as the public seemed to become increasingly sceptical of brand rhetoric. The Commerce Commission received a complaint over misleading labelling that Speight's was brewed in Dunedin, when in fact all canned and bottled beer was brewed and packaged in Christchurch and Auckland. Although the Speight's label stated 'Traditionally brewed at Speight's Brewery Dunedin since 1876', Lion insisted that this did not mean the beer was brewed in Dunedin, but merely that Speight's had been brewed in Dunedin since 1876. The Commerce Commission also warned Lion following its recent acquisition of Mac's, of which more in the next chapter, that it was misleading to still label the beer as being owned by the McCashin family. Lion's corporate affairs director, Graham Seatter, argued that it was ridiculous to require precise point-of-origin details. Cheese manufacturers were not required to say where the herd came from or which grass the cow chewed. Both breweries then insisted that the expense of having to change labels would inevitably be passed on to the consumer.[113] Despite urging from DB that Lion should be prosecuted, the matter was eventually resolved with an out-of-court settlement in which they admitted to breaches of the Fair Trading Act and changed labels accordingly.[114] Aside from the matter of the origin of Monteith's, DB was also scrutinised for claims that Tui was brewed at Mangatainoka when much of it was actually brewed in Auckland. Changes were subsequently made.[115]

Brand controversies of other kinds emerged from the strategies used to target specific audiences. From the 1980s, mainly from alcohol health research, there was increasingly sophisticated data to identify who was drinking what, where and how frequently. The 1995 *Drinking in New Zealand* national survey indicated that 40 per cent of beer was consumed by 12 per cent of those surveyed – eighteen- to twenty-nine-year-old males. Men drank 70 per cent of all alcohol, and the eighteen- to twenty-nine-year-olds drank more than all men over thirty. It was therefore essential and logical to target beer advertising to younger males and to do so with themes that resonated with them.[116] Consequently, while advertising standards dictated the nature of images and messages around responsible drinking and counselled against 'unduly masculine themes', content has repeatedly been criticised for its overwhelmingly masculine archetypes and sexism. A 1996–97 Lion Red slogan declared, 'If you want me to spend time in the kitchen, put more beer in the fridge.'[117] At the end of 1997 Lion conceded that there had been a good deal of sexism in its previous advertising and announced that it wished to 'de-bloke' the image of Lion Red to appeal more to the 'man of the 90s'. But at least one marketing representative suggested that Lion risked alienating loyal male customers. 'If you change their perception of the brand into something else, say Lion Pink, they will switch to something else and won't be back. It's possible to be blokey without being chauvinistic. I think this may be an over-reaction to past criticism.'[118]

But de-bloking clearly had its limits within the industry. In 2003, Lion was forced to abandon a promotion involving young women dressed as 'Lion Red nurses', who were to sell beer and give away Lion Red products at the international rugby sevens tournament in Wellington. Nurses' representatives condemned the campaign for portraying nurses as 'bimbos and dolly-birds in a bid to sell more beer. This is demeaning. Nurses are knowledgeable, skilled and dedicated [and] play an extremely important role in society.'[119] As part of an enduring portrayal of beer drinking as an exclusively masculine activity, a Ranfurly Draught billboard declared 'A group of men is called a pub', while another of its ads for Frontier Lager portrayed a futuristic 'man park', with a narrator explaining in the manner of a nature documentary, 'These are the last real men left. Large shifts towards veganism and vampire based TV series nearly wiped them out. But here they're free to live as they wish. They grow their own vegetables, barley and hops mainly.' Beer is clearly rescuing this endangered species from near-emasculation.[120]

In 2012, Tui was challenged by groups such as Auckland Feminist Action for ads featuring a group of men attempting to infiltrate the Tui Brewery, run by bikini-clad women. As a spokesperson explained, 'The women are depicted as even more stupid than the dorky men who try to steal the beer. The dorky men outwit the women. If you look at the advert in its entirety, it's demeaning women, it's showing women as just a collection of body parts, it's saying that it is OK to value women on how they look – that's the only value women have in that ad. That has a very corrosive effect on women's self-worth.' A spokesman for DB replied, 'Tui Brewery Limited is an equal opportunities employer and we have no grounds for dismissal of the Tui Brewery girls, especially when they're doing such a great job.'[121] A complaint about a DB Export ad under the slogan 'let nothing come between a man and a great beer' also failed to gain traction when the agency responsible argued that 'man' was generic as in 'mankind' and was aesthetically better in the ad than 'person'. The Advertising Complaints Board agreed and also 'further acknowledged that beer had a predominant appeal to men and there was a level of acceptance that advertisements about beer would be more likely to appeal to men'.[122] The next chapter will reveal that such assumptions about women and beer were soon to be challenged more consistently.

Others criticised the beer–sport–male-bonding nexus for subconsciously associating drinking with success and sporting prowess, not least because national sporting heroes such as the All Blacks were sponsored by beer brands. Although breweries were not specifically encouraging heavy drinking, the association through advertising and sponsorship of sport was doing so. During a 2011 debate on alcohol sponsorship, Progressive Party MP and sometime Deputy Prime Minister Jim Anderton, who had been instrumental in the campaign to ban tobacco sponsorship in 1995, captured a popular view:

> Most young Kiwi boys watch those games; they grow up dreaming of being just like the stars they see on TV. So what do they do when they see those games dripping in beer sponsorship? They learn that saturation in beer is part of the game. Beer marketers like Steinlager know that. That is why they sponsor the All Blacks . . . . Administrators, media and talkback callers act surprised when athletes like Jimmy Cowan or Jesse Ryder get into trouble for drinking too much. But I don't blame those players for making mistakes. I blame the culture that associates alcohol with their sporting brilliance as a way of making money out of selling more beer.[123]

As high-profile sportspeople periodically attracted attention for alcohol-fuelled indiscretions, there were calls to ban alcohol sponsorship. There were also periodic complaints about the behaviour of drunken spectators at some major sporting events, and beer cans were eventually banned lest they became missiles. Proposals to serve only 'stadium strength' weaker beer did not gain momentum. For the most part there was a reluctance to challenge the significant revenue stream that brewery sponsorship provided to sport.[124] During the mid-1990s this was estimated at $40–50 million of a total direct sport sponsorship pool of perhaps $90 million. Below the elite level, the two major breweries also competed for the 'tap rights' to sports club bars by paying clubs money for access. The greater the turnover of a club bar, the more money a club was paid.[125]

Ultimately, the enduring critique of beer advertising is a reminder of much that did not change amid the rivalries and rebranding of the late twentieth century. Although the 'premium' beer market increased from 12 per cent in 1996 to perhaps 20 per cent by 2000, the definition of that category was very much dependent on the whims of marketing executives. By the mid-1990s, 70 per cent of production was still brown beer, with most produced by DB and Lion. The main difference between the two was that DB persisted with continuous fermentation for longer. Conventional lager provided most of the remainder of production, but the distinction between 'brown' and 'lager' was more cosmetic than real.[126] During a visit in 1999, renowned international beer expert Michael Jackson was bemused by the tendency for major breweries to pass off draught beers as ales when they were in fact 'bronzed lagers'.[127] In 1996 Lion Red made up about 20 per cent of all beer available for consumption, dominating the greater Auckland region in particular, while DB Draught made up about 10 per cent and was still the most popular beer in the South Island. Third place was split evenly between DB Export Gold, Steinlager, Lion Brown and Speight's. Mac's held perhaps 1 per cent of the market, and fifty or so small breweries less than that between them.[128] While Terry McCashin had undoubtedly instigated an important development in 1981 and overcome multiple challenges to achieve success, the New Zealand brewing industry was still far short of a 'craft' revolution with new ingredients and beer styles.

Rhyme x Reason Brewery, Wānaka, was one of many new establishments to offer tasting paddles and food on the brewery premises. Photograph: Liz Martyn

ELEVEN

The Pursuit of Hoppiness: Craft Beer and Beyond, 2000–2020

Despite the best efforts of Mac's and other smaller breweries of varying influence, it was not until the twenty-first century that New Zealand could make a genuine claim to be a 'craft' beer paradise. Although exact brewery numbers are difficult to determine by 2022, with estimates ranging between 170 and 220, it was claimed that New Zealand had more breweries per capita than the United States, the United Kingdom or Australia. However, from 1994 to 2021 New Zealand fell from tenth to twenty-seventh in global per-capita beer consumption.[1] The emphasis had shifted to quality over quantity, to a broadening of available beer styles, a willingness to experiment, and the emergence of a more sophisticated and diverse beer culture in terms of who was drinking and in what settings. In November 2001, journalist Bruce Holloway drew a neat parallel between beer and politics.

> Mainstream breweries are a bit like mainstream political parties. Because they are always aiming for the middle ground in the hope of not offending anyone, they invariably end up with a wishy-washy end product. Similarly, just as it requires a small ginger group to lead the way in any radical political reform, boutique breweries have become the chief civilising agents of beer culture. It is precisely because their strong uncompromising flavours are 'not for everybody' that boutique breweries can play such a major role in beer education. But the ironic thing is that most microbrewers would like nothing more than to be big brewers.[2]

Perhaps the only elements that did not join the march to sophistication were the promotional strategies for mainstream beer and the general tone of the alcohol harm debate, which continued a long-established habit of not drawing distinctions between the impacts of different beverages.

New Zealand followed the global trend to small breweries from the 1980s and had nearly sixty by the turn of the century. In very broad terms these developments can be seen as part of a global 'artisan' movement reacting against corporate capitalism generally and neo-liberal upheaval in particular. Everything from the slow food movement to farmers' markets promoting local produce and traditions in bread, cheese, chocolate and olive oil, among many others, emphasised 'authentic' alternatives to the homogeneity of industrial mass production.[3]

Figure 11.1: New Breweries, 2000–2021

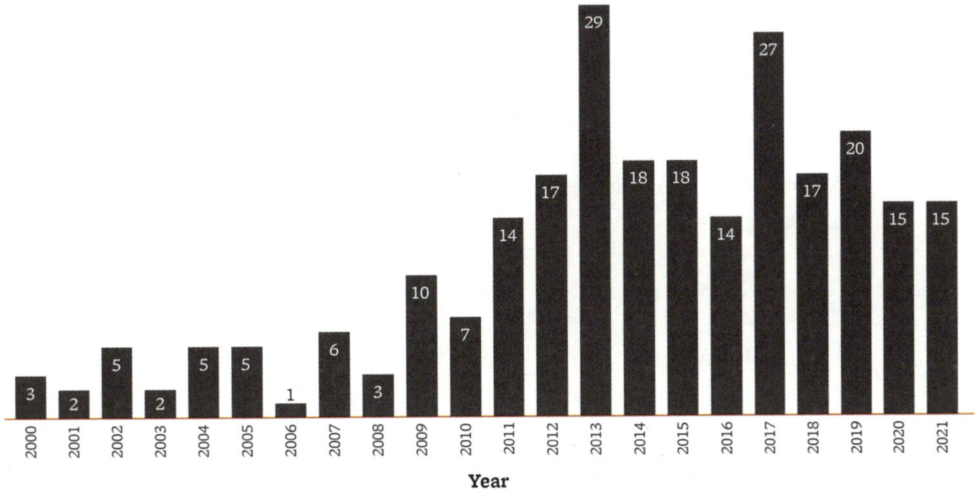

Source: Greg Ryan, New Zealand Brewery Database, doi:10.25400/lincolnuninz.21944714

Figure 11.2 Total Number of Operating Breweries 1835–2022

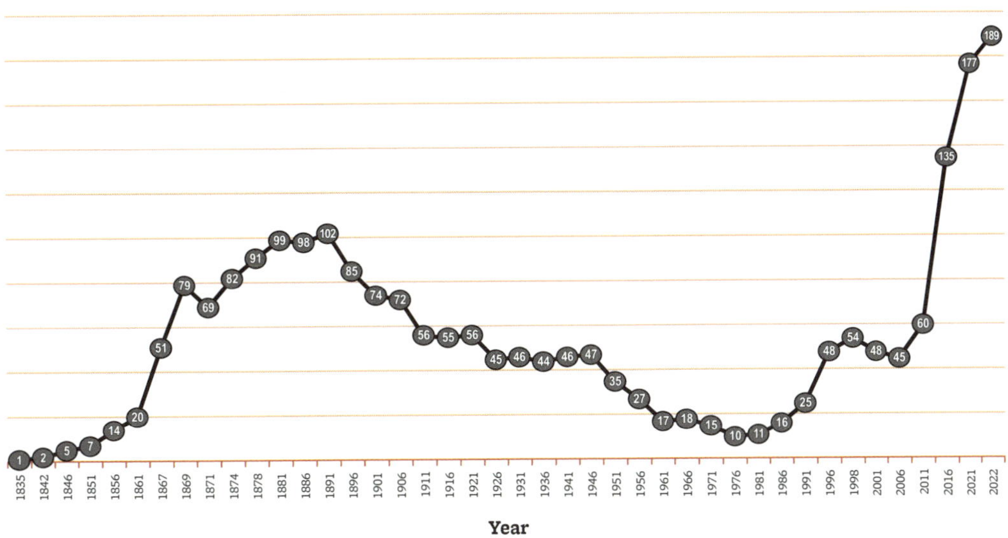

Source: New Zealand Census, various issues. These figures refer to breweries rather than brewing companies, noting that both New Zealand Breweries and Dominion Breweries brew from several different sites.

Figure 11.1 shows steady growth in new breweries during the first decade of the twenty-first century and dramatic growth in the second. Net growth is more difficult to determine as the arrival of new breweries has always been better documented than their departure.

While the majority were in the main and larger provincial centres, locations such as Tūrangi, Ohakune and Kaikōura received their first brewery and there was particular growth in Central Otago. For some the turning point from notions of 'small' or 'boutique' to 'craft' breweries was 2009, when oversupply and recession caused the price of a bottle of Marlborough sauvignon blanc to fall below that of some craft beers, which seemed to remain recession-proof as consumers sought new beer innovations and were willing to pay for them.[4] The period from 2010 saw an explosion of new labels and the estimates of craft beer devotees rose from 10,000 to nearer 100,000. The number of professional breweries trebled.[5] Although the extent of market share is always confounded by questions of definition, to be discussed shortly, by 2015 it was estimated that 'craft' beer had 12–15 per cent of the market and annual growth in the vicinity of 30–40 per cent, with the highest proportions in the upper North Island. Two years later it was reported that 'small breweries' were responsible for 5.8 per cent of total consumption but 10 per cent of total expenditure on beer.[6] In sum, the market was dynamic, but exactly how dynamic was a moot point.

As with the first cohort of independent breweries during the 1980s and 1990s, there were different motivations and approaches. Some, such as West Coast Brewing in Westport, wished to undercut the major breweries with cheaper, more mainstream beer in local markets. In 2007 it sought to coax shareholders with an offer of two dozen bottles of beer for every $8500 investment and a promise to deliver a six-pack of any new beer to all investors.[7] Others, including the Mussel Inn near Tākaka and Galbraith Brewing Company in Auckland, were determinedly small-scale and focused on a specific venue. Parrotdog's roots were in home brewing, with the aim of 'saving money on booze', but exposure to craft beer made them hone their brewing skills.[8] A number of others, such as Yeastie Boys, did not possess an actual brewery and instead arranged to have their recipes contract-brewed at other breweries, especially the Invercargill Brewery, established in 1999. Some were sophisticated, well-financed and ambitious, while others were one- or two-person operations. There were all shades in between. Some lasted only a short time, while others established an enduring national reputation. Some cultivated a specific market niche, such as Founders Brewery in Nelson. This was established by John Duncan in 1999 as New Zealand's first organic brewery. Duncan was a sixth-generation brewer, being the great-great-great-grandson of Joseph Dodson, who had brewed in partnership with George Hooper from the 1850s.[9] In Ōamaru, Scott's Brewing Co. specialised in gluten-free beer, while in another part of town Craftwork Brewery described itself as a 'small idiosyncratic Belgian-style nano brewery'.[10] Irrespective of the motivation or niche, once a critical mass developed it gave more brewers the confidence to take the step from hobby to profession.

Whereas the standard 'yellow, amber and black' beer range of the early years had gained some ground as an alternative to mainstream beer, greater innovation was required to thrive in an increasingly crowded market. The likes of wheat beer, dark lager, Belgian styles and chocolate stouts were new to most beer drinkers at the turn of the century. While Britain and Europe still provided points of inspiration, as they had for the likes of Terry McCashin and Richard Emerson, others took notice of developments in the United States. American pale ale became a relatively common style to brew and there was a general trend to emphasise particular varieties and combinations of hops. Consistent with a growing consumer preference for all things fresh and natural, there was also an emphasis on fresh-hopped beers, in which it was not unknown for helicopters to be used to move hops from vine to brewing kettle as quickly as possible.[11]

Andrew Dixon from the Mussel Inn, near Tākaka, displays the sophisticated bottling technology at his brewery in 2004. Photograph: Neil Macbeth

At other times, the range and innovation of breweries was shaped by the need to keep pace with trends such as the vogue for sour beer that took hold in the 2010s, or that for hazy ale, which appeared around 2020. Seasonal beers such as saison for midsummer and doppelbock or smoked beer in winter also became popular. For others it was a matter of capturing attention with a one-off innovation. Beer was no longer boring. If Kiwifruit and Tequila Gose was an option for the cocktail fraternity, one wonders at the market for the Stag Semen Stout that appeared in 2015. Brick Lane E.1. curry ale appealed to a certain English tradition, while peanut butter, bacon, chilli, pineapple lumps and smoked eel were among other flavours to capture attention.[12] Others were inspired by local and global events – not least being Behemoth Brewing Co., which kept a close eye on the political pulse. In 2018, its Dump the Trump IPA attracted an avalanche of hostile reviews from US Republicans on the company's Facebook page and on beer rating sites, followed by an equally fervent response from Trump opponents. The GOP was similarly unimpressed with Behemoth's ImPEACHment Sour Ale. At the height of the Covid pandemic during 2020–21, Behemoth released 'Pastor of Muppets' pale ale – an open attack on religious fundamentalism and opposition to government vaccine mandates.[13] Still others took their inspiration from the quest for sustainability with beer produced from waste bread, or the likes of Aotea Brewing on Great Barrier Island emphasising zero-waste production and environmental values in its decision-making. In 2019 the New Zealand Beer Awards, of which more shortly, introduced a Brewery Sustainability Award.[14]

By 2019, the New Zealand brewing industry apparently had a value of $2.8 billion – although the meaning of 'value' was not defined. It provided direct and indirect employment for more than 6600 people, and contributed $996 million to government revenue – all as beer consumption continued to decline.[15]

Table 11.1: Alcohol Available for Consumption, 1960–2020 (pure alcohol and liquid volume)

	Beer (L)	PA (%)	LV (%)	Wine (L)	PA (%)	LV (%)	Spirits (L)	PA (%)	LV (%)
1960	237 million	74	95	5 million	6	2	6 million	20	3
1975	390 million	63	91	26 million	15	6	13 million	22	3
1990	371 million	58	86	49 million	20	11	12 million	22	3
2005	315 million	44	69	89 million	33	19	54 million	23	12
2020	293 million	36	59	113 million	32	23	89 million	32	18

LV = each beverage as % of liquid volume of all alcohol available for consumption
PA = each beverage as % of all pure alcohol available for consumption
The figures for 2005 and 2020 include spirit-based drinks as well as spirits.

Source: H. K. Orchard, 'New Zealand Alcohol Consumption Statistics 1955–1989', Wellington: Alcoholic Liquor Advisory Council, 1990, 3–6; 'Alcohol Available for Consumption Year Ended December 2020', Stats NZ, https://www.stats.govt.nz/information-releases/alcohol-available-for-consumption-year-ended-december-2020

The profile of beer in twenty-first-century New Zealand is intertwined with wider patterns of alcohol consumption over a longer period. Up to the late 1980s a growing population was paralleled by increasing beer availability, both local production and imports, which peaked at 399 million litres in 1987 before embarking on a steady decline to a low of 279 million litres in 2012. Obviously, beer as a proportion of all pure alcohol available and of the liquid volume of all alcohol available also declined steadily. Initially the changing proportions were triggered by the growing popularity of wine noted in the previous chapter. But from the 1990s, especially with the rise of Independent Liquor, spirit-based RTDs sharply increased their proportion of both pure alcohol and liquid volume, mostly at the expense of beer, especially among younger drinkers. Moreover, as the actual quantity of beer available decreased and the population increased, per-capita consumption fell dramatically. From a peak of 133.9 litres in 1975 it was scarcely 60 litres by 2020. Within beer consumption there was also a clear shift from quantity to quality.[16] The move from draught beer to cans and bottles was noted in the previous chapter. While the vast bulk of offerings even in the late twentieth century were around 4 per cent alcohol by volume, by 2020 beer under 4.35 per cent made up around 40 per cent; that between 4.35 and 5 per cent, roughly the strength during the inter-war period, comprised about 45 per cent; and the remaining 15 per cent was beer stronger than 5 per cent alcohol by volume – the volume of which had increased fivefold in a decade. The bright future repeatedly predicted during the 1980s and 1990s for low-alcohol beer, below 2.5 per cent alcohol by volume, never really materialised. For example, although its share increased in the three years to 2015, it declined in each of the next five years and only constituted 2 per cent of beer available for consumption in 2020. Of the total volume of beer available in New Zealand in that year, 83 per cent was produced locally and 17 per cent imported. Twenty-four million litres of New Zealand beer was exported.[17]

Strong growth after 2000 both in numbers of breweries and the variety they produced owed much to the sale of beer in supermarkets from December 1999. This development clearly caused concern among the retail chains aligned to the major breweries as the supermarkets offered a 'neutral' distribution channel at very competitive prices. To attract new customers, supermarkets unveiled a wide range of imported beers and later local offerings. Although there was periodic criticism that they used these as loss leaders to draw in customers amid existing supermarket rivalries, in reality the low cost fell on the breweries, which were forced to keep their prices low to compete for access. In turn, the specialist liquor stores responded with their own discounts, with the result that, aside from increases due to excise, the price of bottled and canned beer remained very stable relative to the consumer price index.[18] By 2013, it was estimated that 'craft' beer comprised 10–13 per cent of supermarket beer sales and was increasing rapidly.[19] Yet there remained considerable variation. While supermarkets have always varied their stock in relation to the affluence or otherwise of the particular demographic each serves, the range of beer that some carried compared to others was also a reflection of the enthusiasm of individual purchasing managers rather than uniform company policy. Perhaps the only criticism of supermarkets was that as they were not specialist beer retailers they were inexperienced in treating the variety of beer appropriately, tending to have everything cold or nothing cold, whereas the lager and stout ought to have been treated differently.[20]

Supermarkets contributed to two other significant changes. Firstly, they tended to sell beer in smaller units than the liquor stores. In 1998, 94 per cent of packaged beer was sold in packs of twelve or more, and 69 per cent in packs of twenty-four or more. Gradually, especially as lower blood-alcohol limits to curb drink-driving increased the popularity of the 330-millilitre or similar-sized bottle or can, the purchase of individual bottles and later cans became more common. This had obvious benefits to smaller breweries, which tended not to offer packages, and had implications for older notions of style and brand loyalty as more consumers became likely to purchase a wider variety of beer.[21] Secondly, while beer was mostly consumed by men, women were by far the majority of supermarket shoppers and controllers of household budgets. Many undoubtedly knew the beer preferences of the males for whom they purchased, but there was also scope to influence purchasing decisions and to gradually entice more women to drink beer.[22] Certain supermarkets also established a reputation for their range of beer, and in 2014 New World inaugurated annual Beer & Cider Awards, which simultaneously enhanced their credentials as purveyors of good beer and promoted that beer to a wider market.[23]

Much as supermarkets assisted the growth of new breweries by easing the obstacles to retail access that had prevailed before late 1999, there were other factors that created a feasible environment for aspiring brewers. Globally, the number of new breweries skyrocketed, especially in markets influencing New Zealand, as consumers everywhere came to expect choice and variety. By the start of the twenty-first century there were around 500 independent breweries in Britain. By 2020 there were more than 3000.[24] The number in the United States increased from perhaps 1500 at the turn of the century to around 9000 by 2021.[25] While only a fraction of this new variety reached New Zealand, as importers long remained cautious in testing the market beyond popular lagers, there was certainly more variety available to those determined to look for it. Moreover, as overall beer consumption fell, some drinkers were more inclined to explore their pleasure in a premium product rather than one unlikely to linger in the memory or tastebuds. Consumers more conscious of health and well-being were attracted to smaller breweries with an emphasis on natural, chemical- and sugar-free beers.[26] On the other hand, with the emergence of the internet during the 1990s and various forms of social media during the twenty-first century, the marketing and promotional reach of new breweries became considerably easier than before. The internet also became an important forum for discussion and advice among brewers and drinkers.

As breweries expanded their repertoire, there was demand for new varieties of hops and especially a shift in emphasis from ‘alpha hops’, mainly associated with bittering, to ‘aroma hops’ such as Hallertau Aroma and Pacific Hallertau, which imparted a more subtle taste and smell to the beer. It was no coincidence that the Hop Research Committee named a new cultivar Nelson Sauvin to emphasise that the palate of the beer drinker was as refined as that of the wine drinker.[27] By 2006, with funding from New Zealand Hops Ltd and the Foundation for Research, Science and Technology, plant scientists at HortResearch were analysing global beer trends and developing new plant breeding strategies crossing established New Zealand hybrids with traditional European aroma varieties. In turn, this attracted overseas brewery interest in New Zealand hops, including from the likes of industry giants such as Anheuser-Busch.[28] At the other extreme, from 2008 Christchurch’s Wigram Brewing took a more traditional approach, using wild hops planted on the West Coast during the 1860s. Its Kortegast Sparkling Ale was named in honour of W. C. J. Kortegast of the Crown Brewery, Hokitika, established in 1865.[29] Others chose to emphasise the variety of hop used by including it in the name of their beer – such as Nectaron IPA and Green Bullet Pilsner. The surge in popularity of craft beer triggered a global hop shortage by 2012 that threatened to limit production of some types of beer and required careful management in coming years.[30]

A range of beer produced in the upper South Island in 2004. Photograph: Neil Macbeth

Inevitably, the pool of brewing expertise also expanded. Brewers as a species tended to be collegial and to share ideas and advice. The principle was very much that if any individual was producing a quality product, this benefited the industry as a whole by enhancing the demand for good beer. Both Lincoln and Massey universities offered brewing courses within degrees, and in November 2009 the latter offered a week-long craft brewing course covering everything from malting, hop utilisation, brewery engineering and hygiene to yeasts, fermentation and sensory evaluation.[31] Home brewing, long a means to produce cheap beer of uncertain palatability, also became a space for exploring flavours and styles. For the entry-level home brewer, Mac's offered home brew kits in such styles as Late-Hopped Lager, Creamy Brown Ale and Oatmeal Stout. Serious brewers were able to draw on suppliers of a bewildering variety of local and imported hops, specialty malts, yeasts and equipment. The winner of the annual Society of Beer Advocates (SOBA) Home Brewing Competition also had their beer brewed commercially. More than a few dedicated home brewers eventually took the leap to commercial brewing.[32]

As the number of breweries increased, they spawned a supporting infrastructure. Bottling, which could be messy and time-consuming or required investment in expensive plant, was made easier by the emergence of independent and mobile bottling plants. Eventually mobile canning plants also appeared on the back of a global acknowledgement that canning was more efficient, more sustainable and better for beer as it protected it against deterioration from light strike – although consumer perception, no doubt based on the sort of beer traditionally put in cans, took a little longer to change.[33] Another advantage of cans, especially amid crowded supermarket shelving, was that they provided a larger canvas for artistic and eye-catching labelling. In 2020, the NZ Label Design Awards were inaugurated to recognise the creative talent on display. Packaging awards had been established as early as 2007.[34] Another piece of the puzzle was solved for retailers and individual consumers alike when distribution businesses such as BeerNZ and mail order suppliers such as Beer Hug emerged to source beer from different parts of the country and bring it to new audiences.[35]

Yet not all was unrelenting progress. Some breweries were thwarted by rogue yeast strains or other technical difficulties. Some failed to convince enough palates to gain a toehold in a crowded market. Others were confronted with obstacles to their own identity. In 2001, Duff's Brewery in Dunedin, run by Gavin and Bev Duff, yielded to legal letters from 20th Century Fox after producing a beer with the same name as that drunk by Homer Simpson. The brewery soon became McDuffs so drinkers did not confuse its offering with an imaginary beer drunk by a yellow cartoon character.[36] For Moa Brewing, established in Blenheim in 2003, it was a case of bad marketing decisions. In 2010, they offended the gay community with a campaign implying that low-carb beer was only suitable for homosexuals. T-shirts were printed with the slogan 'Low Carb Beers' with a pink Q superimposed over the B.[37] When Moa was floated on the stock exchange and became a publicly listed company in 2012, its shareholder prospectus was labelled misogynist for its emphasis on 'manly' themes and products, and images of models in short skirts. Similar concerns were voiced by female shareholders in 2016. When the brewery was purchased by new owners in 2021 after several years of substantial losses, its new chief executive observed, 'What they were thinking back then, only they can comment on. I think it was a pretty lazy way to attempt to disrupt and build the brand, personally.'[38] The success of the new brewing wave also created tensions as debate erupted as to whether Nelson or Wellington could claim to be the 'craft beer capital' of New Zealand. While Wellington staged the largest beer festival in the country and had a number of high-profile breweries and bars, Nelsonians, especially John Duncan, pointed to a strong tradition dating to the 1840s, the Mac's revival and the fact that it was the only region growing hops. In 2011, a new Nelson group registered 'craft brewing capital' as part of a strategy to promote local beer and beer tourism. Later in the decade, Wellington responded with 'craft beer capital'.[39]

By around 2017, there were suggestions that the industry had come to something of a crossroads as the proliferation of new breweries and their volume of production had not yet been matched by an increase in market share. Commentators emphasised that beyond the ability to produce good beer, there needed to be greater emphasis on distribution, branding and developing new markets, including exports, as the wine industry had done.[40] Yet even the existing market was problematic. In 2019, Luke Nicholas of Epic Beer described the breweries as being in a 'crazy arms race' to cater to customers who would not drink the same beer twice and equated 'new' with 'good'. Brewers were having to produce many more small batches of new beer to avoid being left behind in the market, but this was costly, time-consuming and ultimately not sustainable. There was a risk that consumers would be overwhelmed with choice. Epic's contribution to the arms race was a brew containing a $2000, 40-kilogram amethyst crystal which had been 'charged' under a full moon. As Nicholas wryly noted, 'I did everything I could to spin this amazing story. People lapped it up.'[41] Perhaps the only advantage was that new beers were often sold at a higher price, thus creating better margins, and helped to promote sales of a brewery's core range.

Although not on a significant scale, some breweries eased the congestion through export opportunities. In 2001, a Nelson export company sent a container of local craft beer to Germany for distribution throughout Europe. The following year Emerson's secured regular export orders to Melbourne. In 2004, North Canterbury's Brew Moon Brewing Company achieved the unusual feat of exporting its stout to Dublin. By 2015, it was estimated that twenty-five breweries were pursuing sales overseas, with another twenty planning to do so.[42] Other international connections took the form of collaborative brewing, such as Andrew Dixon of the Mussel Inn assisting with a Belgium-brewed edition of his Captain Cooker mānuka beer, or Kelly Ryan of Wellington's Fork Brewcorp collaborating with Thornbridge Brewery in Derbyshire where he had previously worked. In 2020, Behemoth secured a contract to brew beer under licence from Vietnam's Heart of Darkness craft brewery.[43]

If market share did not grow at the pace desired, it was not for want of official recognition and strategies to boost public profile. From the early 1990s, small breweries began to win recognition from various national and international beer awards. At the first New Zealand Beer Awards in Wellington in 1997, with renowned beer writer Michael Jackson as chief judge, Emerson's was declared the champion brewery.[44] However, the media sometimes struggled to grasp the significance of the accolades. In July 1998, under the headline 'Tui ale takes prize', the *Dominion* reported that Tui's East India Pale Ale had been voted the best mid-strength beer in the international section of a beer awards in Melbourne. Only as an afterthought was it mentioned that Mac's had won six gold awards at the same event. No mention was made of the successes of Christchurch's Dux de Lux or Nelson's Lighthouse Brewery.[45] By 2002, now sponsored by the Hop Marketing Board and held in Nelson, the New Zealand Beer Awards attracted 200 individual entries from about 45 breweries. At that point, for reasons not clear, Lion and DB boycotted the event and focused on the New Zealand Beer Cup, established in Wellington as part of the BrewNZ festival. By the following year the rival competitions were merged into the Wellington event, which by 2006 was being held in Parliament's banquet hall. From 2007 the awards were under the auspices of the Brewers Guild of New Zealand.[46]

Connoisseurs studying the form at the South Island Beer Festival, Christchurch, in 2015. Photograph: Neil Macbeth

Thereafter brewers trumpeted their award successes on labels and in other publicity. For some, endorsement also came via reviews as beer entered the lifestyle columns of newspapers and magazines.[47]

For many of the drinking public the first encounter beyond mainstream beer may have come from beer festivals, although the first such events in New Zealand experienced mixed fortunes. While the 'Munich beer festival' organised in Wellington in April 1971 to raise funds for the New Zealand Olympic and British Commonwealth Games Association, in anticipation of the 1972 Olympic games in Munich, was apparently successful,[48] the 'Munich style beer festival' held in Hastings later in the year for the Te Mata Kindergarten Establishment Committee 'proved a sobering affair for the organisers'. Although 1500 tickets were printed and sold, gatecrashers swelled attendance to nearly 2000.

> Onions which had been put on tables in the hall were used as missiles by some people, and within a short time, pieces of chicken were being thrown at random over the heads of the crowd. A large flag, one of 12 lent to the organisers by the German Embassy in Wellington, was hauled down near the roof beams by a group of men who used tables and chairs to form a pyramid to reach the flag.[49]

Nevertheless, by the end of 1971 Munich beer festivals as fundraising events were being described as the 'in thing' and NZB frequently supplied a special 'Munich bier'.[50]

By the 1990s civility generally prevailed at such events as the Blues, Brews and Barbecues festivals held throughout the country, and later at specialist beer festivals. The most significant of these was the BrewNZ festival inaugurated in Wellington in 2001, which evolved into Beervana. Others included the Great Kiwi Beer Festival, the Great Australasian Beer SpecTAPular, Auckland's City of Ales, Wellington's Beers at the Basin, the South Island Beer Festival and a variety of smaller regional events. Yet there was a certain stigma attached to beer festival-goers that did not accompany their wine equivalents. As beer writer Bruce Holloway observed in 2003, 'You may have noticed that whenever there is a wine festival, some arty high-brow type gets to report on it in the local media. But when there is a beer festival, for some reason the police reporter is invariably assigned the job. As a result, the newspaper copy always begins "Police were delighted there were only xxx arrests at yesterday's beer festival. . ." and the actual quality of the event gets overlooked.'[51] Four years later beer drinkers were equally offended by the decision of the Marlborough District Council's liquor licensing sub-committee to ban glass mugs at the Blues, Brews and Barbecues festival, when a similar ban had not been placed on the use of glasses at Marlborough's wine festival.[52] There was simply no escaping the fact that beer drinking, although it had moved a long way from the swill, was still perceived as a very masculine and periodically uncouth activity.

From the 1990s there were signs, albeit sometimes clichéd and patronising, that the male dominance of all things beer needed to accommodate women as both drinkers and brewers. Although a few women worked on bottling lines and in administrative roles at the major breweries by the 1980s, none were involved in brewing. The macho advertising campaigns for mainstream brands also confirm that the breweries had no interest in appealing to women, perhaps for fear that doing so risked implying to men that they were now drinking 'girls' beer'.[53] Indeed, as one female journalist observed in 1997:

> Congratulations Lion, your relentless effort to alienate a female drinking population which may be larger than you realise has at least in this case paid off. I will not drink your beer again. I remind you that women who do not enjoy beer themselves, may well purchase for partners, and for entertaining. Lastly, dare I suggest that your portrayal of the loud, chauvinistic, drunken, 'macho' and seemingly cloned beer drinker may be less impressive to many male consumers than you think.[54]

However, the shrinking beer market from the 1980s, in a society that was slowly confronting its entrenched gender stereotypes, also triggered some change as the breweries began to explore lighter and less bitter beers that could appeal to women. In 1994, Lion Nathan considered advertising its Ice Beer in women's magazines during the summer as its research showed that the ratio of male to female drinkers for the product was 70:30, as opposed to the usual 88:12.[55] By 1997, the two major breweries were strategising from different ends of the spectrum. A Lion representative explained, 'Light Ice has given us an easy drinking beer with a moderate alcohol content. As a brand it's more about Ice than low alcohol. It's a beer for people who are not really beer drinkers.' Meanwhile, Monteith's was attempting to attract women to its Black Beer.[56] As noted earlier, the sale of beer in supermarkets from the end of 1999 was also significant as the breweries recognised both the need to influence women as purchasers for men and the potential for more of them to become consumers. The appointment of Nicki Stewart in 2000 as the first female chief executive of the Beer, Wine and Spirits Council, a body representing the non-commercial interests of both Lion and DB, was also timely in breaking down barriers. A qualified chef, Stewart was well placed as a trend to beer and food matching emerged through such events as the Monteith's Wild Food Challenge, established in 1998, and various restaurants seeking to broaden their beer offerings. There was a feeling, in the manner of wine as an accompaniment to food, that women would be more likely to try beer in such settings.[57] Yet not all were convinced that strategies to court the female market were authentic or conducive to good beer. As a 2007 review of Lion's new 'mash' range noted, 'it could badly do with somebody else pushing hop advocacy. These beers sport plenty of curious additives in what appears a determined effort to capture a slice of the adolescent female slapper-juice market. Citrus, lime etc.'[58]

Led by Tracy Banner, women began to enter the industry as brewers from the 1990s. Trained in England and arriving in New Zealand in 1994, Banner worked briefly in technical services for Lion before joining Mac's and enjoying considerable acclaim as head of its brewing team. Among her innovations was Blonde Mac, a less bitter, orange- and coriander-infused beer, designed to entice women from wine. After two years as head brewer at Speight's from 2005, she returned to Nelson and in 2009 secured a shareholding in the Sprig & Fern Brewery, which was beginning to enjoy success through its own small chain of pubs – one of a very few among the new generation of brewers to venture down this path.[59] Following Banner's lead, breweries such as Brew Moon, Beer Baroness, Valkyrie and Abandoned Brewing have been established or led by women, and both Lion and DB have employed more women as brewers.[60] The Pink Boots Society, founded in the United States in 2007 as a brewers' network for women and non-binary people, was extended to New Zealand in 2014 – since when it has enjoyed a strong profile at beer festivals, produced a variety of collaborative brews and engaged in various strands of advocacy.[61] However, as Kathleen Kuehn and Sophie Parker point out, being the product of a female brewer has tended not to be an explicit marketing strategy. Rather, the emphasis has been in established gender-neutral values of authenticity and quality.[62]

A DB promotion for its seventy-fifth anniversary in 2005. DB Breweries

This is not to suggest that the macho character of beer and inherent sexism associated with it was easily dismantled. As discussed in the previous chapter, sexist advertising and promotion of mainstream brands remained a staple in the twenty-first century. Women drinking beer were still treated as a novelty, and there was an enduring assumption that specific lighter beer styles needed to be produced to satisfy them rather than effort being made to broaden their beer knowledge through a full range of styles.[63] As Alice Neville mused in 2018, 'I've never quite understood why it's presumed that girls are the main consumers of tasteless beer. Anyone who's been on Tinder will tell you that every second dude is grasping a green bottle in at least one of his pics, or – if he's really feeling fancy – he might be swigging from that bastion of basicness, a Corona.'[64] While the case of a woman being refused entry to a 'blokes only' Central Otago home brew competition in 2012 was the most visible example,[65] the context of the #MeToo movement and revelations of sexism and sexual harassment within the US craft brewing industry in particular prompted reflection in New Zealand. While there has, so far, been no sign of specific public allegations, familiar themes of belittling and harassment have emerged. However, even with opportunities for anonymous reporting, there is a sense of some reluctance to do so in a relatively small industry.[66] A racist social media outburst by one brewer in May 2021 also focused attention on a wider lack of cultural diversity in an industry dominated by white males as both brewers and drinkers.[67] In sum, no matter how good the beer, there is a clear sense that those behind it are obligated to behave responsibly and inclusively. Indeed, those such as Parrotdog and Yeastie Boys who used PledgeMe or other crowd-funding platforms to expand their operations were acutely aware of the importance in maintaining good public relations.[68]

These and other debates took place in an increasingly organised advocacy and support environment. While brewers' associations in various forms existed since the late nineteenth century, they lacked real purpose as the industry became a duopoly. The Brewers Association, representing Lion and DB, positioned itself as being 'committed to the promotion of responsible consumption of alcohol' and targeted intervention to reduce alcohol harm for specific groups. It has also commissioned various reports on the economic footprint of the industry.[69] On a wider scale, the Brewers Guild of New Zealand was established in 2007 with a vision 'that the quality and creativity of New Zealand beer is celebrated at home and abroad'. The Guild represents all brewing businesses from small to large, organises the New Zealand Beer Awards, and provides such support as employment agreement templates, health and safety guidelines and advocacy amid perennial alcohol-harm debates.[70] In 2006, partly on the model of CAMRA, the Society of Beer Advocates was formed as a consumer-based organisation with a vision that 'People have an appreciation of, and access to, a diverse range of quality beer'.[71] In 2008 SOBA began publishing *Pursuit of Hoppiness* as a free magazine to 'inform, educate, entertain and to spread the word about good beer'.[72]

Ultimately, the growth of small breweries was relative and the publicity they generated was somewhat deceptive. The vast majority of the beer market remained with the two major breweries and the majority of consumption remained with mainstream beer. However, in an ever-shrinking market, Lion and DB, and later Independent Liquor, clearly believed that every mouthful was worth fighting for. Hence the 'premium beer' strategies of the 1990s, described in the previous chapter, gathered pace in the twenty-first century. As Ralph Bungard of Christchurch's Three Boys Brewery suggested in 2009, initial innovation came from the craft brewers who were able to experiment on a small scale and had customers more likely to take risks. 'The micro-breweries are the proving ground for beer styles, and then the big boys will

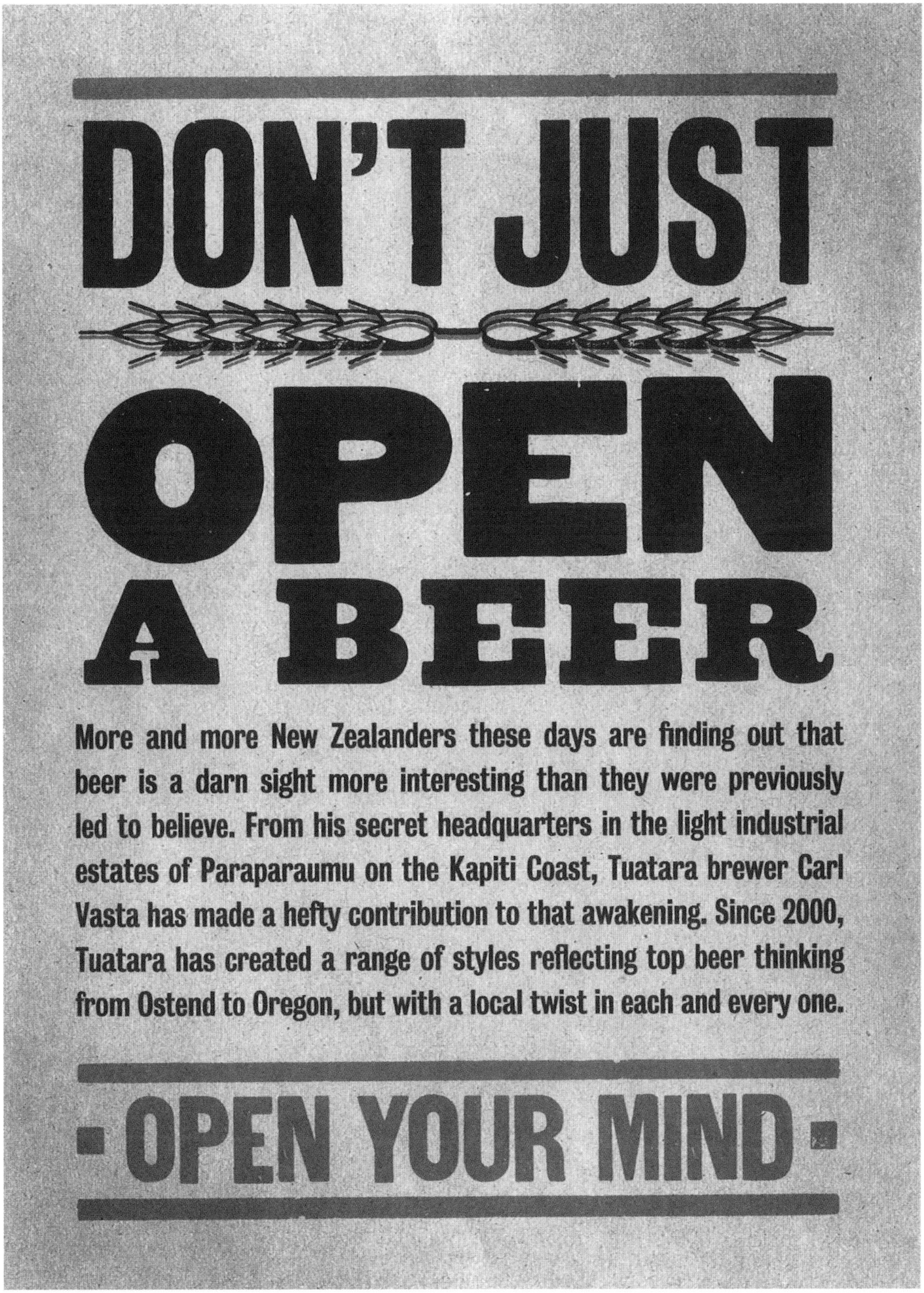

This 2016 Tuatara advertisement emphasised the contribution of the brewery to the broadening of New Zealand beer tastes. Tuatara Brewing Co.

come in and get on the bandwagon.'[73] DB continued to expand its Monteith's repertoire with seasonal offerings. Among Lion's innovations were a Speight's boutique range using traditional recipes and restored open kauri fermenting gyles at the Dunedin brewery, and a short-lived production of New Zealand's first beer to receive carbon-zero certification.[74] Its other major innovation, while not visible to consumers, was nevertheless significant, when in 2010 the brewery moved from Khyber Pass Road to East Tāmaki, thus ending 150 years of brewing on the site. While the original brewery had been on the edge of the Auckland business district, the growth of Auckland throughout the twentieth century had transformed Newmarket into a congested retail, light commercial and residential area in which transport and other restrictions made it unsuited for large-scale brewing. The new $250 million brewery, christened The Pride, with 20 per cent more capacity than its predecessor, emphasised environmental responsibility and sustainability, being largely constructed with recycled glass bottles as an aggregate in the building's concrete.[75]

As well as premium beer, there was a growing emphasis on premium presentation in improved surroundings. From the mid-1970s, especially through its Cobb & Co. chain, which consisted of fifteen outlets by 1979 and was described as 'a national catering institution' by 1982, Lion had consciously targeted family socialising.[76] From the beginning of the new century, the likes of Lion's Shed 22 in Wellington and the nationwide Speight's Ale House chain provided a family-friendly restaurant–bar setting deliberately designed to appeal to women as well as men and to soften the image of beer consumption. Although some criticised the limited beer range at such establishments and suggested that other bar owners 'sell their souls' when accepting financial support from the major breweries, a counter argument was that the two breweries were more interested in outflanking each other and had relaxed somewhat from their earlier efforts to stifle small breweries. This was also driven in part by a recognition that they needed guest beers on tap and a wider variety in bottle stores to attract and retain customers who no longer displayed a fierce loyalty to one beer.[77] Other initiatives included the breweries 'accrediting' certain establishments for the quality of their beer service and staging such events as the Stella Artois Draught Master beer-pouring competition and the New Zealand Guinness Pint Master competition.[78]

The clearest indication that the major breweries recognised the importance of variety and the future of craft beer was in their acquisition of some of its most significant pioneers. The first move came early in 1999 when Mac's sold its beer brands, but not the Nelson brewery, to Lion for an undisclosed but significant sum. As Terry McCashin explained, 'It was the first year we didn't have any growth and that let us know that we had to do something.' In part Mac's was being challenged by the rise of Monteith's, but it had also failed in efforts to develop export markets with Australia and Japan. Investment by Lion gave the brewery more 'muscle' to expand.[79] In 2002, a $1 million bottling plant was installed at the brewery to help it meet growing demand, and nationwide distribution was also improved by Lion as Mac's became a key component of its premium beer range.[80] Lion ceased brewing Mac's in Nelson in 2007, and after the restraint of trade expired early in 2009 members of the McCashin family returned to produce Stoke beer and Rochdale cider.[81]

In 2012, Lion purchased Emerson's for $8 million. For Richard Emerson the offer came at a point where the brewery needed significant investment to expand its production capacity to meet demand. For Lion it was a matter of acquiring and enhancing the brand without undermining its quality. As Danny Phillips, Lion's head of marketing, explained: 'We knew that if we didn't protect the magic, we'd be wasting our money – and we'd be accused of corporatising it or homogenising it. We wanted the richness and depth of the Emerson's story – the reason people were attracted to the beers and brand.'[82] While some were perturbed by the sale, SOBA saw a positive for the future of craft beer in New Zealand. 'Lion have seen a really good brewery and have decided to invest in it. By doing so they have shown that they are intent on building their craft beer business; Emerson's are not a "rival" to be taken over. . . . With any luck, with some capital investment from Lion, New Zealand will soon see Emerson's excellent beer being made available on tap at more outlets across the country and, who knows, even abroad.'[83] Around the same time, Founders in Nelson was acquired by Independent Liquor, which in turn had been acquired by the Japanese brewing giant Asahi in 2011. In a familiar refrain, John Duncan stressed that his family still owned the brewery and made the beer, but now with more access to the market.[84] In early 2017, DB joined the party when they acquired Wellington's Tuatara Brewery, followed almost immediately by Lion paying as much as $25 million for Wellington's Panhead Custom Ales, the success of which they described as a 'runaway train'.[85]

Yet in the eyes of craft beer devotees there was periodic evidence that the big players were corporate bullies who did not really understand beer. In August 2003 DB launched Monteith's Radler, accompanied by a media release acknowledging that radler, meaning 'cyclist' in German, was a beer style originating from Bavaria during the 1920s – an 'easy drinking lager with a citrus tang'. By one estimate, perhaps half of German breweries produced a radler by the twenty-first century. These were generally two parts lager or other lighter beer mixed with one part lemonade or lemon–lime soda.[86] Despite radler being an acknowledged international beer style, DB applied to the Intellectual Property Office of New Zealand (IPONZ) to trademark the word 'radler' in New Zealand. Although objection was raised at the time by an IPONZ examiner who pointed out that radler is defined as 'beer mixed with lemonade' and therefore did not appear to have any distinctive character as a trademark, DB responded that beer mixed with lemonade is known as a shandy in New Zealand, and therefore 'radler' was distinctive to DB here. IPONZ granted the trademark in a move not noticed by anyone else in the brewing industry.[87] Despite being thwarted in the High Court in 2007 when it tried to stop Lion using another generic beer style, summer ale, in 2008 DB acted to enforce its trademark against Dunedin's small Green Man brewery, whose German brewer had produced a radler. It was required to cease production or rename its beer. Unable to finance a complex legal battle with DB, Green Man rebranded its beer 'Green Man Cyclist'. However, in May 2009 SOBA applied to have DB's trademark revoked on the grounds that IPONZ should never have allowed it for a generic beer style. Nonetheless, the best efforts of multiple submissions from beer connoisseurs and SOBA's pro bono legal team did not convince IPONZ to reverse the trademark. Its assistant commissioner noted in 2011, 'I am of the opinion that the applicant has not established that RADLER had any meaning whatsoever to consumers at the application date and, as such, has not satisfied its onus of proof.'[88]

DB's tactics and the subsequent IPONZ decision caused considerable disquiet in the craft beer community. One critic tweeted, 'I think I might start up some coffee brands called latte, cappuccino, espresso & flat white and trademark the names.' Others suggested a boycott of DB products.[89] The West Coast Brewing Company responded with Reldar – Radler spelt backwards – and put a sign up outside its Thirsty Bloke Bottle Shop in Christchurch saying West Coast MP Damien O'Connor 'might have been right when he called DB deceitful bastards' – a reference to the Monteith's controversy in 2001.[90] DB insisted that it had not sought the trademark to prevent competition or restrict consumer access to different types of beer, but to protect its intellectual property and effort in creating the brand. But judiciously DB decided not to claim the hearing costs awarded against SOBA.[91] In 2014, it was suggested that the matter could be one for the Commerce Commission as a breach of the Fair Trading Act, in that DB's 5 per cent alcohol by volume product was not actually a true radler, and New Zealand was the only country in the world where other brewers were not allowed to use the term 'radler'.[92] By that point the Green Man brewery had closed and the campaign failed to gain momentum.

For Lion, less than six months after acquiring Emerson's, it was a matter of seeming to bite the hand that fed it. In March 2013, it released a new 'Crafty Beggars' range with the marketing tagline 'A craft beer you can actually drink'. Although its brand marketing director suggested the campaign was 'tongue in cheek', he also said that 'Lion was catering to traditional beer drinkers who found boutique craft beer too challenging because of the strong hop flavour'. Moreover, the company had as much right to be in the craft beer market as a small brewery. 'Just because you're small doesn't mean you're good.' But one bottle store responded that it would not stock the new range as the marketing was an insult to the wider brewing industry.[93] Craft products continued to emerge from the big breweries, including Independent Liquor launching its Boundary Road Brewery range through a campaign in which 999 volunteer 'tasters' received three test brews and then voted for their preferred option, which was then released as The Chosen One.[94]

Big brewery 'incursions' on the craft market, and the acquisition of Emerson's in particular, prompted sometimes heated reactions of a sort that had not been evident when Mac's was acquired at a time before the 'beer revolution' had gained full momentum. The key debate focused on what was and was not 'craft' beer. Some drinkers and several establishments announced that they would not be drinking or stocking Emerson's again. Amid strands of opposition to global capitalism and foreign ownership, there was a perception that big breweries sacrificed quality to quantity and were now crowding genuine craft beer off supermarket shelves. In June 2013, Josh Scott of Moa organised a forum to discuss what some regarded as fake or faux craft beer. As he explained: 'Bigger breweries have the marketing budgets to just create what they say are craft brands overnight. The rest of us work . . . for years building ours. They're not there to grow the category. They're there to take the market share craft brewers have created. . . . When I started up, there was passion and integrity, whereas now with the big boys, it's just about crunching numbers and mass production.' Scott argued that consumers were being misled and that a clear definition as to what could be described as 'craft' was needed.[95]

Some suggested that the size of a brewery could be a defining factor, in the manner that the American Brewers Association included a volume limit in its definition, and stated that a genuine craft brewer must be 'small, independent and traditional'. Others felt that the best strategy was better public education about beer and especially more emphasis on beer style rather than brand.[96] Both SOBA and DB had earlier pointed out that the size of the brewery was not relevant to the quality of the beer, as the prominent US craft brewer Sam Adams was listed on the New York Stock Exchange and was producing more barrels annually than Lion produced at East Tāmaki. For the Brewers Guild of New Zealand, Ralph Bungard suggested that the big breweries' craft offerings were important in introducing consumers to new kinds of beer with the implication that they may then be more adventurous.[97] By 2017, with some of the higher-profile craft brewers such as Garage Project and Parrotdog increasing production volumes, and the latter signing a distribution deal with an Australian supermarket chain, beer writer Michael Donaldson concluded that 'craft' was more about the ethos of a beer, rather than the mechanics.[98] Of course, such semantics about craft did not concern the majority of beer drinkers, who remained in a mainstream market, or the Carterton correspondent who actively raged against it:

> All this crap about craft beer. I can't stand the weird, expensive stuff. I was given some recently and it wasn't just me that tipped it out over the rose bushes. If you want a nice beer that hits the spot, just get yourself a dozen 4% 750ml Lion Red in a wooden crate. It works out at $2.89 a pint when you take the bottles and crate back to be recycled. . . . And pubs? Well, they should be about real conversation, not a round-circle discussion about an overhopped lager with pinot gris undertones and rays of grapefruit singing out from the bottom of the glass, or whatever the latest fad is that sets you back $15 a pint.[99]

One big brewery would no doubt have been pleased that Lion Red was making ground in what many regarded as a Tui domain.

*

The dynamic change in the brewing industry from the 1990s ran in parallel to renewed disquiet about alcohol harm and calls for legislative reform. In some ways the concern was understandable. Following the liberalisation of the Sale of Liquor Act 1989, the number of liquor outlets more than doubled during the next two decades to around 14,000. In the decade to 2008, there was a 9 per cent increase in per-capita consumption of pure alcohol. The emergence of RTDs, noted earlier, was altering drinking behaviour and bringing renewed focus to the consequences of binge drinking. Falling unemployment and a relatively buoyant economy meant, it was suggested, more surplus income to use on alcohol consumption. There were also periodic claims that supermarkets used alcohol as a loss leader.[100] However, in other respects New Zealand was not unique within a global movement. A more diverse media, especially social media with its tendency to amplify and sensationalise, also fed perceptions that alcohol harm was on the rise. There was a particular emphasis on problematic youth drinking and a more traditional moral indignation at drinking by young women.[101]

Periodically, in an adjunct to the macho advertising campaigns discussed in the previous chapter, the major breweries provided grist to the media mill with products and promotions that reinforced stereotypes of the problematic drinker and attracted criticism from across the alcohol health spectrum. In 2002, DB released its Hopper range of sweet fruit-flavoured beers as an alternative to bitter beer, specifically aimed at drinkers aged eighteen to twenty-nine and women in particular. Health researchers expressed concern that such beverages were known as bridging drinks, marketed to introduce new drinkers, and possibly children, to alcohol. They were no doubt pleased when the beer failed in the market and was withdrawn in August 2004.[102] In the meantime, Lion had run a competition to win your weight in Lion Red. Tui offered packs of beer with 'penalty cards' for watching rugby, requiring the holder to drink in response to such events as the Blues scoring a try or an attractive woman being seen on screen. The chief executive of the Alcohol Advisory Council condemned this 'out of touch' promotion.[103] In 2008, Alcohol Healthwatch attacked Tui for its Tui-mato sauce, which had a small amount of beer in it, as a cynical ploy to get around alcohol advertising guidelines and expose young people to potentially harmful images. In 2010, a rival brewery posted a picture showing an Auckland school bus with a large advertisement for Monteith's Summer Ale, although in this case the brewery was not at fault and the bus company apologised.[104] In the same year, one of the most apparently 'sacred' beer relationships of recent decades was somewhat curtailed when the University of Otago, in a stand against New Zealand's binge-drinking culture and excessive drinking by some university students, banned alcohol advertising and sponsorship on campus, thereby ending Speights's long-standing sponsorship of the Otago University Students' Association Orientation Week.[105]

By contrast, craft beer producers and drinkers were determined to emphasise that they were in a different realm. In a 2008 exhortation to drink local rather than mainstream beer, a Nelson journalist suggested:

> Getting more people drinking beer brewed in their own town brings more than economic benefits. There are environmental advantages and significant cultural ones too. Having things that are unique to your community fosters identity, identity fosters pride and pride fosters civility. I could be dreaming, but I suspect this has something to do with the fact you never see people getting aggro at Founders, the Sprig and Fern, the Mussel Inn or the Beer Fete. Craft beer drinkers are a tidy bunch, too. Over the last couple of months I've picked up more than 100 bottles off the streets of Nelson and not one of them has come from a local brewery.[106]

In 2009, the Brewers Guild called for the first 10,000 litres of beer its members produced to be excise tax-free to encourage the growth of smaller brewers and thus create jobs in the regions. Guild Chairman David Cryer argued that the higher-strength, and therefore higher-taxed, beers produced by some craft brewers were not the products being abused by binge drinkers.[107] While Alcohol Healthwatch noted in 2019 that it was keeping a 'cautious eye' on the increased production of high-strength craft beer but wanted evidence before suggesting any limits, Gemma Smith of Hawke's Bay's Brave Brewing Co. concluded, 'Craft beer drinkers aren't the kind of people to smash a handful of drinks and get hammered. They drink it for the taste and flavour of the beer.'[108]

As the alcohol harm debate escalated during the early twenty-first century, there were some small gestures towards reform. From late 2002 bottles and cans of beer were required to carry a 'best before' date, specify the number of 'standard drinks' contained, and provide the name of the New Zealand importer if not brewed domestically. While the standard drink measure was perhaps a more useful guide to drinkers than alcohol by volume, it caused certain frustrations to importers who were confronted with the time-consuming and expensive task of hand-labelling imported products before selling them to a demographic generally not associated with problem drinking.[109] In 2006, both Lion and DB presented reform proposals to Parliament's Law and Order Select Committee. These included banning teenagers from drinking in public places or driving with any alcohol in their system; tighter restrictions on new liquor licences; and tougher penalties for outlets breaching their licence conditions. But they did not believe that raising the legal drinking age back to twenty or restricting alcohol advertising on television would reduce alcohol harm.[110]

Finally in 2008, the Labour Government directed the Law Commission, under the presidency of former Prime Minister Sir Geoffrey Palmer, to carry out a further comprehensive review of the liquor laws. In doing so, the Commission repeatedly emphasised a March 2009 Business and Economic Research Limited (BERL) report, *Costs of Harmful Alcohol and Other Drug Use*, commissioned by the Ministry of Health and Accident Compensation Commission, which calculated the annual social costs of alcohol consumption at $4.8 billion.[111] In April 2010, the Law Commission released a 587-page report entitled *Alcohol in Our Lives: Curbing the Harm*. Its 153 recommendations included: an Alcohol Harm Reduction Act to replace the Sale of Liquor Act;

returning the alcohol purchase age to twenty; regulating alcohol advertising, sponsorship and promotion; reducing opening hours for licensed premises; streamlining the enforcement of the alcohol laws, including strengthening the grounds upon which licences could be declined and allowing more local input into licensing decisions; and increasing the price of alcohol through a 50 per cent excise tax increase in order to both reduce consumption and increase government revenue to offset more of the social cost of alcohol.[112]

Long before the Commission issued its final report, Nelson's Free House pub launched a nationwide online petition urging the government not to impose a blanket tax increase and indeed to give tax breaks to craft brewers. The petitioners were concerned that taxation would disproportionately penalise craft brewers, who were more dependent on price for profit, whereas larger breweries could push a greater volume of product on to the market without significantly altering price.[113] Whether the petition had any impact is unknown, but in April 2010 the National Government announced that it would not increase the excise tax on alcohol. Martin Townshend of Townshend Brewery in Upper Moutere made the familiar point that the empties littering the street on a Sunday morning were not coming from his more expensive beer. He saw the problem as too much cheap beer and RTDs, and recommended a sliding-scale tax system whereby the more a company produced, the more tax it would pay. By that means, smaller breweries would stay in business, but the makers of the products that contributed to binge drinking would pay more tax, resulting in the price of those products going up. 'I think they should pay more. They have economies of scale and they have about 90 per cent of the market.'[114] The point frequently made by Townshend and other critics of both the Law Commission and the alcohol health lobby more generally was that it lacked nuance and was prone to suggest one-size-fits-all solutions that did not distinguish the range of possibilities behind the generic term 'alcohol'. As shown at the beginning of this chapter, beer consumption had declined dramatically within a generation. Within the remaining beer consumption, premium, more expensive products have grown at the expense of bulk and mainstream. While some beer drinkers have clearly shifted to wine since the 1970s, the single most significant change to drinking patterns during the twenty-first century has been the rise of spirit-based drinks, which are primarily consumed by a younger population repeatedly identified as being at most risk of alcohol harm.

Another robust critique of the Law Commission would be familiar to many who argued against the prohibitionist onslaught that gripped New Zealand from the 1870s. The BERL threshold for harm from alcohol was crossed after 1.8 pints of beer per day – a measure low enough to include one in six adult New Zealanders, who were assumed to be responsible for half of all alcohol consumption. But as Eric Crampton and Matt Burgess pointed out, the 'costs' of alcohol calculated by BERL and the remedies endorsed by the Law Commission needed to be balanced against the overall consumer benefit from moderate drinking, not least in terms of aiding conviviality, sociability and the development of 'human capital'.[115] Moreover, BERL's assumption that alcohol had zero benefits meant that everything associated with it was then seen as a cost – a position which, among other things, took no account of those taxpayers employed in and around the industry. Further, even if one accepted that BERL's 1.8 pints had some deleterious impact on the drinker, there was scope to debate how much of this was a private cost, being one borne by the individual through their own behaviour, or an external cost in terms of one borne by others as a result of the drinker's activity.[116] While it was important to develop proposals to address the very real external costs on victims of drunk drivers and alcohol-related crime, Crampton stressed that proposed solutions needed to be weighed against the majority of New Zealanders who reasonably enjoyed moderate consumption of alcohol. In particular, 'Any per unit tax on alcohol aimed at recouping costs generated by a few hoons will always overcharge moderate drinkers while undercharging the folks who generate most of the problems.'[117]

Eventually the government responded with an Alcohol Reform Bill, which adopted in full or in part 126 of the Law Commission's recommendations and became the Sale and Supply of Alcohol Act in 2012. The drinking age was kept at eighteen and the excise tax increases were rejected. On the other hand, parents were required to give consent for minors to be supplied with alcohol; some restrictions were placed on opening hours; dairies and convenience stores were prohibited from selling liquor; and local councils were empowered to further restrict opening hours and control the location of liquor outlets. Some accused the government of pulling the teeth out of the Law Commission's proposals and pandering to both the powerful alcohol industry and the votes of drinkers.[118] Moderate beer drinkers and smaller breweries perhaps understood more than ever why missionaries on an apparently inebriated colonial frontier had established a temperance society at Kororāreka in 1836 that condemned 'ardent spirits' but found no problem with wine or beer.

This 2015 advertisement for Three Boys Brewery in Christchurch emphasised the link between global influences and local tastes. Three Boys Brewery

Epilogue: The Future of Beer

New Zealand beer and brewing have navigated a multitude of challenges since the first offering at Dusky Sound in 1773 – brewery fires, contaminated water, prohibitionist onslaughts, mergers, weak beer, political inertia, escalating taxation and declining consumption among others. Beer has endured not because it numbed the senses of drinkers, but because it has been accepted since ancient times as a vital component of sociability within communities. Where that role was questioned, especially during the swill era and its legacy in the bulk beer economies of the major breweries, the fault lay less with the drinker than with restrictive liquor legislation that had its misguided origins in ideology rather than evidence. Society will be periodically buffeted by neo-prohibition and new solutions to old legislative shortcomings, delivered at strategic moments within the election cycle, and too many of which reference alcohol with little acknowledgement of its different types and settings. As such debates unfold, there is much in the preceding history, both positive and negative, that ought to inform options for the future. One alcohol and health researcher could perhaps be forgiven for claiming that during six o'clock closing 'we were trained for 60 years to binge drink'. But attributing this to 'our history of relatively lax alcohol laws' would certainly bemuse any historian acquainted with the detail.[1] Similarly, as another addiction specialist argued that we have been lulled into regarding alcohol as 'a necessary recreational tonic that "normal" people consume regularly as part of "normal" life',[2] the long history of beer, if not other alcohol, suggests the need to turn such statements around to an understanding of why, how and with what consequences such a historically entrenched normality was curbed and demonised.

Predicting the future of beer has always been a somewhat fraught exercise. An 1875 account predicted a bright future for 'beer powder' being developed in Frankfurt that simply needed to be added to cold water. 'Shortly, therefore, a great revolution may be expected. With a few grains of Bass in one pocket, and a pinch or two of Ind and Coope in the other, the thirsty but not even bona fide traveller of the future will calmly make for the nearest pump, and there contentedly drink.'[3]

By 1932 chemists had reportedly developed means to colour beer blue, green, purple and other shades of the rainbow without altering the taste. 'The best hostesses will in future be able to order their beer to fit the colour schemes of their flats, and to order a drink in harmony with their frocks.'[4] An 1887 prediction that beer would become the 'drink of the world' was certainly more prophetic. China had ordered English beer for its navy, Japan was also encouraging beer as a 'national drink', and German lager was supplanting whisky in the United States.[5] But by January 1979 *The Press* facetiously suggested that at the present rate of brewery concentration there would be none by 1986. A correspondent presciently replied:

> We are well on our way to guzzling the national beer which has more to do with test tubes than the brewer's art. We will have not the beer we would like to drink, but beer that we have allowed the system to evolve for us. It is desirable that responsible free enterprise swings the pendulum in the opposite direction – by allowing hundreds of brewers to brew something different from their neighbourhood breweries, officially discouraging tied houses, and abolishing price control.[6]

The outcome was somewhere in the middle as the guzzling declined rapidly but the number of breweries soon expanded. It is likely that the per-capita consumption of beer will continue to decline, at least until its spirit-based alternatives are brought under control. But the variety of beer, measured by both styles and the range of producers, will increase, as will the diversity of people drinking it. Perhaps low-alcohol and no-alcohol beer will also enjoy the success frequently predicted for them. In 2021, consumption of beer below 2.5 per cent alcohol by volume increased by 20 per cent following five consecutive falls. On the other hand, the proportion of beer stronger than 5 per cent also continued to increase.[7] While it is too soon to write an obituary for displays of boorish, beer-influenced masculinity and some of the branding and advertising that encourages it, the tolerance for such cultures both within and beyond the brewing industry is certainly in retreat.

Where there is a disconcerting future for beer, it may be less of the industry's own making and more in terms of the impact of global environmental challenges on its raw materials. Deteriorating water quality is one problem, while climate change with droughts and generally more extreme weather conditions poses a significant threat to barley and hop yields. Some modelling for coming decades suggests a 15 per cent decline in global beer consumption and a doubling of price for that which remains.[8] Perhaps such a fundamental threat to one of the staples of human history is grave enough to persuade major powers to bring their emissions under control – or at least to prioritise the development of drought- and heat-tolerant barley cultivars so that long-established traditions of beer and sociability can accompany the demise of the planet.

Notes

INTRODUCTION

1 Caroline Daley, *Leisure and Pleasure: Reshaping and Revealing the New Zealand Body, 1900–1960*, Auckland: Auckland University Press, 2003, 10.

2 Greg Ryan, 'Drink and the Historians: Sober Reflections on Alcohol in New Zealand, 1840–1914', *New Zealand Journal of History*, 44(1), 2010, 35–52.

3 P. F. McKimmey, 'The Temperance Movement in New Zealand, 1835–1894', MA thesis, University of Auckland, 1968; A. R. Grigg, 'The Attack on the Citadels of Liquordom: A Study of the Prohibition Movement in New Zealand, 1894–1914', PhD thesis, University of Otago, 1977; Richard Newman, 'New Zealand's Vote for Prohibition in 1911', *New Zealand Journal of History*, 9(1), 1975, 52–71.

4 Jock Phillips, *A Man's Country? The Image of the Pakeha Male – A History*, Auckland: Penguin, 1987, 43–81; Stevan Eldred-Grigg, *Pleasures of the Flesh: Sex and Drugs in Colonial New Zealand, 1840–1915*, Wellington: Reed, 1984.

5 Paul Christoffel, 'Removing Temptation: New Zealand's Alcohol Restrictions, 1881–2005', PhD thesis, Victoria University of Wellington, 2006.

6 T. R. Gourvish and R. G. Wilson, *The British Brewing Industry, 1830–1980*, Cambridge: Cambridge University Press, 1994, xix–xx.

ONE – 'Inspissated Juice of Wort' and Other Early Alcoholic Forays Before 1850

1 Solomon H. Katz and Fritz Maytag, 'Brewing an Ancient Beer', *Archaeology*, 44(4), 1991, 24–33.

2 Peter Damerow, 'Sumerian Beer: The Origins of Brewing Technology in Ancient Mesopotamia', *Cuneiform Digital Library Journal*, 2, 2012, 1–20; Ian S. Hornsey, *A History of Beer and Brewing*, London: Royal Society of Chemistry, 2003, 1–30.

3 Hornsey, *History*, 79–83; Martyn Cornell, *Beer: The Story of the Pint*, London: Headline Book Publishing, 2003, 3–18.

4 Max Nelson, *The Barbarian's Beverage: A History of Beer in Ancient Europe*, London: Routledge, 2005, 9–38, 67–77.

5 Ibid., 78–110; Hornsey, *History*, 165–275.

6 Cornell, *Beer*, 19–58; John Burnett, *Liquid Pleasures: A Social History of Drinks in Modern Britain*, London: Routledge, 1999, 112.

7 Richard W. Unger, *Beer in the Middle Ages and the Renaissance*, Philadelphia: University of Pennsylvania Press, 2004, xiii.

8 Cornell, *Beer*, 59–88.

9 Judith M. Bennett, *Ale, Beer and Brewsters in England: Women's Work in a Changing World, 1300–1600*, Oxford: Oxford University Press, 1996.

10 Josephine A. Spring and David H. Buss, 'Three Centuries of Alcohol in the British Diet', *Nature*, 270(15), December 1977, 567–68; Burnett, *Liquid Pleasures*, 179; James Nicholls, *The Politics of Alcohol: A History of the Drink Question in England*, Manchester: Manchester University Press, 2009, 81.

11 Burnett, *Liquid Pleasures*, 114–18.

12 Martyn Cornell, *Amber, Gold and Black: The History of Britain's Great Beers*, Stroud: History Press, 2010, 53–78; Roger Protz, *The Taste of Beer: A Guide to Appreciating the Great Beers of the World*, London: Rigel Publications, 1998, 18–21.

13 Ian L. Donnachie, *A History of the Brewing Industry in Scotland*, Edinburgh: Donald, 1979, 1–37; Gourvish and Wilson, *British Brewing Industry*, 99–110.

14 Kenneth J. Carpenter, *The History of Scurvy and Vitamin C*, Cambridge: Cambridge University Press, 1986, 43–74; Brett J. Stubbs, 'Captain Cook's Beer: The Anti-scorbutic Use of Malt and Beer in Late 18th Century Sea Voyages', *Asia Pacific Journal of Clinical Nutrition*, 12(2), 2003, 129–37, at 129.

15 Carpenter, *Scurvy*, 173–97; Stubbs, 'Captain Cook's Beer', 130–37.

16 Carpenter, *Scurvy*, 64–66; Stubbs, 'Captain Cook's Beer', 130–31.

17 Martyn Cornell, 'A Short History of Spruce Beer in Britain', *Brewery History*, 165, 2016, 2–6.

18 Stubbs, 'Captain Cook's Beer', 131–32.

19 Peter Mathias, *The Brewing Industry in England, 1700–1830*, Cambridge: Cambridge University Press, 1959, 204–9.

20 J. C. Beaglehole, ed., *The Journals of Captain James Cook: II The Voyage of the* Resolution *and* Adventure, *1772–1775*, Cambridge: The Hakluyt Society, 1961, 38.

21 Ibid., 114.

22 James Cook, Tobias Furneaux and William Hodges, *A Voyage Towards the South Pole and Round the World: Performed in His Majesty's Ships the* Resolution *and* Adventure, *in the Years 1772, 1773, 1774, and 1775*, London: W. Strahan & T. Cadell, 1777, 99–100.

23 Beaglehole, *Journals*, 114.

24 Stubbs, 'Captain Cook's Beer', 132–34.

25 Ibid., 131; Beaglehole, *Journals*, 909.

26 Carpenter, *Scurvy*, 75–97; Stubbs, 'Captain Cook's Beer', 130; Mathias, *Brewing Industry*, 208–9.

27 Robert McNab, *Murihiku: A History of the South Island of New Zealand and the Islands Adjacent and Lying to the South, from 1642 to 1835*, Christchurch: Whitcombe & Tombs, 1909, 83.

28 Carpenter, *Scurvy*, 95; Stubbs, 'Captain Cook's Beer', 134–35.

29 Brian Harrison, *Drink and the Victorians: The Temperance Question in England, 1815–72*, London: Faber & Faber, 1971, 37–39, 298–302.

30 Ibid., 39, 41, 302–3, 307–8.

31 Ibid., 46–47. See also Peter Clark, *The English Alehouse: A Social History, 1200–1830*, London: Longman, 1983.

32 Paul Jennings, *The Local: A History of the English Pub*, London: The History Press, 2011, 14–15.

33 Harrison, *Drink*, 48–53.

34 Ibid., 42–43.

35 Jennings, *The Local*, 124.

36 Harrison, *Drink*, 302–3, 319–20, 334–39.

37 Nicholls, *Politics of Alcohol*, 5–20, 34–50. See also Patrick Dillon, *Gin: The Much Lamented Death of Madam Geneva*, London: Justin, Charles & Co., 2003.

38 Nicholls, *Politics of Alcohol*, 80–83.

39 Harrison, *Drink*, 87–106; Nicholls, *Politics of Alcohol*, 96–108.

40 Harrison, *Drink*, 114.

41 Nicholls, *Politics of Alcohol*, 85–87.

42 Ibid., 88.

43 1830: 11 Geo. 4 and 1 Will. 4 c. 64, The Beerhouse Act.

44 Nicholls, *Politics of Alcohol*, 88–93; Cornell, *Beer*, 167–69.

45 Burnett, *Liquid Pleasures*, 120; Cornell, *Beer*, 167–68.

46 Peter Haydon, *Beer and Britannia: An Inebriated History of Britain*, Stroud: Sutton Publishing, 2001, 171.

47 Harrison, *Drink*, 107–46, 196–246.

48 Ibid., 66, 312, 317.

49 Murdoch Riley, *Māori Healing and Herbal*, Paraparaumu: Viking Sevenseas, 1994, 16–17; Marten Hutt, *Te Iwi Maori me te Inu Waipiro: He Tuhituhinga Hitori / Maori and Alcohol: A History*, Wellington: Health Services Research Centre for Kaunihera Whakatupato Waipiro o Aotearoa/ Alcohol Advisory Council of New Zealand (ALAC), 1999, 3–14.

50 Ronald Sharp, 'Alcohol in New Zealand 1642–1840', BA Hons research essay, Massey University, 1976, 5–10.

51 David Hughes, 'The Brewing Industry in Early Australia', in R. G. Wilson and T. R. Gourvish, eds, *The Dynamics of the International Brewing Industry Since 1800* (pp. 213–228), London: Routledge, 1998, 213; N. G. Butlin, 'Yo, Ho, Ho and How Many Bottles of Rum?', *Australian Economic History Review*, 23(2), 1983, 1–27.

52 Matthew Allen, 'Alcohol and Authority in Early New South Wales', *History Australia*, 9(3), 2012, 7–26.

53 Robert McNab, *Historical Records of New Zealand*, Wellington: Government Printer, 1908–14, vol.1, 714.

54 Keith Deutsher, *The Breweries of Australia: A History*, Melbourne: Lothian, 1999, 10; David Hughes, 'Australia's First Brewer', *Journal of the Royal Australian Historical Society*, 82(2), 1996, 153–67.

55 Quoted in Sharp, 'Alcohol in New Zealand', 9.

56 Quoted in McKimmey, 'Temperance Movement', 5.

57 Quoted in Sharp, 'Alcohol in New Zealand', 5.

58 R. Darwin Keynes, *Charles Darwin's* Beagle *Diary*, Cambridge: Cambridge University Press, 1988, 384.

59 Quoted in Eldred-Grigg, *Pleasures of the Flesh*, 61.

60 Riley, *Māori Healing and Herbal*, 18.

61 R. J. Barton, ed., *Earliest New Zealand: The Journals and Correspondence of the Rev. John Butler*, Masterton: Palamontain and Petherick, 1927, 142.

62 Jocelyn Chisholm, 'Polack, Joel Samuel 1807–1882', Dictionary of New Zealand Biography, Te Ara – the Encyclopedia of New Zealand, https://teara.govt.nz/en/biographies/1p18/polack-joel-samuel (accessed 5 December 2022).

63 J. S. Polack, *Manners and Customs of the New Zealanders*, London: James Madden & Co., 1840, vol.2, 279.

64 Ibid., 280. There is no evidence for a claim in the *New Zealand Herald* (29 January 1890, 2, supp.) that he brewed until the Māori attack on Kororāreka in March 1845.

65 Sharp, 'Alcohol in New Zealand', 23.

66 Ibid., 18–27.

67 Quoted in ibid., 27.

68 Church Missionary Society, *Report of the Formation and Establishment of the New Zealand Temperance Society*, Paihia: Church Mission Press, 1836.

69 Ibid.

70 Eric Ramsden, *Busby of Waitangi: H.M.'s Resident at New Zealand, 1833–40*, Wellington: A. H. & A. W. Reed, 1942, 193.

71 For example, *New Zealand Advertiser & Bay of Islands Gazette*, 15 June 1840, 3; 29 October 1840, 1.

72 Unless specifically stated, all population, import, export and alcohol consumption statistics are derived from C. G. F. Simkin and B. G. Hardie, *Statistics of New Zealand for the Crown Colony Period 1840–1852*, Auckland: Auckland University College Department of Economics, 1954, and various editions of *Statistics of New Zealand* from 1853 onwards.

73 *Wellington Independent*, 28 December 1850, 4.

74 Eldred-Grigg, *Pleasures of the Flesh*, 58.

75 *New Zealand Gazette & Wellington Spectator*, 1 May 1844, 2.

76 'A Late Resident in the Colony' [G. B. Earp], *The Emigrant's Guide to New Zealand*, London: William S. Orr & Co., 1851, n.p.

77 R. R. Scott, 'Public Control of the Liquor Trade in New Zealand, 1840–1899', MA thesis, Victoria University College, 1952, 10.

78 *New Zealand Gazette & Wellington Spectator*, 24 April 1841; Scott, 'Public Control', 11.

79 1841: 4 Vic. No.5: Distillation Prohibition Ordinance.

80 Eldred-Grigg, *Pleasures of the Flesh*, 63; Scott, 'Public Control', 11–12.

81 1842: 5 Vic. No.12: Licensing Ordinance.

82 Eldred-Grigg, *Pleasures of the Flesh*, 65; Scott, 'Public Control', 12–15; *Otago Witness*, 7 April 1860, 4.

83 *New Zealander*, 14 April 1849, 3. Among many similar examples see also 12 April 1852, 3; *New Zealand Spectator & Cook's Strait Guardian*, 17 April 1850, 2; *Wellington Independent*, 15 May 1852, 3.

84 The most comprehensive treatment of these themes is Jason Wright, 'Company of Strangers: Patea and Wanganui Hotels, 1866–1899', MA thesis, Massey University, 2004.

85 M. N. Pearson, 'Brown and Campbell in Early Auckland, 1840–56', MA thesis, University of Auckland, 1964, 19.

86 Grigg, 'Attack on the Citadels of Liquordom', 11–12.

87 Miles Fairburn, *The Ideal Society and its Enemies*, Auckland: Auckland University Press, 1989, 203–4.

88 *Report of the 1974 Royal Commission on the Sale of Liquor in New Zealand*, Wellington: Government Printer, 1974, 22.

89 Eldred-Grigg, *Pleasures of the Flesh*, 76–77.

90 *Statistics of New Zealand*, 1854, No.19.

91 Pearson, 'Brown and Campbell', 67.

92 For estimates of nineteenth-century beer strength, see: Cornell, *Beer*, 146–49, 154–55; Frank G. Leckie, *Otago's Breweries: Past and Present*, Dunedin: Otago Heritage Books, 1997, 13; Alan V. Stewart and Jimmy K. Stewart, *The Last of the Small Breweries in New Zealand: A History of Stewart's Brewery, Greymouth, 1880–1961*, Christchurch: A. V. & J. K. Stewart, 2004, 18, 45; Brewers Journal: William Dawson, Brewer, City Brewery [Speight's], Dunedin, 1879–91, Lion Technical Library, Auckland. For wine and spirits see Brian Glover, *The Illustrated Encyclopedia of Wine, Beer and Spirits*, London: Lorenz Books, 2006; Jancis Robinson, ed., *The Oxford Companion to Wine*, 3rd ed., Oxford: Oxford University Press, 2006, 10.

93 Eldred-Grigg, *Pleasures of the Flesh*, 94.

94 G. B. Wilson, *Alcohol and the Nation: A Contribution to the Study of the Liquor Problem in the United Kingdom from 1800 to 1935*, London: Nicholson and Watson, 1940, 430.

95 *New Zealand Gazette & Wellington Spectator*, 27 March 1841, 2.

96 *New Zealander*, 12 February 1848, 3.

97 *Nelson Examiner*, 22 July 1848, 2.

98 McKimmey, 'Temperance Movement', 13–15.

99 Gourvish and Wilson, *British Brewing Industry*, 252–53.

TWO – 'It is a Brewer's Paradise': Beer and Settlement, 1840–1860

1 Hughes, 'Brewing Industry in Early Australia', 213.

2 Gourvish and Wilson, *British Brewing Industry*, 109–10, 169–70.

3 Ibid., 40–47, 169–73; Mathias, *Brewing Industry*, 194.

4 Cornell, *Beer*, 132.

5 Ibid., 131–33.

6 Ibid., 133–38.

7 *New Zealand Gazette & Wellington Spectator*, 10 December 1840, 2; 10 April 1841, 1.

8 *New Zealand Gazette & Wellington Spectator*, 16 October 1841, 1.

9 *New Zealand Gazette & Wellington Spectator*, 31 October 1840, 1; 23 December 1843, 1; *Nelson Examiner & New Zealand Chronicle*, 2 March 1844, 1; *New Zealand Spectator & Cook's Strait Guardian*, 31 May 1845, 1; *Daily Southern Cross*, 22 October 1850, 2.

10 Pearson, 'Brown and Campbell', 67.

11 *Daily Southern Cross*, 17 June 1848, 4; 15 December 1854, 2.

12 *Lyttelton Times*, 13 June 1860, 1.

13 Ibid.

14 Martyn Cornell, 'The Mysterious Australian Ale', http://zythophile.wordpress.com/2011/06/10/the-mysterious-australian-ale (accessed 12 June 2011).

15 *Wellington Independent*, 19 May 1858, 3.

16 'Prices of Provisions in Wellington 1840–45', Simkin and Hardie, Table 55.

17 Burnett, *Liquid Pleasures*, 119–20.

18 *Nelson Examiner & New Zealand Chronicle*, 22 July 1848, 82.

19 *Nelson Examiner & New Zealand Chronicle*, 22 July 1848, 82; 27 September 1851, 134; *Taranaki Herald*, 15 September 1852, 1.

20 Alison Drummond, ed., *The Auckland Journal of Vicesimus Lush, 1850–63*, Christchurch: Pegasus Press, 1971, 145.

21 *Lyttelton Times*, 5 December 1857, 5.

22 *Daily Southern Cross*, 22 November 1850, 1.

23 *New Zealand Herald & Auckland Gazette*, 8 January 1842, 1.

24 *New Zealand Gazette & Wellington Spectator*, 20 February 1841, 2; 6 March 1841, 2.

25 *New Zealand Gazette & Wellington Spectator*, 25 June 1842, 2. See also *Nelson Examiner & New Zealand Chronicle*, 30 July 1842, 82.

26 *New Zealand Gazette & Wellington Spectator*, 21 January 1843, 1; *New Zealand Colonist & Port Nicholson Advertiser*, 21 March 1843, 3.

27 *New Zealand Colonist & Port Nicholson Advertiser*, 4 April 1843, 2.

28 Ibid.; *New Zealand Gazette & Wellington Spectator*, 8 April 1843, 2; 22 April 1843, 2.

29 *New Zealand Spectator & Cook's Strait Guardian*, 2 January 1847, 2; 7 April 1847, 1; *Wellington Independent*, 11 July 1849, 4.

30 *Nelson Examiner & New Zealand Chronicle*, 28 October 1843, 341.

31 Rob Packer, *A History of Nelson's Early Manufacturers: Their Wares and Containers*, Nelson: R. V. Packer, 1995, 7.

32 M. D. Lash and D. Smith, *Nelson Notables, 1840–1940: A Dictionary of Regional Biography*, Nelson: Nelson Historical Society, 1992, 84, 121; *Nelson Examiner & New Zealand Chronicle*, 6 January 1844, 1.

33 *New Zealand Gazette & Wellington Spectator*, 28 February 1844, 3.

34 *Nelson Examiner & New Zealand Chronicle*, 23 March 1844, 10; 8 June 1844, 54.

35 *New Zealand Spectator & Cook's Strait Guardian*, 21 June 1845, 4.

36 *Nelson Examiner & New Zealand Chronicle*, 18 August 1858, 2.

37 *Nelson Examiner & New Zealand Chronicle*, 21 March 1846, 10.

38 Ibid.; Colin C. Aikman to Thomas Renwick, 16 October 1846, AG83 A3930 Nelson Breweries Ltd. D-E L527.29.2 box, folder AG 83, E.1, Nelson Provincial Museum.

39 *Nelson Examiner & New Zealand Chronicle*, 30 August 1845, 3; 4 April 1846, 18; 27 March 1847, 14.

40 *Nelson Examiner*, 24 September 1863, 1.

41 R. C. J. Stone, 'Campbell, John Logan 1817–1912', Dictionary of New Zealand Biography, Te Ara – the Encyclopedia of New Zealand, https://teara.govt.nz/en/biographies/1c3/campbell-john-logan (accessed 6 December 2022).

42 *Nelson Examiner & New Zealand Chronicle*, 10 December 1856, 3.

43 *Daily Southern Cross*, 25 May 1858, 2; New Zealand Breweries, *Hosts to the Nation: The First Fifty Years of New Zealand Breweries Limited*, Wellington: New Zealand Breweries, 1973, 8.

44 G. R. Macdonald, Dictionary of Canterbury Biographies, Christchurch: Canterbury Museum, n.d., C803 (Croft), W142 (Ward).

45 John R. Godley, ed., *Letters from Early New Zealand by Charlotte Godley, 1850–53*, Christchurch: Whitcombe & Tombs, 1951, 328.

46 Macdonald, M83/84 (Manning).

47 Leckie, *Otago's Breweries*, 32–35, 49–55.

48 Breweries (historical file) [1940–1970], MS 77-067-6/05: Patrick Anthony Lawlor Papers, Alexander Turnbull Library, Wellington.

49 Cornell, *Beer*, 148–49. The use of X derives from the Latin names for beer used by medieval monastic brewers – *beirra simplex* (single beer) and *bierra duplex* (double beer). As these apparently sounded like 'single X' and 'double X' in English, casks were marked in this fashion. Over time a wider range of Xs evolved.

50 Ibid., 156–58.

51 Ibid., 169–70; Gourvish and Wilson, *British Brewing Industry*, 40–58.

52 *New Zealander*, 14 December 1850, 3.

53 *Nelson Examiner*, 26 January 1856, 2.

54 *New Zealander*, 21 July 1860, 3.

55 Letter dated 20 July 1861, in Henry W. Harper, *Letters from New Zealand 1857–1911 Being Some Account of Life and Work in the Province of Canterbury, South Island*, London: Hugh Rees Ltd, 1914, 67.

56 *Daily Southern Cross*, 17 January 1862, 3.

57 *The Press*, 18 March 1870, 2.

58 *Auckland Star*, 1 December 1870, 2.

59 *Nelson Examiner & New Zealand Chronicle*, 16 December 1843, 369–70.

60 *Nelson Examiner & New Zealand Chronicle*, 6 September 1845, 107.

61 *New Zealander*, 19 July 1845, 2.

62 *Nelson Examiner & New Zealand Chronicle*, 5 August 1848, 90.

63 *Otago Witness*, 10 January 1857, 3.

64 *Lyttelton Times*, 22 April 1857, 7.

65 Charles Hursthouse, *An Account of the Settlement of New Plymouth: In New Zealand, From Personal Observations, During a Residence There of Five Years*, Christchurch: Capper Press, 1975 [1849], 135–36.

66 *First Annual Report of the Agricultural and Horticultural Society of Auckland*, Auckland: Southern Cross, 1843, 18; *New Zealand Gazette & Wellington Spectator*, 31 July 184, 4.

67 *New Zealand Gazette & Wellington Spectator*, 10 July 1844, 2.

68 *Taranaki Herald*, 18 April 1857, 2.

69 *Nelson Examiner & New Zealand Chronicle*, 15 September 1858, 3.

70 For example, *New Zealand Colonist & Port Nicholson Advertiser*, 11 April 1843, 1; *Daily Southern Cross*, 9 October 1847, 4.

71 David Veart, *First Catch Your Weka: A Story of New Zealand Cooking*, Auckland: Auckland University Press, 2008, 37–38.

72 For example, *Taranaki Herald*, 16 February 1853, 1; *Nelson Examiner & New Zealand Chronicle*, 3 November 1858, 2.

73 *Dunedin Leader*, 16 January 1864, quoted in Eldred-Grigg, *Pleasures of the Flesh*, 59.

74 1847: 11 Vic. No.3: Sale of Spirits Ordinance; Hutt, *Maori and Alcohol*, 20, 23–30.

75 *Daily Southern Cross*, 26 May 1849, 2.

76 Quoted in Hutt, *Maori and Alcohol*, 20.

77 *Nelson Examiner & New Zealand Chronicle*, 26 November 1859, 2.

78 *Nelson Examiner & New Zealand Chronicle*, 7 January 1860, 2.

79 *Daily Southern Cross*, 13 December 1859, 3.

80 McKimmey, 'Temperance Movement', 54–55; *Nelson Examiner and New Zealand Chronicle*, 12 August 1843, 299; *New Zealander*, 9 May 1846, 3.

81 *New Zealand Gazette & Wellington Spectator*, 7 February 1844, 3.

82 *Nelson Examiner & New Zealand Chronicle*, 17 August 1850, 98.

83 Jim McAloon, *Nelson: A Regional History*, Whatamango Bay: Cape Catley in association with the Nelson City Council, 1997, 45.

84 McKimmey, 'Temperance Movement', 55–56.

85 R. G. Golledge, 'The New Zealand Brewing Industry', *New Zealand Geographer*, 19(1), 1963, 7–24, at 10; Cornell, *Beer*, 166–67.

86 Packer, *Nelson's Early Manufacturers*, 13.

87 Ibid., 25; *Nelson Examiner & New Zealand Chronicle*, 30 April 1853, 39.

88 *Nelson Examiner & New Zealand Chronicle*, 31 December 1853, 3.

89 *Daily Southern Cross*, 20 April 1852, 3.

90 *Nelson Examiner & New Zealand Chronicle*, 27 August 1853, 6.

91 For example, *Daily Southern Cross*, 3 March 1849, 4; 26 November 1850, 1; 30 January 1855, 2.

92 *Taranaki Herald*, 23 August 1854, 2.

93 *Otago Witness*, 13 February 1858, 1.

94 *New Zealander*, 18 July 1846, 1; *Otago Daily Times*, 7 July 1863, 1.

95 Cornell, *Beer*, 154; Gourvish and Wilson, *British Brewing Industry*, 205–11.

96 *The Press*, 10 November 1865, 1; Jeremy Sutherland, 'A History of Brewing in South Canterbury, New Zealand, 1865–1996, part 1', http://www.timaru.com/aeratedwaters_history/history.htm (accessed 24 July 2011); *Timaru Herald*, 1 September 1879, 4.

97 *Lyttelton Times*, 3 April 1861, 4.

THREE – Golden Ale? The Rush to Brew, 1860–1870

1 *West Coast Times*, 6 December 1876, 2.

2 Erik Olssen, *A History of Otago*, Dunedin: McIndoe, 1984, 51–70.

3 Raewyn Dalziel, *Julius Vogel: Business Politician*, Auckland: Auckland University Press, 1986, 80–115.

4 E. Pawson and N. C. Quigley, 'The Circulation of Information and Frontier Development: Canterbury, 1850–90', *New Zealand Geographer*, 38(2), 1982, 65–76.

5 A. H. Gray, *Aotearoa and New Zealand: A Historical Geography*, Christchurch: Canterbury University Press, 1994, 333–37.

6 Donnachie, *History of the Brewing Industry in Scotland*, 197.

7 Ibid.; Gourvish and Wilson, *British Brewing Industry*, 109–10.

8 *Daily Southern Cross*, 1 December 1870, 3.

9 *New Zealand Gazette & Wellington Spectator*, 14 November 1840, 1. Among numerous other examples, see: *New Zealander*, 11 December 1850, 1; *Daily Southern Cross*, 5 August 1856, 1.

10 *Lyttelton Times*, 5 October 1859, 1; *Otago Daily Times*, 7 September 1863, 1; *West Coast Times*, 11 September 1866, 2.

11 Donnachie, *History of the Brewing Industry in Scotland*, 197, 221–26; *Cyclopedia of New Zealand: Industrial, Descriptive, Historical, Biographical* (hereafter *CNZ*), Wellington & Christchurch: Cyclopedia Company, 1897–1908, vol.2, 94.

12 *Otago Daily Times*, 7 September 1863, 8; *New Zealand Licensee*, 20 January 1974, 15.

13 *Nelson Examiner & New Zealand Chronicle*, 12 April 1851, 29; 12 July 1851, 83; *Otago Daily Times*, 8 August 1862, 3.

14 *Grey River Argus*, 24 November 1868, 3; 9 January 1869, 4.

15 *Nelson Examiner & New Zealand Chronicle*, 27 January 1855, 1.

16 Sutherland, 'South Canterbury'.

17 Macdonald, D167 (Rogers Deacon).

18 Macdonald, M84 (William Manning).

19 Sutherland, 'South Canterbury'.

20 Stewart and Stewart, *Last of the Small Breweries*.

21 Nola Easdale, *Five Gentleman's Residences in Princess Street*, Auckland: Auckland City Council, 1980, 19; Duncan Waterson, Auckland Provincial History Project Notes: Distilling and brewing industries, NZMS 1166, Auckland City Libraries, 3.

22 Macdonald, J44 (Henry Jaggar). I am grateful to Lindley Turnbull for revealing this family network.

23 Donald Gordon, *Speight's: The Story of Dunedin's Historic Brewery*, Dunedin: Avon Publishers, 1993, 10; Donald Gordon, 'Dawson, William', in Jane Thomson, ed., *Southern People: A Dictionary of Otago Southland Biography*, Dunedin: Longacre Press in association with Dunedin City Council, 1998, 122.

24 Leckie, *Otago's Breweries*, 75.

25 Ibid., 32, 66–67; *CNZ*, vol.4, 292–93; New Zealand Breweries, *Hosts to the Nation*, 22.

26 Macdonald, M538 (Augustus Moore); T10 (Richard Taylor).

27 W. Coleman, ed., *A Short Sketch of Some Incidents in the Colonial Life of Mr Thomas Hancock*, Auckland: J. D. Wickham, 1885, 62–63.

28 Michael Bassett and Paul Goldsmith, *The Myers*, Auckland: David Ling, 2007, 46.

29 *The Press*, 10 November 1870, 3; 11 April 1873, 3.

30 Leckie, *Otago's Breweries*, 10; Gordon, *Speight's*, 23, 184.

31 Colin V. Innes, 'Innes, Mary Jane 1852–1941', Dictionary of New Zealand Biography, Te Ara – the Encyclopedia of New Zealand, https://teara.govt.nz/en/biographies/2i1/innes-mary-jane (accessed 8 December 2022).

32 *Evening Post*, 19 December 1865, 2.

33 Lloyd Carpenter, 'The Merchants of the Rush: The Proto-bankers of the Goldfields', in Lloyd Carpenter and Lyndon Fraser, eds, *Rushing for Gold: Life and Commerce on the Goldfields of New Zealand and Australia* (pp. 222–39), Dunedin: Otago University Press, 2016.

34 Philip Ross May, *The West Coast Gold Rushes*, 2nd rev. ed., Christchurch: Pegasus, 1967, 289.

35 Ibid.; Stevan Eldred-Grigg, *Diggers, Hatters and Whores: The Story of New Zealand Gold Rushes*, Auckland: Random House, 2008, 285–87; M. P. McCarthy, 'A City in Transition: Diversification in the Social Life of Dunedin, 1860–1864', MA thesis, University of Otago, 1977, 29–33.

36 May, *West Coast Gold Rushes*, 291.

37 Phillips, *A Man's Country?*, 58.

38 *Daily Southern Cross*, 8 October 1863, 3.

39 Harper, *Letters from New Zealand*, 64–65.

40 Quoted in Eldred-Grigg, *Diggers, Hatters and Whores*, 289.

41 *West Coast Times*, 26 September 1865, quoted in Philip Ross May, *Gold Town, Ross, Westland*, Christchurch: Pegasus, 1970, 19.

42 Harper, *Letters from New Zealand*, 165.

43 See, for example, Jim McAloon, *No Idle Rich: The Wealthy in Canterbury and Otago, 1840–1914*, Dunedin: Otago University Press, 2002, 48–49. Numerous others are to be found in the *Cyclopedia of New Zealand*.

44 Martin R. Howard, 'Red Jackets and Red Noses: Alcohol and the British Napoleonic Soldier', *Journal of the Royal Society of Medicine*, 93(1), 2000, 38–41.

45 Charlotte Macdonald, 'Woolwich to Wellington: From Settler Colony to Garrisoned Sovereignty', *New Zealand Journal of History*, 53(1), 2019, 50–76, at 58.

46 *Otago Daily Times*, 10 December 1863, 10.

47 *New Zealander*, 29 December 1863, 6.

48 *Daily Southern Cross*, 3 January 1866, 3.

49 Quoted in *Hawkes Bay Times*, 4 November 1867, 273.

50 *Wanganui Herald*, 5 November 1867, 2.

51 1868: 32 Vic. No.6: The Public House Act.

52 *Otago Daily Times*, 5 October 1868, 3; *Evening Post*, 8 October 1868, 2.

53 Eldred-Grigg, *Pleasures of the Flesh*, 73–75.

54 Burnett, *Liquid Pleasures*, 169–70.

55 Harrison, *Drink*, 37.

56 May, *West Coast Gold Rushes*, 309–10, 330–31.

57 Darrell Latham, *The Golden Reefs: An Account of the Great Days of Quartz-Mining at Reefton, Waiuta and the Lyell*, Nelson: Nikau Press, 1992, 53. See also W. J. Gardner, *The Amuri: A County History*, 2nd ed., Culverden: Amuri County Council, 1983, 207.

58 Latham, *Golden Reefs*, 65.

59 Leckie, *Otago's Breweries*, 10–11.

60 Ibid., 15–16.

61 *The Press*, 9 April 1863, 1, 7.

62 *Otago Witness*, 14 November 1863, 3.

63 *West Coast Times*, 2 December 1868, 2.

64 *Cromwell Argus*, 3 February 1874, 6.

65 J. E. Coutts, 'Kühtze, Frederick Joseph 1833–1901', Dictionary of New Zealand Biography, Te Ara – the Encyclopedia of New Zealand, https://teara.govt.nz/en/biographies/2k12/kuhtze-frederick-joseph (accessed 8 December 2022); Leckie, *Otago's Breweries*, 118–19; Carpenter, 'Merchants', 235.

66 *Evening Post*, 2 August 1867, 2; *North Otago Times*, 18 February 1868, 2; Leckie, *Otago's Breweries*, 64.

67 *Lyttelton Times*, 25 May 1861, 4–5.

68 *Taranaki Herald*, 22 March 1862, 2; Sutherland, 'South Canterbury'; *Waikato Times*, 30 June 1881, 2.

69 Leckie, *Otago's Breweries*, 48, 60.

70 *New Zealand Mail*, 2 August 1873, 14; *Taranaki Herald*, 3 February 1875, 2.

71 *Hawera & Normanby Star*, 2 January 1884, 2.

72 *New Zealand Police Gazette*, 1 May 1895, 7.

73 *Alexandra Herald & Central Otago Gazette*, 23 July 1903, 5.

74 *Nelson Examiner & New Zealand Chronicle*, 25 December 1858, 3; 22 February 1860, 2; 11 February 1865, 3.

75 *Daily Southern Cross*, 19 June 1863, 5.

76 *Daily Southern Cross*, 20 May 1864, 3.

77 Gourvish and Wilson, *British Brewing Industry*, 47–59; James Sumner, *Brewing Science, Technology and Print, 1700–1880*, London: Pickering & Chatto, 2013.

78 Gourvish and Wilson, *British Brewing Industry*, 59–60; Cornell, *Beer*, 39–40, 141–42.

79 *Otago Daily Times*, 3 September 1872, 2.

80 Leckie, *Otago's Breweries*, 55–56; K. G. Duncan, 'Duncan, George', in Thomson, *Southern People*, 141.

81 *Lyttelton Times*, 2 July 1881, 6.

82 Macdonald, V47 (William Vincent).

83 Leckie, *Otago's Breweries*, 14.

84 J. H. Dwyer, *The Coburg Hotel, the Albert Brewery and the Site They Have Occupied – A Brief History*, Auckland: J. H. Dwyer, 1974, 2.

85 *Timaru Herald*, 20 January 1869, 2.

86 'A Pioneer of Beer: The Story of Thomas Hancock', typed notes, by Mrs H. Carr Rollett, sent to the Brewers' Association of New Zealand, 20 June 1956, Lion Technical Library, Auckland.

87 *Otago Witness*, 23 January 1864, 2.

88 Pamela Wood, *Dirt: Filth and Decay in a New World Arcadia*, Auckland: Auckland University Press, 2005, 71–72; Leckie, *Otago's Breweries*, 113.

89 Leckie, *Otago's Breweries*, 14–15; *The Press*, 21 June 1866, 2; *North Otago Times*, 17 May 1866, 1.

90 Ernest G. Collins, *Malting in Heathcote Valley, 1871–1981*, Christchurch: Canterbury Malting Co., *c*. 1981, 12–16.

91 *Otago Daily Times*, 3 September 1872, 2.

92 *Weekly News* (Auckland), 4 March 1871, Lion Breweries files, Lion Technical Library, Auckland.

93 *The Press*, 1 May 1867, 2.

94 Macdonald, M538 (Augustus Moore).

FOUR – 'Trifling with the Business of the Country': Regulating Beer, 1868–1893

1 Burnett, *Liquid Pleasures*, 122–23; *Daily Southern Cross*, 28 May 1867, 6.

2 1866: 30 Vic. No.55: Adulteration of Food Act.

3 Quoted in Leckie, *Otago's Breweries*, 17.

4 'Committee on Weights and Measures and Adulteration of Food', *Appendix to the Journals of the House of Representatives* (hereafter *AJHR*) 1870, F-5, 4.

5 Martyn Cornell, 'The Bass Red Triangle: Things AB-InBev Won't Tell You', https://zythophile.co.uk/2013/06/28/the-bass-red-triangle-things-ab-inbev-wont-tell-you/ (accessed 18 July 2022).

6 *New Zealand Herald*, 12 July 1873, 1; 14 July 1863, 3; 1 August 1873, 2.

7 1841: 4 Vic: No.5: Distillation Prohibition Ordinance.

8 *Statistics of New Zealand*, 1862, 1866; Hugh R. K. Logan, 'A Business History of Whisky Distillation in Dunedin', Hist 480 research essay, University of Otago, 1993, 4; Stuart Perry, *The New Zealand Whisky Book*, Auckland: Collins, 1980, 70.

9 William Seed, 'Memorandum on Illicit Distillation and Smuggling', *AJHR*, 1867, D-8.

10 *New Zealander*, 7 June 1862, 7.

11 *Dunstan Times*, 9 April 1869, quoted in Leckie, *Otago's Breweries*, 13.

12 Logan, 'Business History', 4–14; P. Hanna, '"Why Not Whisky": The Short History of the "New Zealand Distillery Company"', Postgrad. Dip. essay, University of Otago, 1975, 5–13.

13 1868: 32 Vic: No.72: Distillation Act.

14 *Otago Witness*, 12 September 1868, 1.

15 Logan, 'Business History', 4–14; Perry, *Whisky Book*, 68–73.

16 Logan, 'Business History', 4–14.

17 *New Zealand Herald*, 22 November 1870, 2.

18 'Report of the Select Committee on Excise Duties', *AJHR*, 1874, I-2.

19 Perry, *Whisky Book*, 93–95.

20 'Report of the Select Committee on Excise Duties', 13.

21 Perry, *Whisky Book*, 96–99.

22 Conrad Bollinger, *Grog's Own Country: The Story of Liquor Licensing in New Zealand*, 2nd ed., Auckland: Minerva, 1967, 28–30.

23 Leckie, *Otago's Breweries*, 19.

24 Ian R. Johnstone, *Breweries of Gisborne*, Mosgiel: the author, 1995, 2.

25 *The Press*, 25 November 1863, 2.

26 *Otago Witness*, 4 March 1865, 1.

27 *Otago Daily Times*, 25 July 1866, 6.

28 *Otago Daily Times*, 3 September 1872, 2.

29 *Guide to Wellington and District: With a Complete Map of the City*, Wellington: Walter K. Bishop, 1883, 211.

30 *The Press*, 9 March 1897, 2.

31 *Wanganui Chronicle*, 21 January 1910, 2.

32 *Auckland Weekly News*, 5 November 1870, 6.

33 Hooper and Dodson, Nelson Brewery, to Chairman, Joint Committee on Colonial Industries, 8 August 1870, *AJHR*, 1870, F1, 25–26.

34 *Timaru Herald*, 13 November 1869, 2.

35 *Intercolonial Exhibition 1866: Awards of the Jurors*, Melbourne: Blundell & Co., 1866, n.p.; *Evening Post*, 30 May 1868, 3.

36 *Daily Southern Cross*, 17 August 1871, 2.

37 *Evening Star*, 5 October 1870, 2.

38 Gordon, *Speight's*, 22–23.

39 *Wellington Independent*, 27 September 1859, 6.

40 *Lyttelton Times*, 5 October 1859, 4.

41 *The Press*, 7 January 1870, 3. See also *Otago Witness*, 16 June 1883, 9, for similar views concerning New Zealand water.

42 T. G. Parsons, 'Brewing Beer in the Colonial Days', *Victorian Historical Journal*, 55(2), 1984, 13–19, at 14–15; *Australian Brewers' Journal*, 20 August 1889, 296; Brett Stubbs, 'Brewing with Sugar', BrewsNews, https://www.brewsnews.com.au/2011/09/05/brewing-with-sugar (accessed 1 March 2022).

43 *Timaru Herald*, 26 May 1880, 2.

44 *Evening Star*, 11 November 1881, 4.

45 *Otago Daily Times*, 5 October 1882, 2.

46 Quoted in Alfred Eccles, *An Account of the Brewing Trade of Dunedin and Neighbourhood: Past and Present; Compiled in the Centenary Year of the Province of Otago*, Dunedin: New Zealand Breweries, 1949, n.p.

47 *Evening Post*, 25 February 1897, 6.

48 *The Press*, 28 June 1880, 2.

49 *Wanganui Chronicle*, 22 January 1881, 2. See also: *Evening Post*, 8 June 1871, 2; *Timaru Herald*, 22 March 1876, 4; *Nelson Evening Mail*, 26 March 1878, 2.

50 *Australian Brewers' Journal*, 20 August 1889, 296–97.

51 Brewers Journal: William Dawson.

52 Gordon, *Speight's*, 16.

53 Pat Lawlor, *The Froth-Blowers' Manual: Some Confessions, Some Beer Ballads, a Dash of History, Froth of Personalities and an Encyclopedia*, Wellington: Whitcombe & Tombs, 1965, 16.

54 Stewart and Stewart, *Last of the Small Breweries*, 50.

55 Cornell, *Beer*, 41, 46, 167, 174–75.

56 *New Zealand Parliamentary Debates* (hereafter *NZPD*), 6 August 1878, 92.

57 Ibid.

58 Timothy McIvor, *The Rainmaker: A Biography of John Ballance, Journalist and Politician, 1839–1893*, Auckland: Heinemann Reed, 1989, 78–79.

59 *New Zealand Herald*, 8 August 1878, 2.

60 *Auckland Evening Star*, 9 August 1878, 2.

61 *Grey River Argus*, 8 August 1878, 2.

62 *The Press*, 16 August 1878, 2; 24 August 1878, 2.

63 *Otago Witness*, 17 August 1878, 6.

64 *Nelson Evening Mail*, 22 August 1878, 2.

65 *The Star*, 20 August 1878, 2.

66 Leckie, *Otago's Breweries*, 21.

67 'Anglo-Saxon' [J. W. Jago], *The Economics of Beer*, Dunedin: G. Lodge of New Zealand Independent Order of Good Templars, 1878, 1.

68 Ibid., 7.

69 *Evening Post*, 4 October 1878, 2.

70 *Otago Witness*, 5 October 1878, 14.

71 *Southland Times*, 8 October 1878, 2; McIvor, *The Rainmaker*, 78–79.

72 *Otago Witness*, 12 October 1878, 16.

73 *NZPD*, 8 June 1880, 125.

74 *The Press*, 28 June 1880, 2.

75 *Hawke's Bay Herald*, 14 June 1880, 3; *The Star*, 15 June 1880, 2; *Evening Post*, 17 June 1880, 3; 26 June 1880, 2.

76 *Clutha Leader*, 2 July 1880, 4.

77 *Timaru Herald*, 15 June 1880, 2; 17 June 1880, 3.

78 *Evening Post*, 26 June 1880, 2.

79 *NZPD*, 23 August 1880, 546; 24 August 1880, 603; *Evening Post*, 25 August 1880, 2; *Otago Daily Times*, 27 August 1880, 2.

80 Cornell, *Beer*, 175.

81 1880: 44 Vic. No.44: Beer Duty Act.

82 *Otago Daily Times*, 4 December 1880, 3.

83 *Lyttelton Times*, 29 April 1887, 3.

84 *The Star*, 13 April 1889, 4; 20 April 1889, 3.

85 *Grey River Argus*, 9 March 1897, 2; *Observer*, 24 April 1897, 10.

86 Phillips, *A Man's Country?*, 49.

87 McKimmey, 'Temperance Movement', 30–32; Scott, 'Public Control', 22–25.

88 A. E. Dingle, *The Campaign for Prohibition in Victorian England*, London: Croom Helm, 1980, 16–17; Ian Tyrell, *Sobering Up: From Temperance to Prohibition in Antebellum America, 1809–1860*, Westport, Conn.: Greenwood Press, 1979, 3, 244–45, 254.

89 Eldred-Grigg, *Pleasures of the Flesh*, 65–70.

90 1873: 37 Vic. No.74: An Act to Regulate the Sale of Intoxicating Liquors.

91 Eldred-Grigg, *Pleasures of the Flesh*, 71.

92 1881: 45 Vic. No.21: Licensing Act.

93 Lowther Broad, *The Law of Innkeepers in New Zealand and Licensing Committees Guide*, Nelson: Bond Finney, 1887, 1–2.

94 Christoffel, 'Removing Temptation', 10–31; Grigg, 'Citadels of Liquordom', 15, 35–42.

95 A. R. Grigg, 'Prohibition and Women: The Preservation of an Ideal and a Myth', *New Zealand Journal of History*, 17(2), 1983, 144–65.

96 Grigg, 'Citadels of Liquordom', 16–17, 40–82; Tom Brooking, *Richard Seddon: King of God's Own: The Life and Times of New Zealand's Longest Serving Prime Minister*, Auckland: Penguin, 2014, 117.

97 McKimmey, 'Temperance Movement', 42–46; Brooking, *Seddon*, 118.

98 1893: 57 Vic. No.34: Alcoholic Liquors Sale Control Act; Brooking, *Seddon*, 118–20.

99 Patricia Grimshaw, *Women's Suffrage in New Zealand*, 2nd ed., Auckland: Oxford University Press, 1988, 21–26, 108–10.

100 Linda Moore, 'Gender Counts: Men, Women and Electoral Politics, 1893–1919', MA thesis, University of Canterbury, 2004, 222–46.

101 Grigg, 'Citadels of Liquordom', 102–8, 113.

FIVE – 'Domineering, Purse-Proud, Scheming Brewers': Concentration and Competition, 1890–1914

1 *Australian Brewers' Journal*, 20 August 1889, 296.

2 Gary Hawke, *The Making of New Zealand: An Economic History*, Cambridge: Cambridge University Press, 1985, 38–41, 54–56, 82–83.

3 Ross Fitzgerald and Trevor Jordan, *Under the Influence: A History of Alcohol in Australia*, Sydney: ABC Books, 2009, 90–93.

4 Brett J. Stubbs, 'Tied Houses, Taxes, and Technology: Concentration in the Brewing Industry of New South Wales, 1882 to 1932', *Australian Economic History Review*, 39(2), 1999, 87–113, at 89.

5 S. R. H. Jones, 'The New Zealand Brewing Industry 1840–1995', in Wilson and Gourvish, eds, *Dynamics of the International Brewing Industry* (221–37), at 230–31; S. R. H. Jones and D. R. Paul, 'Concentration and Regulation in the New Zealand Brewing Industry, 1850–1970', *Australian Economic History Review*, 31(2), 1991, 66–94, at 69–71, 74–79.

6 *Tuapeka Times*, 5 May 1870, 5.

7 *Australian Brewers' Journal*, 20 October 1906, 45; Waterson, Auckland Provincial History Project Notes, 4.

8 Leckie, *Otago's Breweries*, 12.

9 Ibid.

10 Jones and Paul, 'Concentration and Regulation', 75.

11 Angus McCallum, *Tui Country. A History of Pahiatua County, 1888–1988*, Pahīatua: Pahiatua County Council, 1988, 281.

12 Ibid., 281–85.

13 *Poverty Bay Herald*, 24 November 1897, 4.

14 *Otago Witness*, 16 January 1875, 16.

15 *The Star*, 16 January 1878, 2; *Lyttelton Times*, 2 July 1881, 6; *Otago Witness*, 25 February 1882, 9.

16 Kevin Bourke, *Kelburn, King Dick and the Kelly Gang: Richard Seddon and Political Patronage*, Wellington: Hit or Miss Publishing, 2008, 145–46.

17 Gordon, *Speight's*, 182; *CNZ*, vol.5, 89.

18 Bassett and Goldsmith, *The Myers*, 32, 36, 39; New Zealand Breweries, *Hosts to the Nation*, 8.

19 R. C. J. Stone, *The Father and His Gift: John Logan Campbell's Later Years*, Auckland: Auckland University Press, 1987, 219–20.

20 R. C. J. Stone, *Makers of Fortune: A Colonial Business Community and its Fall*, Auckland: Auckland University Press, 1973, 158, 165–66.

21 Ibid.

22 Stone, *The Father and His Gift*, 221.

23 Bassett and Goldsmith, *The Myers*, 46; Stone, *The Father and His Gift*, 224–26, 268.

24 *Auckland Industrial & Mining Exhibition: Official Handbook 1898*, Auckland: Geddis & Blomfield, 1898, xix.

25 C. V. Innes, *A Crown for the Lady: The Unravelling of a Pioneer Story*, Tauranga: Moana Press, 1989, 55–56.

26 Golledge, 'The New Zealand Brewing Industry', 10; Jones and Paul, 'Concentration and Regulation', 76. See also *Journal of the Legislative Council* (hereafter *JLC*), 1902, 7, 10–11. I am very grateful to Basil Poff for information on the New Zealand tied-house debates.

27 D. Arbury, *Thames Beer Baron: Louis Ehrenfried*, Thames: Metallum Research, 2003, n.p.

28 Jones and Paul, 'Concentration and Regulation', 6.

29 Christoffel, 'Removing Temptation', 100.

30 Ibid., 184; 1895: 59 Vic. No.54: Alcoholic Liquors Sale Control Act Amendment, s.18.

31 *Waikato Times*, 16 December 1902, 2.

32 Bourke, *Kelburn, King Dick and the Kelly Gang*, 146; *JLC*, 1902, 14–15, 18, 100, 113; Gordon, *Speight's*, 31–32, 49, 61.

33 *Observer*, 29 March 1890, 12.

34 1895: 59 Vic No.45, Alcoholic Liquors Sale Control Act Amendment Act, s.35.

35 *JLC*, 1902, 7–8; Stone, *The Father and His Gift*, 269; Bourke, *Kelburn, King Dick and the Kelly Gang*, 147.

36 *JLC*, 1902, 15.

37 Ibid., 19.

38 Ibid., 15–16.

39 *The Captain Cook Brewery Co. Ltd v. Martin Ryan*, Auckland: J. H. Field, 1901, 3–5.

40 Ibid., 6.

41 Ibid., 10; 'Captain Cook Brewery (Limited) v. Ryan', *New Zealand Law Reports*, XIX, 1903, 595–604.

42 *The Captain Cook Brewery Co. Ltd v. Martin Ryan*, 12.

43 Ibid.

44 Stubbs, 'Tied Houses, Taxes, and Technology', 94.

45 *JLC*, 1902, 98.

46 *Licensed Victuallers Gazette*, 13 May 1897, 13.

47 *JLC*, 1902, 5, 16, 32–33, 55.

48 Ibid., 59.

49 Ibid.

50 Ibid., 101, 103.

51 Ibid.

52 Ibid.; *Evening Post*, 30 July 1903, 2; 4 August 1904, 5.

53 Notes and clippings - Confectionery and brewery workers, 1927–1991, Herbert Roth Papers, MS 94-106-38/10, Alexander Turnbull Library, Wellington; *Otago Witness*, 4 September 1890, 19.

54 *New Zealand Mail*, 22 August 1890, 2.

55 *Auckland Star*, 23 August 1890, 2.

56 Ibid.

57 *Australian Brewers' Journal*, 20 October 1892, 20.

58 Notes and clippings - Confectionery and brewery workers; *Australian Brewers' Journal*, 20 August 1907, 760; 20 February 1911, 289.

59 Notes and clippings - Confectionery and brewery workers; *Australian Brewers' Journal*, 20 October 1892, 20.

60 *Australian Brewers' Journal*, 20 June 1913, 551.

61 Notes and clippings - Confectionery and brewery workers.

62 *Australian Brewers' Journal*, 21 September 1903, 805.

63 *Australian Brewers' Journal*, 20 June 1913, 551.

64 Notes and clippings - Confectionery and brewery workers.

65 N. J. Adamson, 'Hop Culture in Nelson District', *New Zealand Journal of Agriculture*, February 1946, 147–53, at 147.

66 *Nelson Evening Mail*, 29 April 1871, 2. See also *Daily Southern Cross*, 22 February 1871, 2; *Star*, 11 August 1873, 2.

67 *The Colonist*, 6 December 1879, 2; *Report on the Statistics of New Zealand 1889*, Wellington: George Didsbury Govt Printer, 1891, 121.

68 *The Press*, 22 July 1903, 6.

69 *New Zealand Official Yearbook* (*NZOYB*), 1901, 339–40; Adamson, 'Hop Culture', 149.

70 *The Colonist*, 27 March 1882, 3; 26 April 1884, 3. I am grateful to Tim Holt for information in this section.

71 Joanne Benseman, 'She Worked Her Heart Out': Hops Farming Women in the Nelson Region, 1920s–1950s, NPM 998/5/1, 12, Nelson Provincial Museum, 12–13; Mary Boyd, *City of the Plains: A History of Hastings*, Wellington: Victoria University Press, 1984, 90.

72 *Australian Brewers' Journal*, 20 April 1907, 450.

73 For example, *Australian Brewers' Journal*, 20 May 1914, 504.

74 E. Buxton & Co. (1876–1941) Records. AG136 E 10.2 Hop dept Letter book 10, Nelson Provincial Museum.

75 T. G. Parsons, 'Technological Change in the Melbourne Flour-Milling and Brewing Industries, 1870–90', *Australian Economic History Review*, 11(2), 1971, 133–46.

76 *Australian Brewers' Journal*, 20 August 1903, 662.

77 Mervyn Palmer, 'William Soltau Davidson: A Pioneer of New Zealand Estate Management', *New Zealand Journal of History*, 7(2), 1973, 148–64.

78 Ansell & Sons to Henry Duncan, 22 December 1890, in possession of John Duncan, Founders Brewery, Nelson.

79 *CNZ*, vol.5, 89.

80 Gordon, *Speight's*, 33, 35, 51.

81 Lawlor, 'Breweries' (historical file); New Zealand Breweries, *Hosts to the Nation*, 14.

82 New Zealand Breweries, *Hosts to the Nation*, 16; *CNZ*, vol.3, 290.

83 *Australian Brewers' Journal*, 20 October 1893, 15; 20 July 1900, 397.

84 Cornell, *Beer*, 161–63; Protz, *Taste of Beer*, 28–33.

85 Stubbs, 'Tied Houses, Taxes, and Technology', 104–5.

86 *New Zealand Herald*, 14 February 1900, 7.

87 *Observer*, 24 February 1900, 20.

88 *Auckland Weekly News*, 23 February 1900, quoted in Brian Ronson, 'The Capt Cook Brewery', *Brewnews*, August 1971, 41.

89 *Observer*, 24 February 1900, 20; *Brewnews*, October 1975, 4.

90 *New Zealand Herald*, 5 July 1905, 4; 20 November 1905, 8; *Pelorus Guardian and Miners' Advocate*, 8 March 1907, 3.

91 Peter Petchey, 'The Ring Seal Beer/Champagne Bottle', *Archaeology in New Zealand*, 56(4), 2013, 215–24.

92 Arbury, *Thames Beer Baron*, n.p.; *Poverty Bay Herald*, 18 January 1911, 8; Leckie, *Otago's Breweries*, 101.

93 *NZ Truth*, 9 May 1908, 1.

94 *Poverty Bay Herald*, 18 January 1911, 8; *Australian Brewers' Journal*, 20 November 1895, 41; 20 April 1898, 233; Stewart and Stewart, *Last of the Small Breweries*, 73–74.

95 *New Zealand Herald*, 29 May 1878, 2.

96 *Otago Witness*, 12 April 1894, 4; *Ashburton Guardian*, 4 February 1897, 2.

97 R. W. Willett, *Hotels of Dunedin: Historical Record*, Dunedin: Coulls Somerville Wilkie, 1937, n.p.

98 *Otago Daily Times*, 21 December 1892, 4.

99 *Otago Witness*, 17 October 1895, 35.

100 Stewart and Stewart, *Last of the Small Breweries*, 73–74.

101 Gordon, *Speight's*, 31–32.

102 For example, *Wanganui Herald*, 9 December 1908, 2.

103 New Zealand Breweries, *Hosts to the Nation*, 8; Brett J. Stubbs, '"A New Drink for Young Australia": From Ale to Lager Beer in New South Wales, 1880 to 1930', in Robert Dare, ed., *Food, Power and Community: Essays in the History of Food and Drink* (pp. 126–41), Adelaide: Wakefield Press, 1999; Lawlor, 'Breweries (Historical File)'; *Brewnews*, August 1962, n.p.

SIX – 'A Red Hot Nest of Liquor Dealing Intriguers': Beer and Prohibition, 1900–1919

1 Waterson, Auckland Provincial History Project Notes, 20.

2 Grigg, 'Citadels of Liquordom', 292.

3 Ibid., 128–29, 139–40.

4 *The Press*, 23 November 1896, 3.

5 Grigg, 'Citadels of Liquordom', 158–59, 195.

6 Ibid., 128–29, 139–40, 158–59, 195.

7 Christoffel, 'Removing Temptation', 34–35.

8 T. E. Taylor, *The Shadow of Tammany: The Truth About the New Zealand Police Force, Its Condition and Its Needs*, Christchurch: T. E. Fraser, 1898, 43–51.

9 Grigg, 'Citadels of Liquordom', 181; Bollinger, *Grog's Own Country*, 49–51; Richard S. Hill, *The Iron Hand in the Velvet Glove: The Modernisation of Policing in New Zealand, 1886–1917*, Palmerston North: Dunmore Press, 1995, 81–95.

10 Bollinger, *Grog's Own Country*, 61–71.

11 Christoffel, 'Removing Temptation', 34–35.

12 *New Zealand Mail*, 3 June 1903, 57.

13 Ibid.

14 Bollinger, *Grog's Own Country*, 75–77; *New Zealand Times*, 17 May 1904, 5.

15 Judgment of the Lords of the Judicial Committee of the Privy Council on the Appeal of *Smith v. McArthur and Others*, from the Supreme Court of New Zealand; delivered the 14th May 1904.

16 *New Zealand Times*, 17 May 1904, 5; *New Zealand Mail*, 25 May 1904, 74.

17 Grigg, 'Citadels of Liquordom', 136, 193.

18 Ibid., 149, 195; Brooking, *Seddon*, 117.

19 Grigg, 'Citadels of Liquordom', 195; Brooking, *Seddon*, 118–21.

20 Newman, 'New Zealand's Vote', 52–55.

21 *Auckland Star*, 7 December 1909, 5; 11 December 1909, 5.

22 Craig Turney, 'The New Zealand Alliance and Auckland, 1905–1920', MA thesis, University of Auckland, 1996, 204–5.

23 Ibid., 202–3.

24 Ibid., 206–9.

25 Ibid., 209–12; 1910: 1 Geo. V No.46: Licensing Amendment Act.

26 Turney, 'New Zealand Alliance', 206–9; Newman, 'New Zealand's Vote', 59.

27 Grigg, 'Citadels of Liquordom', 100, 278–79; 1893: 57 Vic. No.34: Alcoholic Liquors Sale Control Act, s.12.

28 Susan Upton, *Wanted, a Beautiful Barmaid: Women Behind the Bar in New Zealand, 1830–1976*, Wellington: Victoria University Press, 2013, 112–13; Brewers' Association, AGM Minutes, 6 October 1909, Brewers' and Wine and Spirit Merchants' Association of New Zealand (hereafter Brewers' Association), Records: MS-Papers-1629, Alexander Turnbull Library, Wellington.

29 *Evening Post*, 13 November 1908, 7.

30 1910: 1 Geo. V No.46: Licensing Amendment Act, s.36.

31 Hutt, *Maori and Alcohol*, 42–46.

32 Ibid., 51, 55.

33 *Australian Brewers' Journal*, 20 April 1903, 472.

34 Hutt, *Maori and Alcohol*, 61–62.

35 Bourke, *Kelburn, King Dick and the Kelly Gang*, 158–59.

36 Brewers' Association, *Annual Report*, 1891, Brewers' Association Records.

37 Bourke, *Kelburn, King Dick and the Kelly Gang*, 160.

38 Brewers' Association, *Annual Report*, 1891; *Australian Brewers' Journal*, 20 November 1892, 61; 20 January 1893, 140.

39 Bourke, *Kelburn, King Dick and the Kelly Gang*, 161; Letter Book, 1897–1903, Brewers' Association Records.

40 *The Press*, 12 July 1893, 6.

41 Brewers' Association, Minutes, 27 July 1896.

42 Bourke, *Kelburn, King Dick and the Kelly Gang*, 164; H. J. Williams to ?, 2 July 1900, Letter Book, 1897–1903, Brewers' Association Records.

43 Bourke, *Kelburn, King Dick and the Kelly Gang*, 160, 166.

44 *Licensed Victuallers Gazette*, 26 January 1899, 181; 16 February 1899, 311.

45 Bourke, *Kelburn, King Dick and the Kelly Gang*, 160, 166; ? to Martin Kennedy, 7 June 1901, Letter Book, 1897–1903, Brewers' Association Records.

46 H. J. Williams to Thomas Barnett, 14 January 1902, Letter Book, 1897–1903, Brewers' Association Records.

47 Derived from Brewers' Association Minute Book, 1894–1905, Brewers' Association Records.

48 *The Press*, 15 March 1894, 4.

49 Ibid., 27 August 1895, 8; 28 August 1895, 4.

50 Ibid., 20 January 1899, 3.

51 *CNZ*, vol.2, 334.

52 *CNZ*, vol.2, 335.

53 J. S. Tullett, *The Industrious Heart: A History of New Plymouth*, New Plymouth: New Plymouth City Council, 1981, 263.

54 *Australian Brewers' Journal*, 20 September 1889, 383.

55 Brewing of Hop Beer in Breweries 1908–1977: AALN 847: W4748 202 31/40/-, Archives New Zealand, Wellington.

56 *Mataura Ensign*, 7 July 1903, 2; *Australian Brewers' Journal*, 21 September 1903, 805; *Otago Daily Times*, 28 October 1904, 5.

57 D. R. Paul, 'The Structure of the Brewing Industry and Beer Market to 1980: A Case Study in Industrial Concentration and Market Regulation', MCom thesis, University of Auckland, 1984, 78.

58 Clive A. Lind, *Pubs, Pints and People: 50 Years of the Invercargill Licensing Trust*, Invercargill: The Trust, 1994, 16–17.

59 Beer Duty Act – Prohibition on Breweries Existence: C 1 31/3/3, Archives New Zealand, Wellington.

60 R. C. J. Stone, 'Myers, Arthur Mielziner', Dictionary of New Zealand Biography, Te Ara – the Encyclopedia of New Zealand, https://teara.govt.nz/en/biographies/3m66/myers-arthur-mielziner (accessed 26 August 2021); Diana Beaglehole, 'Gibbons, Hopeful', Dictionary of New Zealand Biography, Te Ara – the Encyclopedia of New Zealand, https://teara.govt.nz/en/biographies/3g5/gibbons-hopeful (accessed 26 August 2021).

61 Leckie, *Otago's Breweries*, 22.

62 Gordon, *Speight's*, 43.

63 *Otago Daily Times*, 17 April 1896, 2.

64 Gordon, *Speight's*, 9, 57, 73.

65 *Australian Brewers' Journal*, 20 March 1900, 247; 20 April 1900, 294; 20 February 1901, 191.

66 *The Press*, 21 May 1897, 5.

67 *Australian Brewers' Journal*, 20 July 1899, 354.

68 *Oamaru Mail*, 14 July 1899, 1.

69 *New Zealand Tablet*, 22 June 1899, 1.

70 *Evening Post*, 5 July 1910, 5; *New Zealand Times*, 12 April 1912, 1.

71 *Grey River Argus*, 13 September 1900, 3; *Kaipara and Waitemata Echo*, 18 June 1913, 2; *New Zealand Herald*, 14 August 1913, 5; Mark Derby, *Rock College: An Unofficial History of Mt Eden Prison*, Palmerston North: Massey University Press, 2020, 132.

72 John McKeague, *Liquor Reform: Prohibition or No License in New Zealand: Its Real Effects*, Sydney: Ross Brothers, 1905, 3.

73 Speight's Brewery, *The History of a Glass of Beer: As Produced at Speight & Co.'s City Brewery, Rattray and Dowling Streets, Dunedin*, Dunedin: Samuel Lister, 1893, 15.

74 Ibid., 61.

75 *Prohibition Fatal to Liberty, Temperance and Morality. Written Specially for Popular Use and Carefully Revised by Three Professors of the University of New Zealand*, Christchurch: Whitcombe & Tombs, 1911. See also *Licensed Victuallers Gazette*, 9 December 1911, 12.

76 W. D. Lysnar, *Opinions and Facts Showing Prohibition to be a Hideous Failure*, Gisborne: Gisborne Times, 1902, 15.

77 McKeague, *Liquor Reform*, 14.

78 *The Press*, 10 June 1899, 9.

79 *Otago Daily Times*, 15 January 1910, 12. See also 'Dunboy', *Two Wrongs, One Right: A Prohibition Maxim*, Dunedin: Otago Daily Times and Witness Newspapers, 1899.

80 W. Thomson, *Anti-Prohibition: A Lecture Delivered . . . in the Zealandia Hall, Invercargill*, Temuka: Temuka Leader, 1905, 6.

81 1 Timothy 5:23.

82 A. Chodowski, *Wine: Its Use and Abuse: The Fermented Wines of the Bible*, Christchurch: Christchurch Press, 1893; J. Allan, *Prohibition and Christianity: A Protest*, Ōamaru: North Otago Times, 1897; John 2:1–11.

83 *New Zealand Tablet*, 23 November 1899, 3.

84 *New Zealand Tablet*, 30 November 1911, 2409; 7 December 1911, 2479.

85 Greg Ryan, 'An Undertaking Worthy Only of Fanatics: Catholic Opinion on Temperance and Prohibition in New Zealand, c1870–1910', *Australasian Journal of Irish Studies*, 10, 2010, 16–36.

86 *Prohibition Fatal to Liberty*, 7.

87 'It Is', *Isitt and His Prohibition Fad Proved Complete Failures*, Dunedin: Otago Workman, 1893, 4.

88 *The Press*, 10 June 1899, 9.

89 *Defence not Defiance*, Wellington: n.p., 1914.

90 W. Salmond, *Prohibition: A Blunder*, Dunedin: R. J. Stark; 1911; Peter Matheson, 'Salmond, William', Dictionary of New Zealand Biography, Te Ara – the Encyclopedia of New Zealand, https://teara.govt.nz/en/biographies/2s3/salmond-william (accessed 27 August 2021).

91 Brewers' Association, AGM Minutes, 28 September 1911; Special Meeting Minutes, 17 May 1912, Brewers' Association Records.

92 Liberty and Regulation, *National Prohibition Opposed to the Old Testament, the New Testament, the Spirit of Christ, Sound Ethics, Common Sense, Enlightened Politics, is a Fallacy, a Tyranny, a Failure, a Cure Worse than the Disease*, Dunedin: NZ Tablet Printing & Publishing Co., 1920.

93 New Zealand Alliance, Executive Minutes, 9 July 1914, 13 August 1914, MS 77-2006-11, Alexander Turnbull Library, Wellington.

94 *The Press*, 12 December 1914, 10.

95 *Evening Post*, 21 August 1914, 8; *Lyttelton Times*, 14 December 1914, 6.

96 *Dominion*, 25 August 1914, 6; 26 May 1915, 8; *Otago Daily Times*, 7 April 1915, 5.

97 *Southland Times*, 1 September 1914, 2.

98 *Australian Brewers' Journal*, 20 November 1913, 61.

99 1915: 6 Geo. V No.39: Finance Act; *Evening Post*, 29 September 1915, 3; *New Zealand Herald*, 10 November 1915, 9.

100 Gordon, *Speight's*, 62–63.

101 Ibid., 63.

102 New Zealand Alliance, Executive Minutes, 25 August 1914.

103 Tim Shoebridge, 'Index of New Zealand Wartime Laws and Regulations, 1914–21', NZ History, https://nzhistory.govt.nz/files/documents/ww1-laws-and-regulations-index.pdf (accessed 4 April 2020).

104 J. Allen to W. F. Massey, 28 October 1916, 1916–1919: Allen, J.: Series 1; Box 9, Archives New Zealand, Wellington.

105 *NZPD*, 12 July 1917, 367.

106 *The Press*, 3 April 1915, 9.

107 *Vanguard*, 8 August 1914, 6; New Zealand Alliance, Executive Minutes, 8 April 1915; 13 September 1915.

108 Steven Early, 'The Six O'clock Closing Debate', Dip. Arts thesis, University of Otago, 1991, 23–26; *Vanguard*, 13 January 1917, 1.

109 Christoffel, 'Removing Temptation', 134–35; *Auckland Star*, 1 August 1891, 5; New Zealand Alliance, Executive Minutes, 8 May 1916; 4 July 1916.

110 *Six O'clock Closing: Auckland Business Men's Deputation to the Right Hon. W.F. Massey*, Auckland: Wright & Jacques, 1916, 6.

111 Ibid., 8.

112 Ibid., 12.

113 *NZ Truth*, 23 October 1915, 1.

114 *Re Petition for Six O'clock Closing of Liquor Bars: Verbatim Report of Addresses Delivered by Counsel Before the M to Z Public Petitions Committee of the House of Representatives on Friday, June 30th, 1916*, Wellington: New Zealand Times Print, 1916, 17.

115 *Dominion*, 6 July 1916, 6.

116 *New Zealand Times*, 13 April 1915, 7.

117 *Evening Star*, 21 July 1916, 2; Susan Butterworth, 'Herdman, Alexander Lawrence', Dictionary of New Zealand Biography, Te Ara – the Encyclopedia of New Zealand, https://teara.govt.nz/en/biographies/3h18/herdman-alexander-lawrence (accessed 1 September 2021).

118 *Dominion*, 29 July 1916, 10; *Evening Post*, 28 July 1916, 6; 2 August 1916, 3.

119 J. E. Martin, 'Blueprint for the Future? "National Efficiency" and the First World War', in J. Crawford and I. McGibbon, eds, *New Zealand's Great War: New Zealand, the Allies and the First World War* (pp. 516–33), Auckland: Exisle Publishing, 2007, 517–19; Paul Baker, *King and Country Call: New Zealanders, Conscription and the Great War*, Auckland: Auckland University

Press, 1988, 42–63, 79–95, 153–210; David Littlewood, 'Striking a Balance: The Military Service Boards', in S. Loveridge, ed., *New Zealand Society at War, 1914–1918* (pp. 112–26), Wellington: Victoria University Press, 2016.

120 Martin, 'Blueprint for the Future?', 520–21.

121 Allen to Massey, 28 April 1917.

122 *NZPD*, 7 August 1917, 30; Baker, *King and Country Call*, 132–35.

123 Ian McGibbon, 'Allen, James', Dictionary of New Zealand Biography, Te Ara – the Encyclopedia of New Zealand, https://teara.govt.nz/en/biographies/3a12/allen-james (accessed 19 March 2020); James Watson, *W.F. Massey: New Zealand*, London: Haus, 2010, 42–60.

124 Martin, 'Blueprint for the Future?', 520–21.

125 'National Efficiency Board Report on the Liquor Question', *AJHR*, H-43A, 1917, 1.

126 Ibid., 2.

127 National Efficiency Board, Memorandum 448, 9 July 1917, Memoranda: Vol.1, qMS-1525, Alexander Turnbull Library, Wellington.

128 New Zealand Alliance, Executive Minutes, 24 August 1917; Martin, 'Blueprint for the Future?', 522.

129 *NZPD*, 18 July 1917, 495.

130 *NZPD*, 19 July 1917, 584.

131 *NZPD*, 19 September 1917, 258.

132 *NZPD*, 20 September 1917, 306; Christoffel, 'Removing Temptation', 139.

133 Allen to Massey, 19 September 1918.

134 *Vanguard*, 12 January 1918, 2.

135 *Vanguard*, 6 April 1918, 2.

136 New Zealand Alliance, Executive Minutes, 25 March 1918; 10 April 1918; 7 June 1918.

137 Christoffel, 'Removing Temptation', 70.

138 William Ferguson to James Allan, 14 August 1918, Memoranda: Vol.3, qMS-1527, Alexander Turnbull Library, Wellington; New Zealand Alliance, Executive Minutes, 26 August 1918.

139 Geoffrey W. Rice, *Black November: The 1918 Influenza Pandemic in New Zealand*, Christchurch: Canterbury University Press, 2005, 85–86, 107; Geoffrey W. Rice, 'Russell, George Warren', Dictionary of New Zealand Biography, Te Ara – the Encyclopedia of New Zealand, https://teara.govt.nz/en/biographies/2r31/russell-george-warren (accessed 2 September 2021).

140 Rice, *Black November*, 246.

141 *Australian Brewers' Journal*, 20 January 1919, 160; *Evening Post*, 6 December 1918, 3.

142 *Otago Daily Times*, 5 December 1918, 4.

143 1918: 9 Geo. V No.12: Licensing Amendment Act.

144 *Australian Brewers' Journal*, 20 June 1905, 488; Christoffel, 'Removing Temptation', 71–72; Bollinger, *Grog's Own Country*, 85.

145 *Australian Brewers' Journal*, 20 February 1919, 297; 21 July 1919, 471–72; *Aberdeen Press and Journal*, 6 March 1919, 2.

146 *Australian Brewers' Journal*, 20 May 1919, 365.

147 *Labour Leader*, 17 April 1919, 7.

148 Paul Christoffel, 'Prohibition and the Myth of 1919', *New Zealand Journal of History*, 42(2), 2008, 154–59.

SEVEN – The Big Ten and a Rival: Mergers, Competition and Depression, 1920–1940

1 Christoffel, 'Removing Temptation', 141; *Report of the Royal Commission on Licensing* (hereafter *RCL*), *AJHR*, H-38, 1946, 99–101.

2 1921: 12 Geo. V No.25: Finance (No.2) Act.

3 *Gisborne Times*, 14 November 1921, 7; *Australian Brewing and Wine Journal* (hereafter *ABWJ*), 20 December 1921, 141–42, 151.

4 Brian Easton, *Not in Narrow Seas: The Economic History of Aotearoa New Zealand*, Wellington: Victoria University of Wellington Press, 2020, 229–32.

5 Bassett and Goldsmith, *The Myers*, 105–9; *New Zealand Herald*, 7 April 1915, 7.

6 *RCL*, 52–55; New Zealand Licensing Reform Association, *The Proposals for Corporate Control of the Liquor Trade as Embodied in the Petition Presented to Parliament in 1923: Together with Proposals for the Immediate Improvement of the Existing System of Private Control*, Wellington: New Zealand Licensing Reform Association, 1923.

7 *ABWJ*, 22 February 1922, 281; 20 June 1923, 490–91.

8 Gordon, *Speight's*, 70–72.

9 *ABWJ*, 20 July 1923, 557–59.

10 Lion Breweries files, typed notes for staff training course, G. C. Reid, 'History and Organisation of the Company', 3 February 1970, Lion Technical Library, Auckland.

11 New Zealand Breweries, *Hosts to the Nation*, 6.

12 Jones and Paul, 'Concentration and Regulation', 82–84.

13 Leckie, *Otago's Breweries*, 12, 94–97, 149; *The Star*, 19 March 1924, 1; *The Press*, 9 June 1936, 13.

14 R. C. J. Stone, '"Sinister" Auckland Business Cliques, 1840–1940', *New Zealand Journal of History*, 21(1), 1987, 29–45, at 38.

15 *ABWJ*, 20 August 1924, 622–23; 20 September 1924, 727; 20 August 1925, 596; 20 November 1925, 92, 96; 20 August 1926, 635.

16 *Hokitika Guardian*, 8 January 1929, 4.

17 *The Press*, 30 June 1928, 12; 31 August 1932, 10.

18 Westland Breweries, *Annual Report*, 1941; *RCL*, 154.

19 Dominion Breweries, *1930–1980: 50th Golden Anniversary: Dominion Breweries*, Auckland: Dominion Breweries, 1980, 3.

20 Stone, '"Sinister" Auckland Business Cliques', 36, n.26.

21 *New Zealand Herald*, 6 November 1929, 13.

22 R. C. J. Stone, 'Kelliher, Henry Joseph', Dictionary of New Zealand Biography, Te Ara – the Encyclopedia of New Zealand, https://teara.govt.nz/en/biographies/4k5/kelliher-henry-joseph (accessed 26 November 2021).

23 Dominion Breweries, *1930–1980*, 6.

24 *New Zealand Herald*, 31 July 1931, 15; Jones and Paul, 'Concentration and Regulation', 87.

25 Jones and Paul, 'Concentration and Regulation', 85.

26 *ABWJ*, 20 November 1930, 127; 20 April 1931, 405; 21 August 1939, 56.

27 *ABWJ*, 20 May 1930, 476.

28 *Evening Post*, 24 April 1936, 13.

29 *ABWJ*, 20 August 1936, 13.

30 Lion Breweries files, typed notes for staff training course.

31 *RCL*, 60–61.

32 *Proceedings of the Royal Commission on Licensing* (hereafter *PRCL*), Wellington: Royal Commission on Licensing, 1945, 2173–74.

33 Nelson Hotels Ltd/Nelson Breweries Ltd – 'Bible', Nelson Provincial Museum.

34 Ibid.

35 *Marlborough Express*, 22 March 1920, 4; *New Zealand Times*, 23 February 1924, 4; *Evening Post*, 24 February 1930, 11.

36 *Otago Daily Times*, 25 July 1921, 5; *The Sun* (Auckland), 27 April 1927, 12; *New Zealand Herald*, 31 January 1928, 8; *The Star*, 9 July 1928, 1.

37 *ABWJ*, 20 April 1922, 431; *Otago Daily Times*, 18 June 1928, 12.

38 *Evening Star*, 26 February 1923, 6; *The Star*, 16 January 1933, 7.

39 *PRCL*, 1223.

40 *Otago Daily Times*, 9 October 1922, 7; *ABWJ*, 20 March 1923, 360.

41 *New Zealand Herald*, 3 December 1927, 16; *ABWJ*, 20 November 1928, 99.

42 *ABWJ*, 20 February 1926, 351.

43 Memorandum from Customs Dept Auckland to Comptroller of Customs, Wellington, 12 October 1932; R. L. Andrew, Assistant Dominion Analyst, to Comptroller of Customs, 4 November 1932; Brewing of Hop Beer in Breweries.

44 *ABWJ*, 21 June 1926, 420.

45 *The Press*, 24 November 1936, 13.

46 Ibid.

47 *New Zealand Journal of Agriculture*, 20 April 1937, 232; 20 September 1938, 227; 20 October 1938, 307.

48 *Brewnews*, March 1957, n.p.

49 Emily Coe, 'The Nelson Hop Industry: Survival and Success', BA Hons long essay, University of Otago, 1998, 38–39.

50 Ibid., 17; *ABWJ*, 20 February 1937, 56; *Nelson Evening Mail*, 22 August 1936, 4.

51 *ABWJ*, 21 August 1939, 56; *New Zealand Journal of Agriculture*, 15 September 1939, 226.

52 Christoffel, 'Myth of 1919', 159, 162.

53 *ABWJ*, 20 August 1921, 663.

54 *Ward and Company Limited v. Commissioner of Taxes* [1922] NZPC 3; [1922] UKPC 90; [1923] AC 145; (1922) NZPCC 625 (16 November 1922).

55 *Cheerio*, 1, June 1922, 1.

56 *Cheerio*, 2, August 1922, 2.

57 Christoffel, 'Removing Temptation', 85.

58 *RCL*, 55.

59 Ibid.; Bollinger, *Grog's Own Country*, 80–81.

60 Graham W. A. Bush, 'Davis, Ernest Hyam', Dictionary of New Zealand Biography, Te Ara – the Encyclopedia of New Zealand, https://teara.govt.nz/en/biographies/4d7/davis-ernest-hyam (accessed 3 December 2021).

61 John Horrocks, 'The Shadowland Manipulator: Sir Ernest "Booze", John A. Lee, and the Labour Party', *Journal of New Zealand Studies*, 24, 2017, 43–56, at 43–48; Stone, '"Sinister" Auckland Business Cliques', 35–36.

62 *NZPD*, 13 August 1930, 226.

63 *New Zealand Worker*, 17 February 1932, 1; 24 February 1932, 1; *Otago Daily Times*, 8 June 1932, 4.

64 Erik Olssen, *John A. Lee*, Dunedin: University of Otago Press, 1977, 55, 81.

65 Horrocks, 'Shadowland Manipulator', 45.

66 Bollinger, *Grog's Own Country*, 89.

67 G. R. Hawke, 'Macarthy, Thomas George', Dictionary of New Zealand Biography, Te Ara – the Encyclopedia of New Zealand, https://teara.govt.nz/en/biographies/2m2/macarthy-thomas-george (accessed 2 December 2021).

68 Ibid.; T. G. Macarthy Trust, https://www.publictrust.co.nz/grants/t-g-macarthy-trust/ (accessed 2 December 2021).

69 T. G. Macarthy Trust; *ABWJ*, 21 October 1928, 8; 20 June 1938, 56; *Evening Post*, 1 August 1928, 11; *The Press*, 30 August 1929, 18.

70 Christoffel, 'Removing Temptation', 72–73; 'Myth of 1919', 162–63, 167.

71 *White Ribbon*, 18 September 1922, 6.

72 New Zealand Alliance, AGM Minutes, 23 May 1928.

73 *White Ribbon*, 18 January 1931, 6; Christoffel, 'Myth of 1919', 164–65.

74 Bollinger, *Grog's Own Country*, 83.

75 New Zealand Alliance, *Annual Report*, 1932.

76 Christoffel, 'Myth of 1919', 165–66; *RCL*, 59; New Zealand Alliance, *Annual Report*, 1936.

77 New Zealand Alliance, *Treasurer's Report*, 1938–39.

78 Christoffel, 'Removing Temptation', 89; Edith Thornton, 'A History of the Prohibition Movement in New Zealand', MA thesis, Canterbury University College, 1936, 74–75.

79 Malcolm McKinnon, *The Broken Decade: Prosperity, Depression and Recovery in New Zealand, 1928–39*, Dunedin: Otago University Press, 2016, 61–152.

80 *New Zealand Herald*, 14 March 1931, 13; 31 July 1931, 13; *Evening Post*, 13 August 1931, 9.

81 *ABWJ*, 20 August 1932, 600; Dominion Breweries, *Annual Report*, 1933.

82 *Evening Post*, 16 July 1931, 15.

83 New Zealand Breweries Ltd to W. Downie Stewart, Minister of Customs, 28 October 1931; NZ Free Trade Brewers' Association to W. Downie Stewart, 11 December 1931; NZ Breweries to E. D. Good, Acting Comptroller of Customs, 18 January 1932; Beer Duty Act – Home Brewing 1931 – 1932: C: 1: 31/7/40, Archives New Zealand, Wellington.

84 New Zealand Alliance, *Annual Report*, 1932.

85 *ABWJ*, 20 November 1931, 132; 21 December 1931, 187.

86 *Mt Benger Mail*, 6 January 1932, 2.

87 *ABWJ*, 21 December 1931, 187.

88 *ABWJ*, 20 April 1932, 409.

89 *ABWJ*, 20 December 1932, 204; 20 April 1933, 489.

90 *New Zealand Police Gazette*, 9 November 1938, 762.

91 *New Zealand Licensee*, 25 July 1933, 2; 25 June 1934, 5; *ABWJ*, 20 October 1934, 22–23.

92 Ian F. Grant, 'Agents of Prosperity: Building an Advertising Industry', in Peter Alsop and Gary Stewart, eds, *Promoting Prosperity: The Art of Early New Zealand Advertising* (pp. 26–32), Nelson: Craig Potton Publishing, 2013.

93 *Otago Daily Times*, 18 December 1936, 18.

94 *Poverty Bay Herald*, 2 July 1932, 4; Marie J. Duncan, 'Cheers! Selling Health and Happiness: Advertising Alcohol in New Zealand, *c.* 1900–1945', MA thesis, Victoria University of Wellington, 2020, 24–28.

95 *The Press*, 6 February 1936, 18; Duncan, 'Cheers!', 38–39.

96 *Dominion*, 30 July 1937, 5.

97 *New Zealand Herald*, 21 September 1936, 26.

98 Duncan, 'Cheers!', 58–60.

99 For example, *Timaru Herald*, 26 September 1935, 4.

100 *Auckland Star*, 25 September 1934, 11.

101 Vita and Mello Stout, 1934–35, New Zealand Alliance Papers, MS 77-206-06/16, Alexander Turnbull Library, Wellington.

102 1884: 48 Vic. No.24, Police Offences Act; 1906: 6 Edw. VII No.32, Motor Regulation Act.

103 *NZ Truth*, 23 October 1909, 5.

104 James Watson, *Links: A History of Transport and New Zealand Society*, Wellington: Ministry of Transport, 1996, 122, 153.

105 *NZOYB*, 1939, 262.

106 1924: 15 Geo. V, No.39, Motor-vehicles Act; *Stratford Evening Post*, 31 August 1926, 6.

107 'Beer . . . The big bluffer' (posters for illustrated talk), New Zealand Alliance, Manawatu Council: Records, 1946–1974, MS 82-217-12, Alexander Turnbull Library, Wellington.

108 *The Menace of the Motor: Prevention of Motor Accidents*, Wellington: Wright & Carman, 1936.

109 *Evening Post*, 24 July 1936, 10.

110 *Evening Post*, 21 August 1936, 11; 1939: 3 Geo. VI No.9: Transport Law Amendment Act.

111 Notes and clippings – Confectionery and brewery workers.

112 *The Press*, 2 August 1939, 10.

113 *Evening Post*, 17 August 1939, 11.

114 *Otago Daily Times*, 18 August 1939, 8.

115 Ibid.

116 *The Press*, 19 August 1939, 8.

117 *Evening Post*, 29 August 1939, 17.

118 *Evening Star*, 30 August 1939, 2.

119 *New Zealand Herald*, 26 August 1939, 14; 28 August 1939, 12; *Evening Post*, 31 August 1939, 5.

120 *Evening Post*, 4 September 1939, 13; *New Zealand Herald*, 21 September 1939, 8; *The Press*, 27 September 1939, 10.

EIGHT – 1036 and All That: War, Swill and Technology, 1942–1957

1 *Behind the Scenes in New Zealand's Premier Brewery: The Waitemata Model Brewery, Otahuhu*, Auckland: Dawson Printing Co., 1949, 11.

2 *Auckland Star*, 23 May 1945, 6.

3 *The Press*, 24 September 1952, 9.

4 *New Zealand Herald*, 4 September 1939, 3; 21 June 1940, 3; 25 July 1940, 10.

5 *New Zealand Herald*, 16 November 1939, 8.

6 *ABWJ*, 21 March 1941, 39.

7 *ABWJ*, 20 November 1939, 83.

8 *Brewnews*, June 1958, n.p.

9 *ABWJ*, 21 March 1941, 39; 21 April 1941, 39.

10 *New Zealand Herald*, 18 June 1941, 6; *ABWJ*, 20 January 1942, 36; 20 June 1942, 20.

11 *ABWJ*, 21 July 1941, 39; 20 November 1942, 32.

12 New Zealand Alliance Papers, 77-206-14/06; 77-206-06/18.

13 Ibid.

14 *White Ribbon*, 18 October 1941, 10.

15 *White Ribbon*, 1 January 1942, 9; 18 April 1942, 3.

16 *Evening Post*, 9 March 1942, 6; 12 March 1942, 6.

17 Brian Glover, *Brewing for Victory: Brewers, Beer and Pubs in World War II*, Cambridge: Lutterworth Press, 1995, 87–90; Alison Painter, *Jolly Good Ale and Old, The History of Coopers Brewery 1862–1987*, Adelaide: Cooper & Son, 1987, 81–83.

18 *New Zealand Herald*, 10 April 1942, 8.

19 *New Zealand Herald*, 20 April 1942, 6.

20 *NZPD*, 30 April 1942, 192.

21 Ibid.

22 Ibid., 196–98, 200; 1942: Geo. VI No.5: Customs Act Amendment Act.

23 *NZPD*, 30 April 1942, 199; Bruce Brown, 'Nordmeyer, Arnold Henry', Dictionary of New Zealand Biography, Te Ara – the Encyclopedia of New Zealand, https://teara.govt.nz/en/biographies/5n12/nordmeyer-arnold-henry (accessed 2 February 2022).

24 *NZPD*, 30 April 1942, 199; 8 May 1942, 346–49.

25 *New Zealand Licensee*, 10 May 1942, 1.

26 *New Zealand Herald*, 11 May 1942, 4; *The Press*, 11 May 1942, 4.

27 *Auckland Star*, 23 June 1942, 4.

28 *Evening Post*, 14 August 1942, 4; *Waihi Daily Telegraph*, 20 July 1942, 3; The Manufacture of Liquor Emergency Regulations 1942.

29 *Gisborne Herald*, 1 May 1942, 1; *Wanganui Chronicle*, 6 June 1942, 1.

30 Bollinger, *Grog's Own Country*, 56.

31 *PRCL*, 1224.

32 *RCL*, 62–68; The Manufacture of Liquor Emergency Regulations, 1942, Amendment No.1.

33 Report of Technical Adviser to Comptroller of Customs, 23 May 1944, Supervision of Breweries General File 1942–1977: AALN 847: W4748 202 31/7/4, Archives New Zealand, Wellington.

34 P. Greenfield, Dunedin Brewery Co. Ltd, to Collector of Customs, Dunedin, 14 December 1945; Customs Department, Wellington, to Collector of Customs, Dunedin, 17 January 1946, Supervision of Breweries General File.

35 Technical Adviser to Comptroller of Customs, 6 March 1947; *PRCL*, 5454.

36 Gordon, *Speight's*, 99–100; *PRCL*, 6410.

37 *PRCL*, 2926–27.

38 *PRCL*, 4534, 6963–64, 6971.

39 *The Press*, 20 October 2007, 20.

40 H. W. Orsman, ed., *The Dictionary of New Zealand English*, Auckland: Oxford University Press, 1997, 736, 800; Tanja Luckins, 'Pigs, Hogs and Aussie Blokes: The Emergence of the Term "Six O'clock Swill"', *History Australia*, 4(1), 2007, 08.1–08.17.

41 D. J. Noble, 'The Origins and Foundation of the Invercargill Licensing Trusts. Account of the First Shaky Years of the First Ever Licensing Trust in N.Z., Founded by the first Labour Government During the War', Postgrad. Dip. essay, University of Otago, 1976, 4–27; Bernard Teahan, *A Great Social Experiment: The Story of Licensing Trusts in New Zealand*, Masterton: Fraser Books, 2017, 11–22.

42 Noble, 'Origins and Foundation', 39–48; Teahan, *Great Social Experiment*, 12–23; 1944: Geo. VI No.4: Invercargill Licensing Act.

43 *The Press*, 18 July 1944, 6; *Otago Daily Times*, 22 March 1945, 4.

44 *RCL*, 197–99; Bollinger, *Grog's Own Country*, 31–32; *New Zealand Licensee*, 31 December 1934, 2.

45 *New Zealand Herald*, 13 June 1944, 4; Jim McAloon, *Judgements of All Kinds: Economic Policy-making in New Zealand, 1945–1984*, Wellington: Victoria University Press, 2013, 63, 94–95.

46 Bollinger, *Grog's Own Country*, 95; G. P. Barton, 'Smith, David Stanley', Dictionary of New Zealand Biography, Te Ara – the Encyclopedia of New Zealand, https://teara.govt.nz/en/biographies/4s30/smith-david-stanley (accessed 5 February 2022).

47 *Evening Post*, 9 March 1945, 7.

48 Ibid.

49 Bollinger, *Grog's Own Country*, 102–3.

50 J. A. Manderson, *Beer Slops*, Auckland: the author, 1946, 5–6; Bollinger, *Grog's Own Country*, 98–99.

51 *Evening Post*, 27 April 1945, 8–9; 20 October 1945, 8–9.

52 Bollinger, *Grog's Own Country*, 106–11; Christoffel, 'Removing Temptation', 111.

53 Christoffel, 'Removing Temptation', 104–6; *RCL*, 279, 311, 317–21.

54 *The Press*, 6 September 1946, 8.

55 *Otago Daily Times*, 6 September 1946, 4.

56 Bollinger, *Grog's Own Country*, 89.

57 *The Press*, 29 September 1947, 6; October 1947, 6; 2 October 1947, 8.

58 *The Press*, 6 October 1947, 8; 7 October 1947, 8; 10 October 1947, 8.

59 *The Press*, 11 October 1947, 10.

60 *The Press*, 14 October 1947, 5.

61 *The Press*, 18 October 1947, 2; 20 October 1947, 3.

62 *The Press*, 27 October 1947, 3.

63 *The Press*, 28 October 1947, 3.

64 *The Press*, 17 November 1947, 8.

65 *The Press*, 25 November 1947, 5.

66 *The Press*, 26 November 1947, 6.

67 *The Press*, 27 November 1947, 3.

68 *The Press*, 28 November 1947, 8; 3 December 1947, 8.

69 *The Press*, 17 December 1947, 3; 18 December 1947, 8; 22 January 1948, 6.

70 *The Press*, 22 December 1947, 3; 10 January 1948, 2.

71 *The Press*, 30 January 1948, 4; 31 January 1948, 6; 6 February 1948, 6.

72 Teahan, *Great Social Experiment*, 99–227.

73 *NZPD*, 30 November 1948, 4193.

74 John D. Prince, 'Look Back in Amber: The General Licensing Poll in New Zealand, 1919–87', *Political Science*, 48(1), 1996, 48–72, at 48–55; 1948: No.74: Licensing Act Amendment Act.

75 *The Press*, 4 September 1946, 8.

76 *Bay of Plenty Times*, 21 June 1947, 2; *The Press*, 16 January 1948, 8.

77 *Otago Daily Times*, 22 August 1947, 6; *The Press*, 25 August 1947, 8.

78 *The Press*, 3 December 1948, 8.

79 *Bay of Plenty Times*, 11 January 1949, 3.

80 *Otago Daily Times*, 22 June 1949, 7; 23 June 1949, 6; *The Press*, 10 June 1952, 11.

81 *Bay of Plenty Beacon*, 31 March 1950, 1.

82 *Otago Daily Times*, 10 March 1949, 6.

83 Tim Mulcare, 'The Political Economy of Six O'clock Closing', working paper, Victoria University of Wellington, 1999, 8–10; Christoffel, 'Removing Temptation', 145–48.

84 Christoffel, 'Removing Temptation', 150–53.

85 *RCL*, 76–78.

86 Ibid., 151–53; Jones and Paul, 'Concentration and Regulation', 89–90.

87 *The Press*, 11 October 1958, 16.

88 *The Press*, 11 October 1958, 16.

89 Jones and Paul, 'Concentration and Regulation', 90; *RCL*, 79–80.

90 New Zealand Breweries, *Hosts to the Nation*, 24, 30.

91 *The Press*, 2 July 1959, 13; *ABWJ*, 20 October 1956, 94.

92 *The Press*, 22 February 1956, 18.

93 *Brewnews*, December 1956, n.p.; Gordon, *Speight's*, 109–11; New Zealand Breweries, *Annual Report*, 1957, 16; Dominion Breweries, *Annual Report*, 1963, 7; *The Press*, 31 December 1958, 13.

94 *The Press*, 17 June 1954, 3; *ABWJ*, 20 July 1955, 56.

95 *The Press*, 24 October 1957, 22.

96 *The Press*, 21 September 1959, 10; 28 December 1959, 7.

97 *The Press*, 4 January 1960, 7.

98 Gourvish and Wilson, *British Brewing Industry*, 538.

99 New Zealand Breweries, *Hosts to the Nation*, 31; Gordon, *Speight's*, 113–14.

100 *Brewnews*, December 1956, n.p.

101 Michael Donaldson, *Beer Nation: The Art and Heart of Kiwi Beer*, Auckland: Penguin, 2012, 73.

102 *Brewnews*, September 1958, n.p.

103 M. J. Kennedy, 'The World's First Brewery Exclusively Designed to Use Continuous Fermentation: Biotechnology History Made in New Zealand', *Australasian Biotechnology*, 6(1), 1996, 13–18.

104 *Brewnews*, March 1960, n.p.

105 Dominion Breweries, *1930–1980*, 18; New Zealand Breweries, *Hosts to the Nation*, 30; *The Press*, 27 July 1957, 15.

106 *Nelson Mail*, 19 June 2007, 14; *Capital Times*, 30 May 2007.

107 *The Press*, 28 July 1960, 20; 3 January 1962, 10; 22 February 1966, 3; *Brewnews*, December 1969, n.p.; February 1972, n.p.; March 1973, n.p.

108 Department of Industries and Commerce to Meshcraft Sales Ltd, Johannesburg, 9 July 1957, Supervision of Breweries General file.

109 *Brewnews*, March 1957, n.p.

110 *ABWJ*, 20 April 1955, 52; Coe, 'Nelson Hop Industry', 19–35, 45, 58; *The Press*, 23 July 1948, 9.

111 Coe, 'Nelson Hop Industry', 30–31.

112 Coe, 'Nelson Hop Industry', 58–59; *ABWJ*, 20 December 1951, 42.

113 Bollinger, *Grog's Own Country*, 57–58.

114 New Zealand Breweries, *Hosts to the Nation*, 17.

115 *The Press*, 20 July 1955, 13; 3 June 1961, 17; 24 September 1968, 10.

116 A. R. D. Fairburn, *We New Zealanders*, Wellington: Progressive Publishing Society, 1944, 8, 25.

117 Rod Derrett, 'Rugby, Racing and Beer', His Master's Voice – 7EGM.6070, 1965.

118 *Evening Post*, 14 June 1952, 8.

119 Robin W. Winks, *These New Zealanders*, Christchurch: Whitcombe & Tombs, 1954, 92.

120 Gerald Durrell, *Two in the Bush*, London: Collins, 1962, 17.

121 David Ausubel, *The Fern and the Tiki: An American View of New Zealand National Character, Social Attitudes and Race Relations*, rev. ed., North Quincy, MA: The Christopher Publishing House, 1977 [1960], 51–52.

122 *New Zealand Listener*, 29 November 1957, 8.

123 *Dominion*, 16 November 1954, 8.

124 Perrin Rowland, *Dining Out: A History of the Restaurant in New Zealand*, Auckland: Auckland University Press, 2010, 2–4; Tanja Luckins, '"Satan Finds Some Mischief"? Drinkers' Responses to the Six O'clock Closing of Pubs in Australia, 1910s–1930s', *Journal of Australian Studies*, 32(3), 2008, 295–307.

125 *Bay of Plenty Times*, 27 April 1949, 6.

126 *New Zealand Herald*, 15 April 1936, 12; *Otago Daily Times*, 15 April 1939, 10.

127 *Otago Daily Times*, 31 March 1948, 6.

128 *The Press*, 2 September 1948, 2.

129 *White Ribbon*, 1 November 1949, 4.

130 *Salient*, 12 April 1956, 2.

131 *The Press*, 15 June 1957, 14.

NINE – Black Budget and False Dawn: Slow Liberalisation, 1958–1980

1 McAloon, *Judgements*, 103–6; *The Press*, 3 January 1958, 8.

2 *The Press*, 16 May 1958, 7.

3 McAloon, *Judgements*, 106–7.

4 Rob Vosslamber, 'Narrating History: New Zealand's "Black Budget" of 1958', *Accounting History*, 17(3–4), 2012, 481–99, at 482–84.

5 *New Zealand Herald*, 28 June 1958, 12.

6 *Brewnews*, September 1958, n.p.

7 Ibid.

8 F. G. Young, report to Federated Hotel Workers conference 1959, Protest re beer prices, MS 96-235-13/08, New Zealand Hotel Workers' Federation. Service Workers Union of Aotearoa: Records, Alexander Turnbull Library, Wellington; Neill Atkinson, 'Young, Frederick George', Dictionary of New Zealand Biography, Te Ara –

the Encyclopedia of New Zealand, https://teara.govt.nz/en/biographies/4y1/young-frederick-george (accessed 4 March 2022).

9 F. G. Young to Tom Skinner, 23 June 1959, Protest re beer prices.

10 F. G. Young to W. Nash, 22 December 1959, Protests re beer prices.

11 Ibid.

12 Atkinson, 'Young'.

13 *Evening Post*, 17 September 1958, 14; New Zealand Breweries, *Annual Report*, 1959.

14 *The Press*, 18 July 1959, 14.

15 Ibid.

16 *ABWJ*, 20 January 1959, 56; 21 September 1959, 62.

17 Department of Industries and Commerce Internal Memorandum, 1 December 1960, Protests re beer prices; *The Press*, 3 September 1959, 16; 7 January 1960, 8.

18 *The Press*, 26 March 1960, 25.

19 *The Press*, 28 July 1962, 12.

20 *The Press*, 4 May 1967, 3.

21 *The Press*, 5 May 1967, 3; 25 February 1968, 10.

22 *The Press*, 13 November 1972, 2.

23 *The Press*, 19 December 1975, 1; McAloon, *Judgements*, 155–63.

24 'Complaint 10/707 – DB Export Beer Advertisements', Scoop, www.scoop.co.nz/stories/BU1102/S00269/complaint-10707-db-export-beer-advertisements.htm (accessed 8 March 2022); *New Zealand Herald*, 11 February 2011.

25 *New Zealand Listener*, 13 December 1957, 36.

26 *The Press*, 22 July 1960, 12.

27 Notes at opening of Lucky Brewery, 23 July 1960, Lion Breweries files, Lion Technical Library, Auckland.

28 *Brewnews*, September 1960, n.p.

29 New Zealand Breweries, *Annual Report*, 1961, 11.

30 Gordon, *Speight's*, 126.

31 *Brewnews*, August 1962, n.p.; September 1963, n.p.; November 1963, n.p.; August 1966, n.p.

32 *Brewnews*, May 1973, n.p.; October 1973, n.p.

33 Diane Kirkby, *Barmaids: A History of Women's Work in Pubs*, Cambridge: Cambridge University Press, 1997, 179–82.

34 Report of the Licensing Control Commission for the year ending 31 March 1952, *AJHR*, 1952, H-3, 6–7.

35 *Dominion*, 16 November 1954, 8.

36 *Dominion*, 23 May 1955, 10.

37 New Zealand Alliance, MS 77-206-06-06.

38 Ibid.

39 *The Press*, 1 September 1959, 16; 5 February 1960, 10.

40 New Zealand Alliance, MS 77-206-06-06.

41 Ibid.

42 Rowland, *Dining Out*, 150–55.

43 *ABWJ*, 20 August 1959, 60; *The Press*, 17 January 1961, 17.

44 Christoffel, 'Removing Temptation', 112–13.

45 Ibid., 157.

46 Margaret McClure, *The Wonder Country: Making New Zealand Tourism*, Auckland: Auckland University Press, 2004, 203–35.

47 Geoffrey Troughton, 'Redeeming the Demon Drink: New Zealand Presbyterians and the Question of Alcohol', *The Social History of Alcohol and Drugs*, 27(2), 2013, 133–155, at 140–42.

48 *The Press*, 25 July 1966, 1; 8 November 1966, 3; 12 November 1966, 14; Steven Ferguson, 'Class at the Margins: The Canterbury Hotel Workers Union, 1908 to 1970', MA thesis, University of Canterbury, 1985, 213; Christoffel, 'Removing Temptation', 160–62.

49 Quoted in *Brewnews*, February 1967, n.p.

50 *The Press*, 26 May 1967, 1; 29 June 1967, 1; 20 July 1967, 1.

51 *Brewnews*, June 1967, n.p.

52 *Auckland Star*, 22 September 1967, 6; Christoffel, 'Removing Temptation', 163.

53 *Evening Post*, 25 September 1967, 14.

54 Christoffel, 'Removing Temptation', 165.

55 Ferguson, 'Class at the Margins', 214–19; *The Press*, 9 October 1967, 1.

56 *The Press*, 10 October 1967, 12.

57 *Brewnews*, October 1967, n.p.

58 R. H. Wheeler, 'The City and Ten O'clock Closing: A Letter to Those Abroad; A Sick Story', *Comment*, April 1968, 8.

59 Mulcare, 'Political Economy', 6; *The Press*, 1 August 1968, 15.

60 *New Zealand Listener*, 15 November 1971, 7.

61 *The Press*, 22 May 1967, 12.

62 *Dominion*, 9 October 1967, 2; 10 October 1967, 1.

63 *NZ Licensee*, November 1973, 39; *The Press*, 2 November 1973, 12.

64 Bollinger, *Grog's Own Country*, 68–69; *The Press*, 28 March 1966, 8.

65 Dominion Breweries, *1930–1980*, 22; *The Press*, 23 January 1969, 14; 26 February 1969, 20; 17 September 1969, 21; 12 February 1970, 18.

66 *The Press*, 5 February 1971, 1.

67 Dominion Breweries, *1930–1980*, 24; *The Press*, 5 June 1970, 1; 16 July 1971, 16.

68 Dominion Breweries, *1930–1980*, 27–30, 40, 47.

69 *New Zealand Licensee*, 20 June 1974, 3; New Zealand Breweries, *Hosts to the Nation*, 61.

70 Christoffel, 'Removing Temptation', 124, 166–67.

71 Mulcare, 'Political Economy', 11.

72 *The Sale of Liquor in New Zealand: Report of the Royal Commission of Inquiry*, Wellington: Government Printer, 1974, 27.

73 Ibid., 31–32; Christoffel, 'Removing Temptation', 127.

74 Christoffel, 'Removing Temptation', 125.

75 *The Press*, 26 March 1970, 11.

76 Nick Bollinger, *Goneville: A Memoir*, Wellington: Awa Press, 2016, 143.

77 *New Zealand Listener*, 19 January 1974, 14.

78 *New Zealand Listener*, 16 November 1974, 14.

79 Ibid.

80 *New Zealand Listener*, 2 February 1974, 4.

81 *The Press*, 6 December 1979, 3.

82 'Death on NZ roads since 1921', Ministry of Transport, https://www.transport.govt.nz/statistics-and-insights/safety-road-deaths/sheet/death-on-nz-roads-since-1921 (accessed 22 March 2022).

83 Nancy Swarbrick, 'Road Accidents – Promoting Road Safety', Te Ara – the Encyclopedia of New Zealand, http://www.TeAra.govt.nz/en/road-accidents/page-3 (accessed 24 March 2022); *The Press*, 22 July 1971, 13.

84 *The Press*, 12 March 1969, 12.

85 *Otago Daily Times*, 4 July 1969, 4; Spectrum 067, 'The Day the Pub Burnt Down', 1973, Ngā Taonga Sound & Vision, https://www.ngataonga.org.nz/collections/catalogue/catalogue-item?record_id=204972 (accessed 22 March 2022).

86 Swarbrick, 'Road Accidents'.

87 'The Day the Pub Burnt Down'.

88 Bollinger, *Goneville*, 147–55, 185, 242.

89 Ibid., 149–50.

90 *Brewnews*, March 1974, n.p.

91 Bollinger, *Goneville*, 151–55.

92 Ibid., 77–78, 142.

93 'The Day the Pub Burnt Down'.

94 *The Sale of Liquor in New Zealand*, 28; Christoffel, 'Removing Temptation', 128; 1976: No.53, Sale of Liquor Amendment Act.

95 *The Press*, 8 December 1961, 16.

96 *The Press*, 24 June 1965, 1.

97 *The Press*, 9 July 1965, 1.

98 Beer prices, MS 1403-210/5, Sir John Ross Marshall, Political Papers, Alexander Turnbull Library, Wellington.

99 *The Press*, 12 March 1966, 3; 26 July 1966, 3.

100 *The Press*, 22 December 1969, 1.

101 *The Press*, 27 January 1970, 1; 10 February 1970, 26; 18 February 1970, 10; 8 April 1970, 9; 1 May 1970, 14; 2 May 1970, 12.

102 *The Press*, 4 February 1970, 1.

103 *The Press*, 29 May 1970, 1; *Evening Post*, 18 June 1970, 8; M. J. Moriarty, Secretary of Industries and Commerce, Report to Minister of Industries and Commerce, 11 May 1970, Beer Price Files.

104 *The Press*, 26 June 1970, 24.

105 *The Press*, 5 August 1970, 28; 8 August 1970, 1.

106 *The Press*, 11 August 1970, 16.

107 *Dominion*, 2 July 1975, 1.

108 *Evening Post*, 9 July 1975, 8; 31 July 1975, 1.

109 *The Press*, 3 July 1968, 1; 9 July 1968, 1; 10 July 1968, 1; 17 July 1968, 1.

110 Easton, *Not in Narrow Seas*, 360–70; Notes and clippings – Confectionery and brewery workers; *The Press*, 12 August 1971, 2; 14 August 1971, 1.

111 Notes and clippings – Confectionery and brewery workers.

112 *The Press*, 6 June 1960, 16.

113 *The Press*, 25 July 1960, 12; New Zealand Breweries, *Annual Report*, 1959, 15.

114 *The Press*, 17 August 1962, 12.

115 *The Press*, 11 December 1962, 16.

116 *The Press*, 9 October 1963, 11.

117 Ibid.

118 *The Press*, 23 October 1963, 14.

119 *The Press*, 19 September 1964, 14; *Brewnews*, October 1964, n.p.

120 Rod Smith, *Guinness Down Under: The Famous Brew and the Family Comes to Australia and New Zealand*, Tauranga: Eyeglass Press, 2018, 297–310.

121 Ibid., 316–27; *The Press*, 30 November 1966, 3; 7 August 1967, 13; 11 August 1967, 5.

122 *The Press*, 11 August 1967, 5; *Dominion*, 24 July 1969, 4.

123 *Waikato Times*, 14 November 1934, 10.

124 *ABWJ*, 20 August 1959, 60; *The Press*, 8 July 1959, 16.

125 *The Press*, 10 September 1959, 14.

126 *The Press*, 11 February 1960, 12.

127 *The Press*, 8 March 1960, 17.

128 Gordon, *Speight's*, 62, 133.

129 *The Star*, 21 July 1973, *Otago Daily Times*, 24 July 1973, Hop Beer File, Lion Technical Library, Auckland.

130 *The Press*, 10 August 1973, 1.

131 Brewing of Hop Beer in Breweries.

132 *The Press*, 11 August 1973, *The Star*, 15 August 1973, Hop Beer File.

133 *New Zealand Licensee*, 20 July 1967, 27; *The Press*, 31 July 1969, 10.

134 D. B. S. Maxwell, 'Beer Cans: A Guide for the Archaeologist', *Historical Archaeology*, 27(1), 1993, 95–113; *Waikato Independent*, 23 May 1936, 4; *Timaru Herald*, 7 November 1936, 21; *Dominion*, 18 March 1936, 6.

135 S. Jones, 'Brewing', in Michael Pickford and Alan Bollard, eds, *The Structure and Dynamics of New Zealand Industries* (pp. 121–48), Palmerston North: Dunmore Press, 1998, 131; New Zealand Breweries, *Annual Report*, 1963; Dominion Breweries, *Annual Report*, 1964.

136 *The Brewing Industry in New Zealand*, Wellington: R. A. Jarden & Co., 1977, 2, 5; *New Zealand Licensee*, 20 August 1973, 49.

137 *Brewnews*, December 1975, n.p.; May 1976, n.p.; March 1977, n.p.

138 *The Press*, 20 May 1970, 18; 10 May 1972, 10; 27 November 1972, 17.

139 'About Us', Super Liquor, https://www.superliquor.co.nz/about-us (accessed 17 December 2022); 'About Liquorland', Liquorland, https://www.liquorland.co.nz/about (accessed 4 November 2022).

140 1942: 6 Geo. VI No.11: Medical Advertisements Act; Dominion Breweries, *1930–1980*, 61.

141 *Gisborne Herald*, 14 October 1949, 3.

142 'Hori', *The Half Gallon Jar*, Auckland: Sporting Publications, 1962.

143 Hutt, *Maori and Alcohol*, 74–81; *The Press*, 6 July 1965, 1; 14 July 1965, 3.

144 'National Party 1975: The Cities', Ngā Taonga Sound & Vision, https://www.ngataonga.org.nz/collections/catalogue/catalogue-item?record_id=147251 (accessed 4 November 2022).
145 *New Zealand Licensee*, October 1975, 21.
146 *The Press*, 19 February 1963, 19; 10 April 1964, 14.
147 *The Press*, 24 October 1964, 13; 31 May 1968, 13.
148 *The Press*, 24 July 1971, 15.
149 'Our Brands: Beer', Lion, https://lionco.com/our-brands/new-zealand/beer/ (accessed 5 April 2022).
150 Smith, *Guinness*, 320–24.
151 Jock Phillips, 'Men's Clubs – Workingmen's Clubs', Te Ara – the Encyclopedia of New Zealand, http://www.teara.govt.nz/en/mens-clubs/page-3 (accessed 7 April 2022).
152 *Salient*, 12 June 1975, 4.
153 *Otago Daily Times*, 27 April 1978, 1.
154 *Brewnews*, June 1963, n.p.; August 1968, n.p.; May 1974, n.p.
155 Quoted in Doug Munro, 'The New Zealand Opera Company: Notes Towards a History', *Turnbull Library Record*, 48(1), 2016, 36–49, at 38.
156 *The Press*, 23 September 1959, 6.
157 *Brewnews*, October 1961, n.p.
158 Warren Feeney, 'Rugby, Racing and Beer and Art', in Alsop and Stewart, eds, *Promoting Prosperity* (pp. 48–55), at 52.
159 Ibid., 48–52.
160 *Brew It Yourself: Want to Make a Better, Stronger, Beer Than You Can Buy – At One-sixth the Cost?*, Dunedin: Otago Daily Times, *c.* 1960.
161 Malcolm Gramaphone, *The Brewers' Bible*, Wellington: Alister Taylor, 1972, 10.

TEN – Challenging the Big Two: New Beer and New Rules, 1980–1990s

1 *NZPD*, 6 June 1979, 481.
2 *New Zealand Listener*, 10 March 1984, 23.
3 Jeff Hayward, 'What's Brewing? Boutique Beer', *North and South*, January 1988, 104.
4 John McCrystal and Simon Farrell-Green, *The McCashin's Story: The First Craft Beer and the Kiwi Brewing Revolution It Sparked*, Auckland: Random House, 2013, 12.
5 *The Press*, 26 January 1979, 1; *NZOYB*, 1996, 163.
6 Easton, *Not in Narrow Seas*, 372–82, 445–55, 477–86, 527–36.
7 H. K. Orchard, 'New Zealand Alcohol Consumption Statistics, 1955–1989', Wellington: Alcoholic Liquor Advisory Council, 1990, 3–6.
8 Bronwyn Dalley, 'Wine', Te Ara – the Encyclopedia of New Zealand, http://www.TeAra.govt.nz/en/wine (accessed 2 May 2022); 'Alcohol Consumption Statistics', 1.
9 *Brewnews*, July 1979, n.p.; July 1982, n.p.
10 Jones, 'Brewing', 125, 131.
11 Donaldson, *Beer Nation*, 76–80; Bassett and Goldsmith, *The Myers*, 248–52, 254–59.
12 Quoted in *Stuff*, 15 April 2017; Bassett and Goldsmith, *The Myers*, 275–82.
13 Quoted in *Stuff*, 15 April 2017.
14 Bryan Gaynor, 'Looking Back: The Hugely Controversial Lion–LD Nathan Merger', *BusinessDesk*, 11 April 2021, https://businessdesk.co.nz/article/the-life/looking-back-the-hugely-controversial-lion-ld-nathan-merger (accessed 2 May 2022); Donaldson, *Beer Nation*, 81; Janice C. Mogford, 'Nathan, David 1816–1886', Dictionary of New Zealand Biography, Te Ara – the Encyclopedia of New Zealand, https://teara.govt.nz/en/biographies/1n1/nathan-david (accessed 3 May 2022).
15 'Our History', Lion Corporate, https://lionco.com/who-we-are/our-history/ (accessed 2 May 2022).
16 The University of Auckland Business History Project, http://www.businesshistory.auckland.ac.nz/db_breweries/key_products_and_services.html (accessed 2 May 2022); Donaldson, *Beer Nation*, 83–85.
17 Donaldson, *Beer Nation*, 135–36.
18 Tom Hyde, 'The Beer War', *Auckland Metro*, March 1987, 78.
19 *National Business Review*, 22 March 1996, 48.
20 Hyde, 'Beer War', 78.
21 *National Business Review*, 28 October 1985, 25.
22 *National Business Review*, 20 April 1988, 8.
23 *National Business Review*, 3 April 1992, 1–2.
24 *New Zealand Herald*, 30 June 2000.
25 *DB Breweries Limited v. Lion Nathan Limited*, Auckland High Court, 9 November 2007, Harrison J, https://andrewbrown.co.nz/recent-case-law/db-breweries-limited-v-lion-nathan-limited-9-november-2007/ (accessed 2 May 2022).
26 Hyde, 'Beer War', 78–79.
27 *National Business Review*, 6 June 1989, 17; 24 October 1990, 4.
28 Chris Mirams, *Beleaguered! The Warriors from Dream to Nightmare*, Auckland: Hodder Moa Beckett, 2001, 38–39.
29 *National Business Review*, 10 July 1992, 5.
30 *New Zealand Herald*, 20 December 2000.
31 *Dominion Post*, 27 September 2004, 10.
32 1976: No.143, Alcohol Advisory Council Act.
33 Sally Casswell, *Drinking by New Zealanders*, Wellington: Alcoholic Liquor Advisory Council, 1980; Allan Wyllie and Sally Casswell, *Drinking in New Zealand: A Survey*, Auckland: Alcohol Research Unit, University of Auckland, 1989; Allan Wyllie, *Drinking in New Zealand: A National Survey*, Auckland: Alcohol and Public Health Research Unit, University of Auckland, 1996.
34 'History', Alcohol Healthwatch, https://www.ahw.org.nz/About-Us/History (accessed 4 May 2022); Donaldson, *Beer Nation*, 84.
35 *Food Technology in New Zealand*, June 1988, 32–33; March 1993, 20–21.
36 *Food Technology in New Zealand*, March 1993, 20–21; *Hospitality*, November 1993, 36.
37 *National Business Review*, 2 July 1993, 7.

38 *Dominion*, 2 February 1995, 3.

39 *Brewnews*, May 1982, n.p.

40 *Brewnews*, March 1980, n.p.; Lindsey Dawson, 'The Imported Beer Boom', *Auckland Metro*, January 1982, 96–98.

41 Eion Scott et al., 'Can Theme Bars Rescue the Pub Trade?', *Marketing Magazine*, March 1994, 22–29.

42 Smith, *Guinness*, 328–31.

43 Ale – Beer – General file re beer 1954–1961: IC 1 13/1/- 1; 1961–1964: IC 1 13/1/- 2; 1965–1966: IC 1 13/1/- 3, Archives New Zealand, Wellington; *ABWJ*, 20 December 1962, 54.

44 *Brewnews*, June 1963, n.p.

45 *Brewnews*, June 1966, n.p.; August 1966, n.p.; August 1967, n.p.

46 *The Press*, 15 August 1967, 26; 23 March 1971, 17; 6 July 1971, 12.

47 *The Press*, 17 October 1979, 6.

48 Hyde, 'Beer War', 78; Warren Berryman, 'DB Thirst for Lion's Share', *Marketing*, June 1992, 18–19.

49 Donaldson, *Beer Nation*, 108, 110; *The Independent*, 28 April 1995, 31; *Evening Post*, 10 October 1995, 4.

50 Christoffel, 'Removing Temptation', 128–31, 195–98; *The Sale of Liquor in New Zealand*.

51 Prince, 'Look Back in Amber', 55–60, 66.

52 *Sale of Liquor in New Zealand*, 42.

53 Ibid., 43; *National Business Review*, 11 February 1994, 2.

54 *National Business Review*, 11 February 1994, 2.

55 *Dominion*, 5 October 1995, 1.

56 *National Business Review*, 27 October 1995, 15; 3 November 1995, 15.

57 *Evening Post*, 24 January 1995, 13.

58 'The History of Alcohol Advertising on Radio and Television', https://www.hpa.org.nz/sites/default/files/imported/field_research_publication_file/HistoryofalcoholAdverts.pdf (accessed 15 April 2022).

59 *Evening Post*, 6 May 1997, 9; Christoffel, 'Removing Temptation', 199–201; Sale of Liquor Amendment Act 1999 (1999 No.92).

60 McCrystal and Farrell-Green, *McCashin's Story*, 54–55.

61 Cornell, *Beer*, 217–20; Campaign for Real Ale, https://camra.org.uk/50-years/ (accessed 6 May 2022).

62 McCrystal and Farrell-Green, *McCashin's Story*, 75–77.

63 Ibid., 82–84, 108.

64 Ibid., 100.

65 *Nelson Mail*, 30 May 2008, in Nelson Provincial Museum: Wine-making and brewing subject file.

66 McCrystal and Farrell-Green, *McCashin's Story*, 95–96, 106, 122–24, 131, 142; Chris Hutching, 'The Mouse That Roared at Lion', *National Business Review*, 20 December 1991, 24.

67 Luke Nicholas and Tony Rutherford, 'Craft Brewing Takes Flight in N.Z. (part III)', https://realbeer.co.nz/category/nz-brewing-history/ (accessed 10 May 2022).

68 Ibid.

69 Michael Donaldson, *The Hopfather: Richard Emerson*, Auckland: Penguin, 2019, 110.

70 Ibid.

71 *The Press*, 28 April 1993, 19; Hayward, 'What's Brewing?', 104.

72 Donaldson, *The Hopfather*, 87–122.

73 John Shaw and Chris Hutching, 'It's Small Beer but Potent', *New Zealand Business*, September 1989, 50.

74 Ibid., 54.

75 *National Business Review*, 17 June 1994, 63; Jones, 'Brewing', 129; Peter Allison, 'Worts 'n' All: The Rise of Home Brewing', *Metro*, April 1990, 100–110.

76 Shaw and Hutchings, 'It's Small Beer', 52.

77 Ibid.

78 Ibid.; *National Business Review*, 7 January 1988, 5; 14 October 1988, 5.

79 *National Business Review*, 20 October 1995, 27.

80 Ibid.

81 Quoted in Jo Mackay, 'Microbreweries Take Off', *NZ Business*, December 1996/January 1997, 46–50.

82 Donaldson, *Beer Nation*, 73, 81; Gifford Lee, 'Celebrating Chuck Hahn's 50 Years in Brewing', *Beer & Brewer*, 17 June 2021, https://www.beerandbrewer.com/celebrating-chuck-hahns-50-years-in-brewing/ (accessed 12 May 2022).

83 *The Independent*, 25 March 1994, 24.

84 *National Business Review*, 23 September 1994, 5.

85 M. Casey, 'Michael and Goliath: Independent', *New Zealand Business*, June 1992, 18–23; 'The Early Years', Asahi Beverages New Zealand, https://www.asahibeverages.co.nz/the-history-of-asahi-beverages-nz (accessed 20 April 2022).

86 Quoted in *Sunday Star Times*, 3 March 1996, 14.

87 Hyde, 'Beer War', 77.

88 Donaldson, *Beer Nation*, 113; Steven J. Jackson, Sarah Gee and Jay Scherer, 'Producing and Consuming Masculinity: New Zealand's (Speight's) "Southern Man"', in Lawrence A. Wenner and Steven J. Jackson, eds, *Sport, Beer, and Gender: Promotional Culture and Contemporary Social Life* (pp. 181–202), New York: Peter Lang, 2009, 184.

89 Jackson et al., 'Producing and Consuming Masculinity', 189–92.

90 Greg Ryan and Geoff Watson, *Sport and the New Zealanders: A History*, Auckland: Auckland University Press, 2018, 264–66.

91 Melissa Kennedy, 'Marketing Our Heritage: Dynamics of Commercial Enterprise and Nostalgia by DB Breweries Ltd in the DB Draught Clydesdales', *Deep South*, 3(3), 1997, http://www.otago.ac.nz/DeepSouth/vol3no3/db.html (accessed 5 March 2022).

92 *National Business Review*, 27 March 1991, 11.

93 Donaldson, *Beer Nation*, 115–17, 128.

94 *Bay of Plenty Times*, 28 May 2010, A6.

95 *New Zealand Herald*, 4 June 2010, A4.

96 *Stuff*, 8 June 2014; 'Beer Plumber', https://www.youtube.com/watch?v=HG_wfMK7dko (accessed 16 May 2022).

97 Press release: Tui Brewery, 23 March 2012.

98 Adele Gautier, 'Battle Brewing', *Marketing*, November 1996, 30.

99 *Stuff*, 6 June 2018.

100 *Grey River Argus*, 2 July 1928, 8; *The Press*, 20 October 1936, 11.

101 *The Press*, 23 March 2001, 1; Sara Walton, Shayne Grice and Bevan Catley, 'The Monteith's Affair: Bitter to the Loyal End?', *Journal of the Australian and New Zealand Academy of Management*, 9(2), 2003, 69–74, at 70.

102 *The Press*, 23 March 2001, 1.

103 *Waikato Times*, 30 March 2001, 6.

104 *National Business Review*, 4 May 2001, 18.

105 *Dominion*, 7 March 2002.

106 *Evening Post*, 28 March 2001, 1; Walton et al., 'Monteith's Affair', 69–71.

107 Walton et al., 'Monteith's Affair', 71–73.

108 Quoted in *The Press*, 28 March 2001, 1.

109 *The Press*, 28 March 2001, 4.

110 Walton et al., 'Monteith's Affair', 73.

111 *Dominion*, 26 April 2001, 3.

112 *Dominion*, 7 March 2002.

113 *Evening Post*, 1 May 2001, 11; *National Business Review*, 27 April 2001, 10.

114 *Evening Post*, 19 November 2001, 4; *New Zealand Herald*, 10 August 2002.

115 'Commerce Commission Settlement for DB Breweries', *Scoop*, 12 October 2001.

116 Linda Hill, 'What it Means to be a Lion Red Man: Alcohol Advertising and Kiwi Masculinity', *Women's Studies Journal*, 15(1), 1999, 65–85, at 65–67.

117 Ibid., 70.

118 *National Business Review*, 24 October 1997, 48; *The Press*, 4 December 1997, 37.

119 *Dominion Post*, 5 February 2003, 1.

120 *The Press*, 30 April 2011, C6–7; 'Ranfurly Man Park – Frontier Lager Beer Ad', https://www.youtube.com/watch?v=_Osp9dzxbk8 (accessed 24 May 2022).

121 'Campaign Launched Against Tui Beer Ads', Scoop, https://www.scoop.co.nz/stories/PO1202/S00209/campaign-launched-against-tui-beer-ads.htm (accessed 13 May 2022); *Otago Daily Times*, 21 February 2012, 5.

122 *The Press*, 30 April 2011, C6–7.

123 Quoted in Kieran Cody and Steve Jackson, 'The Contested Terrain of Alcohol Sponsorship of Sport in New Zealand', *International Review for the Sociology of Sport*, 51(4), 2014, 375–93, at 384.

124 *The Independent*, 22 April 1994, 27; *Sunday Star Times*, 14 February 1999, 2; *The Press*, 25 February 2000, 20; 2 January 2006, 4.

125 *Evening Post*, 23 September 1995, 10.

126 Jones, 'Brewing', 124–27.

127 *Waikato Times*, 1 July 1999, 9.

128 *Sunday Star Times*, 3 March 1996, 14.

ELEVEN – The Pursuit of Hoppiness: Craft Beer and Beyond, 2000–2020

1 'Beer Consumption per Person by Country', https://www.kirinholdings.com/en/investors/library/databook/beer_country/ (accessed 17 November 2022).

2 *Waikato Times*, 6 November 2001.

3 Kathleen M. Kuehn and Sophie Parker, 'One of the Blokes: Brewsters, Branding and Gender (In)Visibility in New Zealand's Craft Beer Industry', *Journal of Consumer Culture*, 21(3), 2021, 519–38, at 520–22.

4 *Marlborough Express*, 24 December 2009, 18.

5 *Hawke's Bay Today*, 24 March 2018, A12.

6 'New Zealand Craft Beer Industry Insights 2015', ANZ, https://comms.anz.co.nz/businsights/article/report.html?industry=Craft%20Beer (accessed 20 June 2022).

7 *New Zealand Herald*, 7 November 2007, 5.

8 *The Spinoff*, 13 August 2021.

9 *Nelson Mail*, 12 June 2003, 3.

10 Scotts Brewing Co., https://scottsbrewing.co.nz (accessed 24 July 2022); Craftwork Brewery, https://www.craftworkbrewery.co.nz (accessed 24 July 2022).

11 'Fresh Hops – The Most Anticipated Release on the Beer Calendar', https://www.finewinedelivery.co.nz/blog-post/fresh-hops-the-most-anticipated-release-on-the-beer-calendar/570 (accessed 20 June 2022).

12 *Daily Post*, 1 November 2014, B7; 'The Best, Weirdest and Most Controversial Kiwi Craft Beers to Try in 2016', Concrete Playground, https://concreteplayground.com/auckland/food-drink/drink/the-best-weirdest-and-most-controversial-kiwi-craft-beers-to-try-in-2016 (accessed 20 June 2022); 'Brick Lane E.1.', Untappd, https://untappd.com/b/valkyrie-brewing-co-brick-lane-e-1/738565 (accessed 20 June 2022).

13 *New Zealand Herald*, 29 May 2018; 7 November 2020; Behemoth Brewing Company, https://shop.behemothbrewing.co.nz/products/pastor-of-muppets-2-mt-eden-prison-hooch-24x330ml (accessed 24 July 2022).

14 'Turning Bread into Beer and Beer into Bread – Using Kiwi Innovation to Reduce Food Waste', Love Food Hate Waste, https://lovefoodhatewaste.co.nz/turning-bread-into-beer-and-beer-into-bread-using-kiwi-innovation-to-reduce-food-waste/ (accessed 20 June 2022); 'NZ Beer Awards

Sustainability Award Winner', New World, https://www.newworld.co.nz/who-we-are/news/nz-beer-awards-sustainability-award-winner (accessed 20 June 2022).

15 'New Zealand Boasts More Breweries Per Capita than the US, UK and Australia', Beveragedaily.com, https://www.beveragedaily.com/Article/2019/03/19/New-Zealand-boasts-more-breweries-per-capita-than-the-US-UK-and-Australia (accessed 20 May 2022).

16 *New Zealand Herald*, 24 November 2012, A24.

17 'Alcohol Available for Consumption Year Ended December 2020', Stats NZ, https://www.stats.govt.nz/information-releases/alcohol-available-for-consumption-year-ended-december-2020.

18 *New Zealand Herald*, 2 December 1999; Mackay, 'Microbreweries Take Off', 46–50; Donaldson, *Beer Nation*, 159–60.

19 *Taranaki Daily News*, 5 July 2013, 6.

20 *FMCG*, August 2004, 39.

21 Martin Craig, 'The Battle for Beer', *New Zealand Retail*, December 1998, 14; Donaldson, *Hopfather*, 221.

22 *New Zealand Herald*, 2 December 1999.

23 'Inaugural New World Beer & Cider Awards Results Announced', Real Beer, https://realbeer.co.nz/2014/11/10/new-world-beer-cider-awards-2014/ (accessed 2 June 2022).

24 Cornell, *Beer*, 233–34; 'UK brewery numbers increase 7.5%, surpassing 3,000', *Brewers Journal*, https://www.brewersjournal.info/uk-brewery-numbers-increase-7-5-surpassing-3000/ (accessed 2 June 2022).

25 'U.S. Craft Brewery Count Reaches 2000', Brewers Association, https://www.brewersassociation.org/brewing-industry-updates/u-s-craft-brewery-count-reaches-2000/ (accessed 2 June 2022); Maureen Ogle, *Ambitious Brew: The Story of American Beer*, New York: Harvest Books, 2007, 314–26.

26 *Marketing Magazine*, February 2000, 25–28.

27 *Evening Post*, 4 April 2000, 13; Peter Darby, 'The History of Hop Breeding and Development', *Brewery History*, 121, 2005, 94–112, at 111.

28 *RuralNews*, 9 May 2006.

29 *Waikato Times*, 12 July 2008, 10.

30 *Waikato Times*, 10 October 2012, 1.

31 *New Zealand Herald*, 5 April 2008, A2; *Manawatu Standard*, 2 September 2009, 3.

32 *Dominion Post*, 4 September 2004, 9; *Nelson Mail*, 19 November 2010, 17.

33 *Stuff*, 15 April 2015; *Dominion Post*, 21 August 2014.

34 'GABS Label Design Awards', https://www.gabsfestival.com/label-design-awards-nz-2022 (accessed 8 January 2023); 'The 2007 BrewNZ Results THE BAR HAS BEEN RAISED!', https://www.brewersguild.org.nz/assets/Awards/2007-Catalogue-of-Results.pdf (accessed 20 June 2022).

35 *Waikato Times*, 20 October 2008, 11.

36 *Evening Post*, 6 December 2001, 2.

37 *Nelson Mail*, 10 December 2010, 4.

38 *The Spinoff*, 25 February 2021, 5 March 2021.

39 *Nelson Mail*, 29 March 2008, 13; 27 May 2011, 13; 'Wellington, the Craft Beer Capital', https://www.craftbeercapital.com (accessed 20 June 2022).

40 *BusinessDesk*, 9 August 2017.

41 *National Business Review*, 13 September 2019, 4; Epic Beer, Twitter, 12 December 2019, https://mobile.twitter.com/epicbeer/status/1205064580243841024 (accessed 24 June 2022).

42 *Nelson Mail*, 23 February 2001, 3; *Evening Post*, 26 January 2002, 3; *The Press*, 17 August 2004, 1; *BusinessDesk*, 13 August 2015.

43 *Nelson Mail*, 30 December 2006, 15; *Stuff*, 7 August 2019; *BusinessDesk*, 26 August 2020.

44 *Evening Post*, 7 February 1997, 2; Donaldson, *Hopfather*, 133–35.

45 *Dominion*, 4 July 1998, 3; *The Press*, 6 July 1998, 4.

46 *Nelson Mail*, 12 March 2002, 8; *The Press*, 5 June 2003, 3; *Dominion Post*, 16 September 2006, 9.

47 For example, *The Press*, 20 January 2003, 8; *Nelson Mail*, 9 December 2004, 15.

48 *The Press*, 18 February 1971, 5.

49 *The Press*, 7 September 1971, 2.

50 *Brewnews*, August 1971, n.p.; December 1971, n.p.

51 *Waikato Times*, 4 February 2003, 24.

52 *Marlborough Express*, 5 February 2007.

53 Donaldson, *Beer Nation*, 176, 178.

54 *Sunday Star Times*, 22 April 1997, 10. See also *New Zealand Retail*, December 1998, 10–17.

55 *The Independent*, 26 August 1994, 28.

56 *National Business Review*, 11 April 1997, 51.

57 *Sunday Star Times*, 12 March 2000, 3; Donaldson, *Beer Nation*, 181.

58 *Waikato Times*, 21 April 2007, 6.

59 *Dominion*, 25 May 2002, 6; *Marlborough Express*, 9 September 2010, 16.

60 *Stuff*, 18 April 2021; 4 August 2021.

61 Pink Boots Society NZ, https://www.pinkbootsnz.co.nz (accessed 24 June 2022).

62 Kuehn and Parker, 'One of the Blokes', 527–30.

63 *Manawatu Standard*, 7 July 2011, 7.

64 Alice Neville, 'Smash the Beer Patriarchy! How Women are Putting Brews Before Bros', *The Spinoff*, 4 July 2018.

65 *Scoop News*, 13 January 2012.

66 *The Spinoff*, 11 July 2021.

67 'NZ Beer Industry Condemns Racism Following Brewer's Comments', BrewsNews, https://www.brewsnews.com.au/2021/05/17/nz-beer-industry-condemns-racism/ (accessed 26 June 2022).

68 *The Spinoff*, 13 August 2021.

69 Brewers Association of New Zealand, https://www.brewers.org.nz (accessed 26 June 2022).

70 Brewers Guild of New Zealand, https://www.brewersguild.org.nz (accessed 26 June 2022).

71 Society of Beer Advocates, https://www.soba.org.nz/content.aspx?page_id=0&club_id=992196 (accessed 26 June 2022).
72 Pursuit of Hoppiness, https://hoppiness.co.nz/product/hoppiness-home-delivered (accessed 18 November 2022).
73 Quoted in *Sunday Star Times*, 26 April 2009, 3.
74 *New Zealand Herald*, 24 October 2002; *Stuff*, 18 December 2021.
75 *New Zealand Herald*, 8 November 2008, C15; 28 December 2010.
76 *Brewnews*, March 1975, n.p.; October 1979, n.p.; March 1972, n.p.
77 *The Press*, 26 March 2002, 21; *Sunday Star Times*, 26 April 2009, 3; 29 August 2010.
78 *Southland Times*, 18 December 2002, 12; *Waikato Times*, 12 September 1994, 6; *Scoop News*, 3 December 2008.
79 *National Business Review*, 22 January 1999; McCrystal and Farrell-Green, *McCashin's Story*, 179–83.
80 *Nelson Mail*, 23 November 2002, 3.
81 McCrystal and Farrell-Green, *McCashin's Story*, 193–200.
82 Donaldson, *Hopfather*, 198.
83 Press release: Society of Beer Advocates, 7 November 2012.
84 *Nelson Mail*, 25 January 2013, 13.
85 *BusinessDesk*, 2 February 2017; 17 February 2017.
86 *Manawatu Standard*, 13 February 2014, 11.
87 *National Business Review*, 20 March 2009; *The Independent*, 20 May 2010, 8.
88 *Manawatu Standard*, 13 February 2014, 11.
89 *The Press*, 15 July 2011, 10.
90 *The Press*, 25 May 2009, 7.
91 *National Business Review*, 20 March 2009; *The Independent*, 20 May 2010, 8; *The Press*, 15 July 2011, 10.
92 *Manawatu Standard*, 13 February 2014, 11.
93 *New Zealand Herald*, 4 March 2013; 'Lion Pushes the Art of "Popular Craft" With Crafty Beggars Range', https://stoppress.co.nz/news/lions-art-crafty-beggars/ (accessed 30 June 2022).
94 *Marlborough Express*, 25 August 2011.
95 *Marlborough Express*, 20 June 2013, 14.
96 *New Zealand Herald*, 3 January 2012, B1.
97 Press release: Society of Beer Advocates, 7 November 2012; *New Zealand Herald*, 4 March 2013.
98 *Stuff*, 15 February 2017.
99 *Dominion Post*, 29 March 2019, 6.
100 New Zealand Law Commission, *Alcohol in Our Lives: Curbing the Harm*, Wellington: The Commission, 2010, 5.
101 Linda Hill and Liz Stewart, 'The Sale of Liquor Act, 1989: Reviewing Regulatory Practices', *Social Policy Journal of New Zealand*, 7, 1996.
102 Radio New Zealand Newswire, 27 June 2002; *Sunday Star Times*, 29 August 2004, 12.
103 *NZ Truth*, 17 January 2003, 2; *Sunday Star Times*, 11 May 2003, 3.
104 Radio New Zealand Newswire, 12 May 2008; *Sunday Star Times*, 31 October 2010, 6.
105 *New Zealand Herald*, 6 January 2010, A5.
106 *Nelson Mail*, 4 October 2008, 15.
107 *Sunday Star Times*, 10 May 2009, 1.
108 *Hawke's Bay Today*, 12 March 2019, A5.
109 *Waikato Times*, 5 November 2002, 24.
110 *New Zealand Herald*, 26 May 2006.
111 Business and Economic Research Limited (BERL), *Costs of Harmful Alcohol and Other Drug Use*, Wellington: BERL, 2009.
112 Law Commission, *Alcohol in Our Lives*.
113 Press release: The Free House, 10 June 2009.
114 Quoted in *Nelson Mail*, 1 May 2010.
115 Eric Crampton and Matt Burgess, 'The Price of Everything, The Value of Nothing: A (Truly) External Review of BERL's Study of Harmful Alcohol and Drug Use', Working Paper No.10/2009, Christchurch: Department of Economics and Finance, University of Canterbury, 2009, 5–7, 37.
116 Ibid., 5–6, 26–27.
117 *The Press*, 6 May 2009; *Stuff*, 23 August 2019.
118 'Where Are We at With Alcohol Reform?', NZ Drug Foundation, https://www.drugfoundation.org.nz/matters-of-substance/archive/may-2016/alcohol-law-reform/ (accessed 30 June 2022).

EPILOGUE

1 *Stuff*, 18 December 2019.
2 Ibid.
3 *Westport Times*, 26 October 1875, 3.
4 *Dominion*, 9 January 1932, 4.
5 *Thames Star*, 14 December 1887, 3.
6 *The Press*, 23 January 1979, 17; 2 February 1979, 12.
7 'Spirits Up, Beer and Wine Down', StatsNZ, 24 February 2022, https://www.stats.govt.nz/news/spirits-up-beer-and-wine-down/ (accessed 2 May 2023).
8 *BusinessDesk*, 10 October 2018; Tariq Ali et al., 'Global: Beer Crisis: How We Discovered Climate Change Could Cause a Global Barley Shortage', The Conversation, https://theconversation.com/beer-crisis-how-we-discovered-climate-change-could-cause-a-global-barley-shortage-104962 (accessed 18 November 2022); 'How Climate Change is Shaking Up the Hops that Give Beer Its Flavour', *New Scientist*, https://www.newscientist.com/article/mg25233654-100-how-climate-change-is-shaking-up-the-hops-that-give-beer-its-flavour/ (accessed 18 November 2022).

Bibliography

ARCHIVAL SOURCES

'A Pioneer of Beer: The Story of Thomas Hancock', typed notes, by Mrs H. Carr Rollett, sent to the Brewers' Association of New Zealand, 20 June 1956, Lion Technical Library, Auckland.

Ale – Beer – General file re beer 1954–1961: IC 1 13/1/- 1; 1961–1964: IC 1 13/1/- 2; 1965–1966: IC 1 13/1/- 3, Archives New Zealand, Wellington.

Beer Duty Act – Home Brewing 1931–1932: C: 1: 31/7/40, Archives New Zealand, Wellington.

Beer prices, MS 1403-210/5, Sir John Ross Marshall, Political Papers, Alexander Turnbull Library, Wellington.

Benseman, Joanne, 'She Worked Her Heart Out': Hops Farming Women in the Nelson Region, 1920s–1950s, NPM 998/5/1, 12, Nelson Provincial Museum.

Breweries (historical file) [1940–1970], MS 77-067-6/05: Patrick Anthony Lawlor Papers, Alexander Turnbull Library, Wellington.

Brewers' and Wine and Spirit Merchants' Association of New Zealand, Records: MS-Papers-1629, Alexander Turnbull Library, Wellington.

Brewers Journal: William Dawson, Brewer, City Brewery [Speight's], Dunedin, 1879–91, Lion Technical Library, Auckland.

Brewing of Hop Beer in Breweries 1908–1977: AALN 847: W4748 202 31/40/-, Archives New Zealand, Wellington.

Buxton, E. & Co. (1876 – 1941) Records. AG136 E 10.2 Hop dept Letter book 10, Nelson Provincial Museum.

G. R. Macdonald, Dictionary of Canterbury Biographies, Christchurch: Canterbury Museum, n.d.

James Allen Papers, 1916–1919: Allen, J.: Series 1; Box 9, Archives New Zealand, Wellington.

Lion Breweries files, typed notes for staff training course, G. C. Reid, 'History and Organisation of the Company', 3 February 1970, Lion Technical Library, Auckland.

National Efficiency Board, Memorandum 448, 9 July 1917, Memoranda: Vol.1, qMS-1525, Alexander Turnbull Library, Wellington.

Nelson Hotels Ltd/Nelson Breweries Ltd, 'Bible', Nelson Provincial Museum.

New Zealand Alliance, Manawatu Council: Records. 1946–1974, MS 82-217-12, Alexander Turnbull Library, Wellington.

New Zealand Alliance Papers, MS 77-206, Alexander Turnbull Library, Wellington.

Notes and clippings – Confectionery and brewery workers, 1927–1991, Herbert Roth Papers, MS 94-106-38/10, Alexander Turnbull Library, Wellington.

Protest re beer prices, MS 96-235-13/08, New Zealand Hotel Workers' Federation. Service Workers Union of Aotearoa: Records, Alexander Turnbull Library, Wellington.

Supervision of Breweries General File 1942–1977: AALN 847: W4748 202 31/7/4, Archives New Zealand, Wellington.

Waterson, Duncan, Auckland Provincial History Project Notes: Distilling and brewing industries, NZMS 1166, Auckland City Libraries.

PERIODICALS AND REPORTS

Australian Brewers' Journal

Australian Brewing and Wine Journal (*ABWJ*)

Brewnews

Journal of the Legislative Council (*JLC*)

New Zealand Journal of Agriculture

New Zealand Parliamentary Debates (*NZPD*)

UNPUBLISHED THESES AND RESEARCH ESSAYS

Christoffel, Paul, 'Removing Temptation: New Zealand's Alcohol Restrictions, 1881–2005', PhD thesis, Victoria University of Wellington, 2006.

Coe, Emily, 'The Nelson Hop Industry: Survival and Success', BA Hons long essay, University of Otago, 1998.

Duncan, Marie J., 'Cheers! Selling Health and Happiness: Advertising Alcohol in New Zealand, *c.* 1900–1945', MA thesis, Victoria University of Wellington, 2020.

Early, Steven, 'The Six O'clock Closing Debate', Dip. Arts thesis, University of Otago, 1991.

Ferguson, Steven, 'Class at the Margins: The Canterbury Hotel Workers Union, 1908 to 1970', MA thesis, University of Canterbury, 1985.

Grigg, A. R., 'The Attack on the Citadels of Liquordom: A Study of the Prohibition Movement in New Zealand, 1894–1914', PhD thesis, University of Otago, 1977.

Hanna, P., '"Why Not Whisky": The Short History of the "New Zealand Distillery Company"', Postgrad. Dip. essay, University of Otago, 1975.

Logan, Hugh R. K., 'A Business History of Whisky Distillation in Dunedin', Hist 480 research essay, University of Otago, 1993.

McCarthy, M. P., 'A City in Transition: Diversification in the Social Life of Dunedin, 1860–1864', MA thesis, University of Otago, 1977.

McKimmey, P. F., 'The Temperance Movement in New Zealand, 1835–1894', MA thesis, University of Auckland, 1968.

Moore, Linda, 'Gender Counts: Men, Women and Electoral Politics, 1893–1919', MA thesis, University of Canterbury, 2004.

Mulcare, Tim, 'The Political Economy of Six O'clock Closing', working paper, Victoria University of Wellington, 1999.

Noble, D. J., 'The Origins and Foundation of the Invercargill Licensing Trusts. Account of the First Shaky Years of the First Ever Licensing Trust in N.Z., Founded by the First Labour Government During the War', Postgrad. Dip. essay, University of Otago, 1976.

Paul, D. R., 'The Structure of the Brewing Industry and Beer Market to 1980: A Case Study in Industrial Concentration and Market Regulation', MCom thesis, University of Auckland, 1984.

Pearson, M. N., 'Brown and Campbell in Early Auckland, 1840–56', MA thesis, University of Auckland, 1964.

Scott, R. R., 'Public Control of the Liquor Trade in New Zealand, 1840–1899', MA thesis, Victoria University College, 1952.

Sharp, Ronald, 'Alcohol in New Zealand 1642–1840', BA Hons research essay, Massey University, 1976.

Thornton, Edith, 'A History of the Prohibition Movement in New Zealand', MA thesis, Canterbury University College, 1936.

Turney, Craig, 'The New Zealand Alliance and Auckland, 1905–1920', MA thesis, University of Auckland, 1996.

Wright, Jason, 'Company of Strangers: Patea and Wanganui Hotels, 1866–1899', MA thesis, Massey University, 2004.

PUBLISHED SOURCES

'A Late Resident in the Colony' [G. B. Earp], *The Emigrant's Guide to New Zealand*, London: William S. Orr & Co., 1851.

Adamson, N. J., 'Hop Culture in Nelson District', *New Zealand Journal of Agriculture*, February 1946, 147–53.

'Anglo-Saxon' [J. W. Jago], *The Economics of Beer*, Dunedin: G. Lodge of New Zealand Independent Order of Good Templars, 1878.

Allan, J., *Prohibition and Christianity: A Protest*, Ōamaru: North Otago Times, 1897.

Allen, Matthew, 'Alcohol and Authority in Early New South Wales', *History Australia*, 9(3), 2012, 7–26.

Arbury, D., *Thames Beer Baron: Louis Ehrenfried*, Thames: Metallum Research, 2003.

Auckland Industrial & Mining Exhibition: Official Handbook 1898, Auckland: Geddis & Blomfield, 1898.

Ausubel, David, *The Fern and the Tiki: An American View of New Zealand National Character, Social Attitudes and Race Relations*, rev. ed., North Quincy, MA: The Christopher Publishing House, 1977 [1960].

Baker, Paul, *King and Country Call: New Zealanders, Conscription and the Great War*, Auckland: Auckland University Press, 1988.

Barton, R. J., ed., *Earliest New Zealand: The Journals and Correspondence of the Rev. John Butler*, Masterton: Palamontain and Petherick, 1927.

Bassett, Michael and Paul Goldsmith, *The Myers*, Auckland: David Ling, 2007.

Beaglehole, J. C., ed., *The Journals of Captain James Cook: II The Voyage of the* Resolution *and* Adventure, *1772–1775*, Cambridge: The Hakluyt Society, 1961.

Behind the Scenes in New Zealand's Premier Brewery: The Waitemata Model Brewery, Otahuhu, Auckland: Dawson Printing Co., 1949.

Bennett, Judith M., *Ale, Beer and Brewsters in England: Women's Work in a Changing World, 1300–1600*, Oxford: Oxford University Press, 1996.

Bollinger, Conrad, *Grog's Own Country: The Story of Liquor Licensing in New Zealand*, 2nd ed., Auckland: Minerva, 1967.

Bollinger, Nick, *Goneville: A Memoir*, Wellington: Awa Press, 2016.

Bourke, Kevin, *Kelburn, King Dick and the Kelly Gang: Richard Seddon and Political Patronage*, Wellington: Hit or Miss Publishing, 2008.

Boyd, Mary, *City of the Plains: A History of Hastings*, Wellington: Victoria University Press, 1984.

Brew It Yourself: Want to Make a Better, Stronger, Beer Than You Can Buy – At One-sixth the Cost?, Dunedin: Otago Daily Times, *c.* 1960.

Brewing Industry in New Zealand, Wellington: R. A. Jarden & Co., 1977.

Broad, Lowther, *The Law of Innkeepers in New Zealand and Licensing Committees Guide*, Nelson: Bond Finney, 1887.

Brooking, Tom, *Richard Seddon: King of God's Own: The Life and Times of New Zealand's Longest Serving Prime Minister*, Auckland: Penguin, 2014.

Burnett, John, *Liquid Pleasures: A Social History of Drinks in Modern Britain*, London: Routledge, 1999.

Business and Economic Research Limited (BERL), *Costs of Harmful Alcohol and Other Drug Use*, Wellington: BERL, 2009.

Butlin, N. G., 'Yo, Ho, Ho and How Many Bottles of Rum?', *Australian Economic History Review*, 23(2), 1983, 1–27.

Captain Cook Brewery Co. Ltd v Martin Ryan, Auckland: J. H. Field, 1901.

Carpenter, Kenneth J., *The History of Scurvy and Vitamin C*, Cambridge: Cambridge University Press, 1986.

Carpenter, Lloyd, 'The Merchants of the Rush: The Proto-bankers of the Goldfields', in Lloyd Carpenter and Lyndon Fraser, eds, *Rushing for Gold: Life and Commerce on the Goldfields of New Zealand and Australia* (pp. 222–39), Dunedin: Otago University Press, 2016.

Casswell, Sally, *Drinking by New Zealanders*, Wellington: Alcoholic Liquor Advisory Council, 1980.

Chodowski, A., *Wine: Its Use and Abuse: The Fermented Wines of the Bible*, Christchurch: Christchurch Press, 1893.

Christoffel, Paul, 'Prohibition and the Myth of 1919', *New Zealand Journal of History*, 42(2), 2008, 154–59.

Church Missionary Society, *Report of the Formation and Establishment of the New Zealand Temperance Society*, Paihia: Church Mission Press, 1836.

Clark, Peter, *The English Alehouse: A Social History, 1200–1830*, London: Longman, 1983.

Cody, Kieran and Steve Jackson, 'The Contested Terrain of Alcohol Sponsorship of Sport in New Zealand', *International Review for the Sociology of Sport*, 51(4), 2014, 375–93.

Coleman, W., ed., *A Short Sketch of Some Incidents in the Colonial Life of Mr Thomas Hancock*, Auckland: J. D. Wickham, 1885.

Collins, Ernest G., *Malting in Heathcote Valley, 1871–1981*, Christchurch: Canterbury Malting Co., *c.* 1981.

Cook, James, Tobias Furneaux and William Hodges, *A Voyage Towards the South Pole and Round the World: Performed in His Majesty's Ships the* Resolution *and* Adventure, *in the Years 1772, 1773, 1774, and 1775*, London: W. Strahan & T. Cadell, 1777.

Cornell, Martyn, *Beer: The Story of the Pint*, London: Headline Book Publishing, 2003.

Cornell, Martyn, *Amber, Gold and Black: The History of Britain's Great Beers*, Stroud: The History Press, 2010.

Cornell, Martyn, 'The Mysterious Australian Ale', http://zythophile.wordpress.com/2011/06/10/the-mysterious-australian-ale/ (accessed 12 June 2011).

Cornell, Martyn, 'The Bass Red Triangle: Things AB-InBev Won't Tell You', https://zythophile.co.uk/2013/06/28/the-bass-red-triangle-things-ab-inbev-wont-tell-you/ (accessed 18 July 2022).

Cornell, Martyn, 'A Short History of Spruce Beer in Britain', *Brewery History*, 165, 2016, 2–14.

Crampton, Eric and Matt Burgess, 'The Price of Everything, The Value of Nothing: A (Truly) External Review of BERL's Study of Harmful Alcohol and Drug Use', Working Paper No.10/2009, Christchurch: Department of Economics and Finance, University of Canterbury, 2009.

Cyclopedia of New Zealand: Industrial, Descriptive, Historical, Biographical, Wellington & Christchurch: Cyclopedia Company, 1897–1908.

Daley, Caroline, *Leisure and Pleasure: Reshaping and Revealing the New Zealand Body, 1900–1960*, Auckland: Auckland University Press, 2003.

Dalziel, Raewyn, *Julius Vogel: Business Politician*, Auckland: Auckland University Press, 1986.

Damerow, Peter, 'Sumerian Beer: The Origins of Brewing Technology in Ancient Mesopotamia', *Cuneiform Digital Library Journal*, 2, 2012, https://cdli.mpiwg-berlin.mpg.de/articles/cdlj/2012-2.pdf.

Darby, Peter, 'The History of Hop Breeding and Development', *Brewery History*, 121, 2005, 94–112.

Darwin Keynes, R., *Charles Darwin's* Beagle *Diary*, Cambridge: Cambridge University Press, 1988.

Derby, Mark, *Rock College: An Unofficial History of Mt Eden Prison*, Palmerston North: Massey University Press, 2020.

Deutsher, Keith, *The Breweries of Australia: A History*, Melbourne: Lothian, 1999.

Dillon, Patrick, *Gin: The Much Lamented Death of Madam Geneva*, London: Justin, Charles & Co., 2003.

Dingle, A. E., *The Campaign for Prohibition in Victorian England*, London: Croom Helm, 1980.

Dominion Breweries, *1930–1980: 50th Golden Anniversary: Dominion Breweries*, Auckland: Dominion Breweries, 1980.

Donaldson, Michael, *Beer Nation: The Art and Heart of Kiwi Beer*, Auckland: Penguin, 2012.

Donaldson, Michael, *The Hopfather: Richard Emerson*, Auckland: Penguin, 2019.

Donnachie, Ian L., *A History of the Brewing Industry in Scotland*, Edinburgh: Donald, 1979.

Drummond, Alison, ed., *The Auckland Journal of Vicesimus Lush, 1850–63*, Christchurch: Pegasus Press, 1971.

'Dunboy', *Two Wrongs, One Right: A Prohibition Maxim*, Dunedin: Otago Daily Times and Witness Newspapers, 1899.

Durrell, Gerald, *Two in the Bush*, London: Collins, 1962.

Dwyer, J. H., *The Coburg Hotel, the Albert Brewery and the Site They Have Occupied – A Brief History*, Auckland: J. H. Dwyer, 1974.

Easdale, Nola, *Five Gentleman's Residences in Princess Street*, Auckland: Auckland City Council, 1980.

Easton, Brian, *Not in Narrow Seas: The Economic History of Aotearoa New Zealand*, Wellington: Victoria University of Wellington Press, 2020.

Eccles, Alfred, *An Account of the Brewing Trade of Dunedin and Neighbourhood: Past and Present; Compiled in the Centenary Year of the Province of Otago*, Dunedin: New Zealand Breweries, 1949.

Eldred-Grigg, Stevan, *Pleasures of the Flesh: Sex and Drugs in Colonial New Zealand, 1840–1915*, Wellington: Reed, 1984.

Eldred-Grigg, Stevan, *Diggers, Hatters and Whores: The Story of New Zealand Gold Rushes*, Auckland: Random House, 2008.

Fairburn, A. R. D., *We New Zealanders*, Wellington: Progressive Publishing Society, 1944.

Fairburn, Miles, *The Ideal Society and its Enemies*, Auckland: Auckland University Press, 1989.

Feeney, Warren, 'Rugby, Racing and Beer and Art', in Peter Alsop and Gary Stewart, eds, *Promoting Prosperity: The Art of Early New Zealand Advertising* (pp. 48–55), Nelson: Craig Potton Publishing, 2013.

Fitzgerald, Ross and Trevor Jordan, *Under the Influence: A History of Alcohol in Australia*, Sydney: ABC Books, 2009.

Gardner, W. J., *The Amuri: A County History*, 2nd ed., Culverden: Amuri County Council, 1983.

Glover, Brian, *Brewing for Victory: Brewers, Beer and Pubs in World War II*, Cambridge: Lutterworth Press, 1995.

Glover, Brian, *The Illustrated Encyclopedia of Wine, Beer and Spirits*, London: Lorenz Books, 2006.

Godley, John R., ed., *Letters from Early New Zealand by Charlotte Godley, 1850–53*, Christchurch: Whitcombe & Tombs, 1951.

Golledge, R. G., 'The New Zealand Brewing Industry', *New Zealand Geographer*, 19(1), 1963, 7–24.

Gordon, Donald, *Speight's: The Story of Dunedin's Historic Brewery*, Dunedin: Avon Publishers, 1993.

Gourvish, T. R. and R. G. Wilson, *The British Brewing Industry, 1830–1980*, Cambridge: Cambridge University Press, 1994.

Gramaphone, Malcolm, *The Brewers' Bible*, Wellington: Alister Taylor, 1972.

Grant, Ian F., 'Agents of Prosperity: Building an Advertising Industry', in Peter Alsop and Gary Stewart, eds, *Promoting Prosperity: The Art of Early New Zealand Advertising* (pp. 26–32), Nelson: Craig Potton Publishing, 2013.

Gray, A. H., *Aotearoa and New Zealand: A Historical Geography*, Christchurch: Canterbury University Press, 1994.

Grigg, A. R., 'Prohibition and Women: The Preservation of an Ideal and a Myth', *New Zealand Journal of History*, 17(2), 1983, 144–65.

Grimshaw, Patricia, *Women's Suffrage in New Zealand*, 2nd ed., Auckland: Oxford University Press, 1988.

Guide to Wellington and District: With a Complete Map of the City, Wellington: Walter K. Bishop, 1883.

Harper, Henry W., *Letters from New Zealand 1857–1911 Being Some Account of Life and Work in the Province of Canterbury, South Island*, London: Hugh Rees Ltd, 1914.

Harrison, Brian, *Drink and the Victorians: The Temperance Question in England, 1815–72*, London: Faber & Faber, 1971.

Hawke, Gary, *The Making of New Zealand: An Economic History*, Cambridge: Cambridge University Press, 1985.

Haydon, Peter, *Beer and Britannia: An Inebriated History of Britain*, Stroud: Sutton Publishing, 2001.

Hill, Linda, 'What it Means to be a Lion Red Man: Alcohol Advertising and Kiwi Masculinity', *Women's Studies Journal*, 15(1), 1999, 65–85.

Hill, Linda and Liz Stewart, 'The Sale of Liquor Act, 1989: Reviewing Regulatory Practices', *Social Policy Journal of New Zealand*, 7, 1996.

Hill, Richard S., *The Iron Hand in the Velvet Glove: The Modernisation of Policing in New Zealand, 1886–1917*, Palmerston North: Dunmore Press, 1995.

'Hori', *The Half Gallon Jar*, Auckland: Sporting Publications, 1962.

Hornsey, Ian S., *A History of Beer and Brewing*, London: Royal Society of Chemistry, 2003.

Horrocks, John, 'The Shadowland Manipulator: Sir Ernest "Booze", John A. Lee, and the Labour Party', *Journal of New Zealand Studies*, 24, 2017, 43–56.

Howard, Martin R., 'Red Jackets and Red Noses: Alcohol and the British Napoleonic Soldier', *Journal of the Royal Society of Medicine*, 93(1), 2000, 38–41.

Hughes, David, 'Australia's First Brewer', *Journal of the Royal Australian Historical Society*, 82(2), 1996, 153–67.

Hughes, David, 'The Brewing Industry in Early Australia', in R. G. Wilson and T. R. Gourvish, eds, *The Dynamics of the International Brewing Industry Since 1800* (pp. 213–228), London: Routledge, 1998.

Hursthouse, Charles, *An Account of the Settlement of New Plymouth: In New Zealand, From Personal Observations, During a Residence There of Five Years*, Christchurch: Capper Press, 1975 [1849].

Hutt, Marten, *Te Iwi Maori me te Inu Waipiro: He Tuhituhinga Hitori / Maori and Alcohol: A History*, Wellington: Health Services Research Centre for Kaunihera Whakatupato Waipiro o Aotearoa/Alcohol Advisory Council of New Zealand (ALAC), 1999.

Innes, C. V., *A Crown for the Lady: The Unravelling of a Pioneer Story*, Tauranga: Moana Press, 1989.

Intercolonial Exhibition 1866: Awards of the Jurors, Melbourne: Blundell & Co., 1866.

'It Is', *Isitt and His Prohibition Fad Proved Complete Failures*, Dunedin: Otago Workman, 1893.

Jackson, Steven J., Sarah Gee and Jay Scherer, 'Producing and Consuming Masculinity: New Zealand's (Speight's) "Southern Man"', in Lawrence A. Wenner and Steven J. Jackson, eds, *Sport, Beer, and Gender: Promotional Culture and Contemporary Social Life* (pp. 181–202), New York: Peter Lang, 2009.

Jennings, Paul, *The Local: A History of the English Pub*, London: The History Press, 2011.

Johnstone, Ian R., *Breweries of Gisborne*, Mosgiel: the author, 1995.

Jones, S., 'Brewing', in Michael Pickford and Alan Bollard, eds, *The Structure and Dynamics of New Zealand Industries* (pp. 121–48), Palmerston North: Dunmore Press, 1998.

Jones, S. R. H., 'The New Zealand Brewing Industry 1840–1995', in R. G. Wilson and T. R. Gourvish, eds, *The Dynamics of the International Brewing Industry Since 1800* (pp. 221–237), London: Routledge, 1998.

Jones, S. R. H. and D. R. Paul, 'Concentration and Regulation in the New Zealand Brewing Industry, 1850–1970', *Australian Economic History Review*, 31(2), 1991, 66–94.

Katz, Solomon H. and Fritz Maytag, 'Brewing an Ancient Beer', *Archaeology*, 44(4), 1991, 24–33.

Kennedy, M. J., 'The World's First Brewery Exclusively Designed to Use Continuous Fermentation: Biotechnology History Made in New Zealand', *Australasian Biotechnology*, 6(1), 1996, 13–18.

Kennedy, Melissa, 'Marketing Our Heritage: Dynamics of Commercial Enterprise and Nostalgia by DB Breweries Ltd in the DB Draught Clydesdales', *Deep South*, 3(3), 1997, https://www.otago.ac.nz/deepsouth/vol3no3/db.html.

Kirkby, Diane, *Barmaids: A History of Women's Work in Pubs*, Cambridge: Cambridge University Press, 1997.

Kuehn, Kathleen M. and Sophie Parker, 'One of the Blokes: Brewsters, Branding and Gender (In) Visibility in New Zealand's Craft Beer Industry', *Journal of Consumer Culture*, 21(3), 2021, 519–38.

Lash, M. D. and D. Smith, *Nelson Notables, 1840–1940: A Dictionary of Regional Biography*, Nelson: Nelson Historical Society, 1992.

Latham, Darrell, *The Golden Reefs: An Account of the Great Days of Quartz-Mining at Reefton, Waiuta and the Lyell*, Nelson: Nikau Press, 1992.

Lawlor, Pat, *The Froth-Blowers' Manual: Some Confessions, Some Beer Ballads, a Dash of History, Froth of Personalities and an Encyclopedia*, Wellington: Whitcombe & Tombs, 1965.

Leckie, Frank G., *Otago's Breweries: Past and Present*, Dunedin: Otago Heritage Books, 1997.

Liberty and Regulation, *National Prohibition Opposed to the Old Testament, the New Testament, the Spirit of Christ, Sound Ethics, Common Sense, Enlightened Politics, is a Fallacy, a Tyranny, a Failure, a Cure Worse than the Disease*, Dunedin: NZ Tablet Printing & Publishing Co., 1920.

Lind, Clive A., *Pubs, Pints and People: 50 Years of the Invercargill Licensing Trust*, Invercargill: The Trust, 1994.

Littlewood, David, 'Striking a Balance: The Military Service Boards', in S. Loveridge, ed., *New Zealand Society at War, 1914–1918* (pp. 112–26), Wellington: Victoria University Press, 2016.

Luckins, Tanja, 'Pigs, Hogs and Aussie Blokes: The Emergence of the Term "Six O'clock Swill"', *History Australia*, 4(1), 2007, 08.1–08.17.

Luckins, Tanja, '"Satan Finds Some Mischief"?: Drinkers' Responses to the Six O'clock Closing of Pubs in Australia, 1910s–1930s', *Journal of Australian Studies*, 32(3), 2008, 295–307.

Lysnar, W. D., *Opinions and Facts Showing Prohibition to be a Hideous Failure*, Gisborne: Gisborne Times, 1902.

Macdonald, Charlotte, 'Woolwich to Wellington: From Settler Colony to Garrisoned Sovereignty', *New Zealand Journal of History*, 53(1), 2019, 50–76.

Manderson, J. A., *Beer Slops*, Auckland: the author, 1946.

Martin, J. E., 'Blueprint for the Future? "National Efficiency" and the First World War', in J. Crawford and I. McGibbon, eds, *New Zealand's Great War: New Zealand, the Allies and the First World War* (pp. 516–33), Auckland: Exisle Publishing, 2007.

Mathias, Peter, *The Brewing Industry in England, 1700–1830*, Cambridge: Cambridge University Press, 1959.

Maxwell, D. B. S., 'Beer Cans: A Guide for the Archaeologist', *Historical Archaeology*, 27(1), 1993, 95–113.

May, Philip Ross, *The West Coast Gold Rushes*, 2nd rev. ed., Christchurch: Pegasus, 1967.

May, Philip Ross, *Gold Town, Ross, Westland*, Christchurch: Pegasus, 1970.

McAloon, Jim, *Nelson: A Regional History*, Whatamango Bay: Cape Catley in association with the Nelson City Council, 1997.

McAloon, Jim, *No Idle Rich: The Wealthy in Canterbury and Otago, 1840–1914*, Dunedin: Otago University Press, 2002.

McAloon, Jim, *Judgements of All Kinds: Economic Policy-making in New Zealand, 1945–1984*, Wellington: Victoria University Press, 2013.

McCallum, Angus, *Tui Country. A History of Pahiatua County, 1888–1988*, Pahīatua: Pahiatua County Council, 1988.

McClure, Margaret, *The Wonder Country: Making New Zealand Tourism*, Auckland: Auckland University Press, 2004.

McCrystal, John and Simon Farrell-Green, *The McCashin's Story: The First Craft Beer and the Kiwi Brewing Revolution It Sparked*, Auckland: Random House, 2013.

McIvor, Timothy, *The Rainmaker: A Biography of John Ballance, Journalist and Politician, 1839–1893*, Auckland: Heinemann Reed, 1989.

McKeague, John, *Liquor Reform: Prohibition or No License in New Zealand: Its Real Effects*, Sydney: Ross Brothers, 1905.

McKinnon, Malcolm, *The Broken Decade: Prosperity, Depression and Recovery in New Zealand, 1928–39*, Dunedin: Otago University Press, 2016.

McNab, Robert, *Historical Records of New Zealand*, Wellington: Government Printer, 1908–14.

McNab, Robert, *Murihiku: A History of the South Island of New Zealand and the Islands Adjacent and Lying to the South, from 1642 to 1835*, Christchurch: Whitcombe & Tombs, 1909.

Mirams, Chris, *Beleaguered! The Warriors from Dream to Nightmare*, Auckland: Hodder Moa Beckett, 2001.

Mitchell, B. R., *Abstract of British Historical Statistics*, Cambridge: Cambridge University Press, 1962.

Munro, Doug, 'The New Zealand Opera Company: Notes Towards a History', *Turnbull Library Record*, 48(1), 2016, 36–49.

Nelson, Max, *The Barbarian's Beverage: A History of Beer in Ancient Europe*, London: Routledge, 2005.

New Zealand Breweries, *Hosts to the Nation: The First Fifty Years of New Zealand Breweries Limited*, Wellington: New Zealand Breweries, 1973.

New Zealand Law Commission, *Alcohol in Our Lives: Curbing the Harm*, Wellington: The Commission, 2010.

New Zealand Licensing Reform Association, *The Proposals for Corporate Control of the Liquor Trade as Embodied in the Petition Presented to Parliament in 1923: Together with Proposals for the Immediate Improvement of the Existing System of Private Control*, Wellington: New Zealand Licensing Reform Association, 1923.

Newman, Richard, 'New Zealand's Vote for Prohibition in 1911', *New Zealand Journal of History*, 9(1), 1975, 52–71.

Nicholls, James, *The Politics of Alcohol: A History of the Drink Question in England*, Manchester: Manchester University Press, 2009.

Ogle, Maureen, *Ambitious Brew: The Story of American Beer*, New York: Harvest Books, 2007.

Olssen, Erik, *John A. Lee*, Dunedin: University of Otago Press, 1977.

Olssen, Erik, *A History of Otago*, Dunedin: McIndoe, 1984.

Orchard, H. K., 'New Zealand Alcohol Consumption Statistics, 1955–1989', Wellington: Alcoholic Liquor Advisory Council, 1990.

Orsman, H. W., ed., *The Dictionary of New Zealand English*, Auckland: Oxford University Press, 1997.

Packer, Rob, *A History of Nelson's Early Manufacturers: Their Wares and Containers*, Nelson: R. V. Packer, 1995.

Painter, Alison, *Jolly Good Ale and Old, The History of Coopers Brewery, 1862–1987*, Adelaide: Cooper & Son, 1987.

Palmer, Mervyn, 'William Soltau Davidson: A Pioneer of New Zealand Estate Management', *New Zealand Journal of History*, 7(2), 1973, 148–64.

Parsons, T. G., 'Technological Change in the Melbourne Flour-Milling and Brewing Industries, 1870–90', *Australian Economic History Review*, 11(2), 1971, 133–46.

Parsons, T. G., 'Brewing Beer in the Colonial Days', *Victorian Historical Journal*, 55(2), 1984, 13–19.

Pawson, E. and N. C. Quigley, 'The Circulation of Information and Frontier Development: Canterbury, 1850–90', *New Zealand Geographer*, 38(2), 1982, 65–76.

Perry, Stuart, *The New Zealand Whisky Book*, Auckland: Collins, 1980.

Petchey, Peter, 'The Ring Seal Beer/Champagne Bottle', *Archaeology in New Zealand*, 56(4), 2013.

Phillips, Jock, *A Man's Country? The Image of the Pakeha Male – A History*, Auckland: Penguin, 1987.

Polack, J. S., *Manners and Customs of the New Zealanders*, London: James Madden & Co., 1840.

Prince, John D., 'Look Back in Amber: The General Licensing Poll in New Zealand, 1919–87', *Political Science*, 48(1), 1996, 48–72.

Prohibition Fatal to Liberty, Temperance and Morality. Written Specially for Popular Use and Carefully Revised by Three Professors of the University of New Zealand, Christchurch: Whitcombe & Tombs, 1911.

Protz, Roger, *The Taste of Beer: A Guide to Appreciating the Great Beers of the World*, London: Rigel Publications, 1998.

Ramsden, Eric, *Busby of Waitangi: H.M.'s Resident at New Zealand, 1833–40*, Wellington: A. H. & A. W. Reed, 1942.

Re Petition for Six O'clock Closing of Liquor Bars: Verbatim Report of Addresses Delivered by Counsel Before the M to Z Public Petitions Committee of the House of Representatives on Friday, June 30th, 1916, Wellington: New Zealand Times Print, 1916.

Report of the 1974 Royal Commission on the Sale of Liquor in New Zealand, Wellington: Government Printer, 1974.

Rice, Geoffrey W., *Black November: The 1918 Influenza Pandemic in New Zealand*, Christchurch: Canterbury University Press, 2005.

Riley, Murdoch, *Māori Healing and Herbal*, Paraparaumu: Viking Sevenseas, 1994.

Robinson, Jancis, ed., *The Oxford Companion to Wine*, 3rd ed., Oxford: Oxford University Press, 2006.

Rowland, Perrin, *Dining Out: A History of the Restaurant in New Zealand*, Auckland: Auckland University Press, 2010.

Ryan, Greg, 'Drink and the Historians: Sober Reflections on Alcohol in New Zealand, 1840–1914', *New Zealand Journal of History*, 44(1), 2010, 35–52.

Ryan, Greg, 'An Undertaking Worthy Only of Fanatics: Catholic Opinion on Temperance and Prohibition in New Zealand, c1870–1910', *Australasian Journal of Irish Studies*, 10, 2010, 16–36.

Ryan, Greg and Geoff Watson, *Sport and the New Zealanders: A History*, Auckland: Auckland University Press, 2018.

Sale of Liquor in New Zealand: Report of the Working Party on Liquor: Presented to the Minister of Justice Rt Hon Geoffrey Palmer, Wellington: Department of Justice, 1986.

Salmond, W., *Prohibition: A Blunder*, Dunedin: R. J. Stark, 1911.

Shoebridge, Tim, 'Index of New Zealand Wartime Laws and Regulations, 1914–21', NZ History, https://nzhistory.govt.nz/files/documents/ww1-laws-and-regulations-index.pdf (accessed 4 April 2020).

Simkin, C. G. F. and B. G. Hardie, *Statistics of New Zealand for the Crown Colony Period 1840–1852*, Auckland: Auckland University College Department of Economics, 1954.

Six O'clock Closing: Auckland Business Men's Deputation to the Right Hon. W.F. Massey, Auckland: Wright & Jacques, 1916.

Smith, Rod, *Guinness Down Under: The Famous Brew and the Family Comes to Australia and New Zealand*, Tauranga: Eyeglass Press, 2018.

Speight's Brewery, *The History of a Glass of Beer: As Produced at Speight & Co.'s City Brewery, Rattray and Dowling Streets, Dunedin*, Dunedin: Samuel Lister, 1893.

Spring, Josephine A. and David H. Buss, 'Three Centuries of Alcohol in the British Diet', *Nature*, 270(15), December 1977, 567–72.

Stewart, Alan V. and Jimmy K. Stewart, *The Last of the Small Breweries in New Zealand: A History of Stewart's Brewery, Greymouth, 1880–1961*, Christchurch: A. V. & J. K. Stewart, 2004.

Stone, R. C. J., *Makers of Fortune: A Colonial Business Community and its Fall*, Auckland: Auckland University Press, 1973.

Stone, R. C. J., *The Father and His Gift: John Logan Campbell's Later Years*, Auckland: Auckland University Press, 1987.

Stone, R. C. J., '"Sinister" Auckland Business Cliques, 1840–1940', *New Zealand Journal of History*, 21(1), 1987, 29–45.

Stubbs, Brett, 'Brewing With Sugar', BrewsNews, https://www.brewsnews.com.au/2011/09/05/brewing-with-sugar/ (accessed 1 March 2022).

Stubbs, Brett J., '"A New Drink for Young Australia": From Ale to Lager Beer in New South Wales, 1880 to 1930', in Robert Dare, ed., *Food, Power and Community: Essays in the History of Food and Drink* (pp. 126–41), Adelaide: Wakefield Press, 1999.

Stubbs, Brett J., 'Tied Houses, Taxes, and Technology: Concentration in the Brewing Industry of New South Wales, 1882 to 1932', *Australian Economic History Review*, 39(2), 1999, 87–113.

Stubbs, Brett J., 'Captain Cook's Beer: The Anti-scorbutic Use of Malt and Beer in Late 18th Century Sea Voyages', *Asia Pacific Journal of Clinical Nutrition*, 12(2), 2003, 129–37.

Sumner, James, *Brewing Science, Technology and Print, 1700–1880*, London: Pickering & Chatto, 2013.

Sutherland, Jeremy, 'A History of Brewing in South Canterbury, New Zealand, 1865–1996, part 1', http://www.timaru.com/aeratedwaters_history/history.htm

Taylor, T. E., *The Shadow of Tammany: The Truth About the New Zealand Police Force, Its Condition and Its Needs*, Christchurch: T. E. Fraser, 1898.

Teahan, Bernard, *A Great Social Experiment: The Story of Licensing Trusts in New Zealand*, Masterton: Fraser Books, 2017.

The Menace of the Motor: Prevention of Motor Accidents, Wellington: Wright & Carman, 1936.

Thomson, Jane, ed., *Southern People: A Dictionary of Otago Southland Biography*, Dunedin: Longacre Press in association with Dunedin City Council, 1998.

Thomson, W., *Anti-prohibition: A Lecture Delivered . . . in the Zealandia Hall, Invercargill*, Temuka: Temuka Leader, 1905.

Troughton, Geoffrey, 'Redeeming the Demon Drink: New Zealand Presbyterians and the Question of Alcohol', *The Social History of Alcohol and Drugs*, 27(2), 2013, 133–55.

Tullett, J. S., *The Industrious Heart: A History of New Plymouth*, New Plymouth: New Plymouth City Council, 1981.

Tyrell, Ian, *Sobering Up: From Temperance to Prohibition in Antebellum America, 1809–1860*, Westport, Conn.: Greenwood Press, 1979.

Unger, Richard W., *Beer in the Middle Ages and the Renaissance*, Philadelphia: University of Pennsylvania Press, 2004.

Upton, Susan, *Wanted, a Beautiful Barmaid: Women Behind the Bar in New Zealand, 1830–1976*, Wellington: Victoria University Press, 2013.

Veart, David, *First Catch Your Weka: A Story of New Zealand Cooking*, Auckland: Auckland University Press, 2008.

Vosslamber, Rob, 'Narrating History: New Zealand's "Black Budget" of 1958', *Accounting History*, 17(3–4), 2012, 481–99.

Walton, Sara, Shayne Grice and Bevan Catley, 'The Monteith's Affair: Bitter to the Loyal End?', *Journal of the Australian and New Zealand Academy of Management*, 9(2), 2003, 69–74.

Watson, James, *Links: A History of Transport and New Zealand Society*, Wellington: Ministry of Transport, 1996.

Watson, James, *W.F. Massey: New Zealand*, London: Haus, 2010.

Willett, R. W., *Hotels of Dunedin: Historical Record*, Dunedin: Coulls Somerville Wilkie, 1937.

Wilson, G. B., *Alcohol and the Nation: A Contribution to the Study of the Liquor Problem in the United Kingdom from 1800 to 1935*, London: Nicholson and Watson, 1940.

Winks, Robin W., *These New Zealanders*, Christchurch: Whitcombe & Tombs, 1954.

Wood, Pamela, *Dirt: Filth and Decay in a New World Arcadia*, Auckland: Auckland University Press, 2005.

Wyllie, Allan, *Drinking in New Zealand: A National Survey*, Auckland: Alcohol and Public Health Research Unit, University of Auckland, 1996.

Wyllie, Allan and Sally Casswell, *Drinking in New Zealand: A Survey*, Auckland: Alcohol Research Unit, University of Auckland, 1989.

Index

Page numbers of illustrations are in **bold** type. A page number in *italics* indicates a graph or a table. An endnote is indicated by a page number followed by 'n' and then the note number.

B

C

D

E

F

G

H

I

J

N

Q

R

S

T

U

V

W

X

Y

Greg Ryan is a history professor and proctor at Lincoln University. He is a fellow of the Australian Society for Sports History, and author of: *Sport and the New Zealanders: A History* (with Geoff Watson, Auckland University Press, 2018), winner of the Australian Society for Sports History Book Award 2019; *The Contest for Rugby Supremacy: Accounting for the 1905 All Blacks* (Canterbury University Press, 2005); *The Making of New Zealand Cricket: 1832–1914* (Frank Cass, 2003), winner of the 2005 Ian Wards Prize; and *Forerunners of the All Blacks: The 1888–89 New Zealand Native Football Team in Britain, Australia and New Zealand* (Canterbury University Press, 1993).

SUICIDE SQUAD